Fundamentals of
Ecological Modelling

Developments in Environmental Modelling

Fundamentals of Ecological Modelling
Applications in Environmental Management and Research

4th edition

by

Sven Erik Jørgensen

Brian D. Fath

ELSEVIER

AMSTERDAM • BOSTON • HEIDELBERG • LONDON • NEW YORK
OXFORD • PARIS • SAN DIEGO • SAN FRANCISCO • SYDNEY
TOKYO

Elsevier
Radarweg 29, PO Box 211, 1000 AE Amsterdam, The Netherlands
Linacre House, Jordan Hill, Oxford OX2 8DP, UK

Library of Congress Cataloging-in-Publication Data
A catalog record for this book is available from the Library of Congress.

British Library Cataloguing-in-Publication Data
A catalogue record for this book is available from the British Library.

978-0-444-53567-2

For information on all Elsevier Publications
visit our Web site at elsevierdirect.com

Dedication

To the memory of G. Bendoricchio

Contents

Author Biography

Dr. Jørgensen is Professor Emeritus at the University of Copenhagen and specializes in systems ecology, ecological modelling, and ecological engineering. Dr. Jørgensen has published 66 books and more than 350 papers. He has served as Editor-In-Chief of *Ecological Modelling: International Journal on Ecological Modelling and Systems Ecology* for 34 years. He is also editor-in-chief of *Encyclopedia of Ecology*. He has received several prizes (The Prigoine Award, The Pascal Medal, The Einstein Profesorship of Chinese Academy of Sciences) and the very prestigious Stockholm Water Prize. He is honorable doctor of Coimbra University, Portugal and Dar es Salaam University, Tanzania. He is an elected member of the European Academy of Sciences. He is president of ISEM (International Society of Ecological Modelling).

Dr. Fath is an Associate Professor in the Department of Biological Sciences at Towson University (Maryland, USA) and is a research scholar in the Dynamic Systems Program at the International Institute for Applied Systems Analysis (Laxenburg, Austria). He has published almost 100 journal articles, reports, and book chapters. Dr. Faith first book, *A New Ecology*, was published with S.E. Jørgensen in June 2007 and in 2008 they co-edited a 5-volume Encyclopedia of Ecology. Dr. Fath has been Editor-in-Chief of the journal *Ecological Modelling* since January 2009. He teaches regular courses in ecosystem ecology, environmental biology, networks, and human ecology and sustainability at Towson and has given short courses in China, Croatia, Denmark, France, Germany, and Portugal. Dr. Fath is currently the chair of the Baltimore County Commission on Environmental Quality.

Preface

This is the fourth edition of *Fundamentals of Ecological Modelling*, and we have given it a longer title: *Fundamentals of Ecological Modelling: Application in Environmental Management and Research*. This was done to emphasize that models, applied in environmental management and ecological research, are particularly considered in the model illustrations included in this book.

Giuseppe Bendoricchio, co-author of the third edition published in 2001, passed away in 2005. We would therefore like to dedicate this book to his memory and his considerable contributions in the 1980s and 1990s to the development of ecological modelling.

The first two editions of this book (published in 1986 and 1994) focused on the roots of the discipline — the four main model types that dominated the field 30-40 years ago: (1) dynamic biogeochemical models, (2) population dynamic models, (3) ecotoxicological models, and (4) steady-state biogeochemical and energy models. Those editions offered the first comprehensive textbook on the topic of ecological modelling. The third edition, with substantial input from Bendoricchio, focused on the mathematical formulations of ecological processes that are included in ecological models. In the third edition, the chapter called Ecological Processes encompasses 118 pages. The same coverage of this topic today would probably require 200 pages, and is better covered in the *Encyclopedia of Ecology*, which was published in the fall of 2008.

This fourth edition uses the four model types previously listed as the foundation and expands the latest model developments in spatial models, structural dynamic models, and individual-based models. As these seven types of models are very different and require different considerations in the model development phase, we found it important for an up-to-date textbook to devote a chapter to the development of each of the seven model types. Throughout the text, the examples given from the literature emphasize the application of models for environmental management and research. Therefore the book is laid out as follows:

Chapter 1: Introduction to Ecological Modelling provides an overview of the topic and sets the stage for the rest of the book.
Chapter 2: Concepts of Modelling covers the main modelling elements of compartments (state variables), connections (flows and the mathematical equations used to represent biological, chemical, and physical processes), controls (parameters, constants), and forcing functions that drive the systems. It

also describes the modelling procedure from conceptual diagram to verification, calibration, validation, and sensitivity analysis.

Chapter 3: An Overview of Different Model Types critiques when each type should or could be applied.

Chapter 4: Mediated or Institutionalized Modelling presents a short introduction to using the modelling process to guide research questions and facilitate stakeholder participation in integrated and interdisciplinary projects.

Chapter 5: Modelling Population Dynamics covers the growth of a population and the interaction of two or more populations using the Lotka-Volterra model, as well as other more realistic predator–prey and parasitism models. Examples include fishery and harvest models, metapopulation dynamics, and infection models.

Chapter 6: Steady-State Models discusses chemostat models, Ecopath software, and ecological network analysis.

Chapter 7: Dynamic Biogeochemical Models are used for many applications starting with the original Streeter-Phelps model up to the current complex eutrophication models.

Chapter 8: Ecotoxicological Models provides a thorough investigation of the various ecotoxicological models and their use in risk assessment and environmental management.

Chapter 9: Individual-based Models discusses the history and rise of individual-based models as a tool to capture the self-motivated and individualistic characteristics individuals have on their environment.

Chapter 10: Structurally Dynamic Models presents 21 examples of where model parameters are variable and adjustable to a higher order goal function (typically thermodynamic).

Chapter 11: Spatial Modelling covers the models that include spatial characteristics that are important to understanding and managing the system.

This fourth edition is maintained as a textbook with many concrete model illustrations and exercises included in each chapter. The previous editions have been widely used as textbooks for past courses in ecological modelling, and it is the hope of the authors that this edition will be an excellent basis for today's ecological modelling courses.

Sven Erik Jørgensen
Copenhagen, Denmark

Brian D. Fath
Laxenburg, Austria

July 2010

1

Introduction

CHAPTER OUTLINE

1.1. Physical and Mathematical Models

Humans have always used models — defined as a simplified picture of reality — as tools to solve problems. The model will never be able to contain all the features of the real system, because then it would be the real system itself, but it is important that the model contains the characteristic features essential in the context of the problem to be solved or described.

The philosophy behind the use of a model is best illustrated by an example. For many years we have used physical models of ships to determine the profile that gives a ship the smallest resistance in water. Such a model has the shape and the relative main dimensions of the real ship, but does not contain all the details such as the instrumentation, the layout of the cabins, and so forth. Such details are irrelevant to the objectives of that model. Other models of the ship serve other purposes: blueprints of the electrical wiring, layout of the various cabins, drawings of pipes, and so forth.

Correspondingly, the ecological model we wish to use must contain the features that will help us solve the management or scientific problem at hand. An ecosystem is a much more complex system than a ship; it is a far more complicated matter to ascertain the main features of importance for an ecological problem. However, intense research during the last three decades has made it possible to set up many workable and applicable ecological models.

Fundamentals of Ecological Modelling. DOI: 10.1016/B978-0-444-53567-2.00001-6

1

Ecological models may also be compared with geographical maps (which are models, too). Different types of maps serve different purposes. There are maps for airplanes, ships, cars, railways, geologists, archaeologists, and so on. They are all different because they focus on different objects. Maps are also available in different scales according to application and underlying knowledge. Furthermore, a map never contains all of the details for a considered geographical area, because it would be irrelevant and distract from the main purpose of the map. If a map contained every detail, for instance, the positions of all cars at a given moment, then it would be rapidly invalidated as the cars move to new positions. Therefore, a map contains only the knowledge relevant for the user of the map, so there are different maps for different purposes.

An ecological model focuses similarly on the objects of interest for a considered well-defined problem. It would disturb the main objectives of a model to include too many irrelevant details. There are many different ecological models of the same ecosystem, as the model version is selected according to the model goals.

The model might be physical, such as the ship model used for the resistance measurements, which may be called microcosm, or it might be a mathematical model, which describes the main characteristics of the ecosystem and the related problems in mathematical terms.

Physical models will be touched on only briefly in this book, which will instead focus entirely on the construction of mathematical ecological models. The field of ecological modelling has developed rapidly during the last 30 years due essentially to three factors:

1. The development of computer technology, which has enabled us to handle very complex mathematical systems.
2. A general understanding of environmental problems, including that a complete elimination of pollution is not feasible (denoted zero discharge). Instead, a proper pollution control with limited economical resources requires serious consideration of the influence of pollution impacts on ecosystems.
3. Our knowledge of environmental and ecological systems has increased significantly; in particular we have gained more knowledge of the quantitative relations in the ecosystems and between the ecological properties and the environmental factors.

Models may be considered a synthesis of what we know about the ecosystem with reference to the considered problem in contrast to a statistical analysis, which only reveals the relationships between the data. A model is able to include our entire knowledge about the system such as:

1. Which components interact with which other components, for instance, that zooplankton grazes on phytoplankton
2. Our knowledge about the processes often formulated as mathematical equations, which have been shown to be generally valid
3. The importance of the processes with reference to the problem

This is a list of a few examples of knowledge that may often be incorporated in an ecological model. It implies that a model can offer a deeper understanding of the system than a statistical analysis. Therefore, it is a stronger research tool that can result in a better management plan for solving an environmental problem. This does not mean that statistical analytical results are not applied in the development of models. On the contrary, models are built on all available knowledge, including that gained by statistical analyses of data, physical-chemical-ecological knowledge, the laws of nature, common sense, and so on. That is the advantage of modelling.

1.2. Models as a Management Tool

The idea behind the use of ecological management models is demonstrated in Figure 1.1. Urbanization and technological development have had an increasing impact on the environment. Energy and pollutants are released into ecosystems where they can cause more rapid growth of algae or bacteria, damage species, or alter the entire ecological structure. An ecosystem is extremely complex, therefore it is an overwhelming task to predict the environmental effects that such emissions may have. It is here that the model is introduced into the picture. With sound ecological knowledge, it is possible to extract the components and processes of the ecosystem involved in a specific pollution problem to form the basis of the ecological model (see also the discussion in Chapter 2, Section 2.3). As indicated in Figure 1.1, the resulting model can be used to select the environmental technology eliminating the emission most effectively.

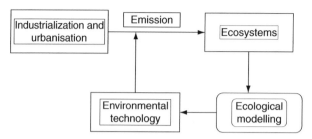

FIGURE 1.1 The environmental problems are rooted in the emissions resulting from industrialization and urbanization. Sound ecological knowledge is used to extract the components and processes of the ecosystem that are particularly involved in a specific pollution problem to form the ecological model applied in environmental management.

Figure 1.1 represents the idea behind the introduction of ecological modelling, which has been a management tool since about 1970. Now environmental management is more complex and is applied to a wider spectrum of tools. Today we have alternatives and supplements to environmental technology such as cleaner technology, ecotechnology, environmental legislation, international agreements, and sustainable management plans. Ecotechnology is mainly applied to solve the problems of nonpoint or diffuse pollution often originated from agriculture. The significance of nonpoint pollution was hardly acknowledged before 1980. Furthermore, the global environmental problems play a more important role today than 20 or 30 years ago; for instance, the reduction of the ozone layer and the climatic changes due to the greenhouse effect. The global problems cannot be solved without international agreements and plans. Figure 1.2 attempts to illustrate the current complex picture of environmental management.

1.3. Models as a Research Tool

Models are widely used instruments in science. Scientists often use physical models to carry out experiments in situ or in the laboratory to eliminate disturbance from processes irrelevant to an investigation: Thermostatic chambers are used to measure algal growth as a function of nutrient concentrations, sediment cores are examined in the laboratory to investigate sediment-water interactions without disturbance from other ecosystems components, reaction chambers are used to find reaction rates for chemical processes, and so on.

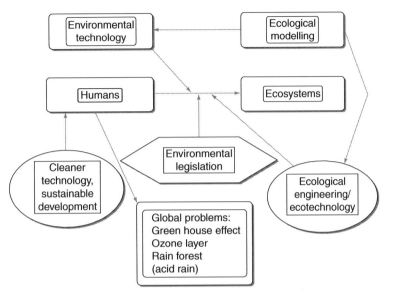

FIGURE 1.2 The idea behind the use of environmental models in environmental management. Environmental management today is very complex and must apply environmental technology, alternative technology, and ecological engineering or ecotechnology. In addition, the global environmental problems play an increasing role. Environmental models are used to select environmental technology, environmental legislation, and ecological engineering.

Mathematical models are widely applied in science as well. For example, Newton's laws are just relatively simple mathematical models of the influence of gravity on bodies, but they do not account for frictional forces, influence of wind, and so forth. Ecological models do not differ essentially from other scientific models except in their complexity, as many models used in nuclear physics may be even more complex than ecological models. The application of models in ecology is almost compulsory if we want to understand the function of such a complex system as an ecosystem. It is simply not possible to survey the many components and their reactions in an ecosystem without the use of a model as holistic tool. The reactions of the system might not necessarily be the sum of all the individual reactions, which implies that the properties of the ecosystem cannot be revealed without the use of a model of the entire system.

It is therefore not surprising that ecological models have been used increasingly in ecology as an instrument to understand the properties of ecosystems as systems. This application has clearly revealed the advantages of models as a useful tool in ecology, which may be summarized in the following:

1. Models are useful instruments in *survey* of complex systems.
2. Models can be used to reveal *system properties*.
3. Models reveal the weakness in our knowledge and can therefore be used to set up *research priorities*.
4. Models are useful in tests of *scientific hypotheses*, as the model can simulate ecosystem reactions that can be compared with observations.

As it will be illustrated several times throughout this volume, models can used to test the hypothesis of ecosystem behavior such as the principle of maximum power presented by H.T. Odum (1983), the ascendency propositions presented by Ulanowicz (1986), the various proposed thermodynamic principles of ecosystems, and the many hypothesis of ecosystem stability.

The certainty of the hypothesis test by using models is, however, not on the same level as the tests used in the more reductionistic disciplines of science. If a relationship is found between two or more variables by the use of statistics on available data, then the relationship is tested on several additional cases to increase the scientific certainty. If the results are accepted, then the relationship is ready to be used to make predictions, and it is again examined to prove whether the predictions are right or wrong in a new context. If the relationship still holds, then we are satisfied and a wider scientific use of the relationship is made possible.

When we are using models as scientific tools to test hypotheses, we have a "double doubt." We anticipate that the model is correct in the problem context, but the model is a hypothesis of its own. We therefore have four cases instead of two (acceptance/nonacceptance):

1. The model is correct in the problem context, and the hypothesis is correct.
2. The model is not correct, but the hypothesis is correct.
3. The model is correct, but the hypothesis is not correct.
4. The model is not correct and the hypothesis is not correct.

To omit cases 2 and 4, only very well-examined and well-accepted models should be used to test hypotheses on system properties, but, unfortunately, our experience in modelling ecosystems is limited. We do have some well-examined models, but we are not completely certain they are correct in the problem context and a wider range of models is needed. A wider experience in modelling may therefore be the prerequisite for further development in ecosystem research.

The use of models as a scientific tool as described earlier is not only known from ecology; other sciences use the same technique when complex problems and complex systems are under investigation. There are simply no other possibilities when dealing with irreducible systems (Wolfram l984a,b). Nuclear physics has used this procedure to find several new nuclear particles. The behavior of protons and neutrons has inspired models of smaller particles, the so-called quarks. These models have been used to predict the results of planned cyclotron experiments, which have inspired further changes of the model.

The idea behind the use of models as scientific tools may be described as an iterative development of a pattern. Each time we can conclude that case 1 (see the earlier list for the four cases) is valid, that is, both the model and the hypothesis are correct, we can add another "piece to the pattern." That provokes the question: Does the piece fit into the general pattern? This signifies an additional test of the hypothesis. If not, we can go back and change the model and/or the hypothesis, or we may be forced to change the pattern, which will require more comprehensive investigations. If the answer is "yes," then we can use the piece at least temporarily in the pattern — which is then used to explain other observations, improve our models, and make other predictions — for further testing. This procedure is used repeatedly to proceed stepwise toward a better understanding of nature on the system level. Figure 1.3 is a conceptual diagram of the procedure applied to test hypotheses by using models.

The application of this procedure in ecosystem theory is still relatively new. We need, as already mentioned, much more modelling experience. We also need a more comprehensive application of our ecological models in this direction and context.

1.4. Models and Holism

Biology (ecology) and physics developed in different directions until about 30 to 50 years ago, when there was more parallel development, which has its roots in the more general trends in science that have been observed in the last 20 years.

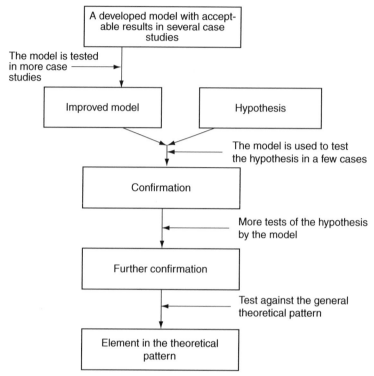

FIGURE 1.3 This diagram shows how it is required to use several test steps, if a model is used to test a hypothesis about ecosystems, as a model may be considered a hypothesis of its own.

The basic philosophy or thinking regarding science is currently changing with other facets of our culture such as the arts and fashion. The driving forces behind such developments are often very complex and are very difficult to explain in detail, but we will attempt to show at least some of tendencies in the development.

1. The sciences have realized that the world is more complex than previously thought. In nuclear physics several new particles have been found. In ecology we have seen new environmental problems. Now we realize how complex nature is and how much more difficult it is to cope with problems occurring in nature than in laboratories. Computations in sciences were often based on the assumption of so many simplifications that they became unrealistic.

2. Ecosystem ecology — we call it the science of (the very complex) ecosystems or systems ecology — has developed very rapidly and has evidently shown the need for systems sciences as well as interpretations, understandings, and implications of the results obtained in other sciences.

3. In the sciences, many systems are so complex that it is impossible to know all the details of every system. In nuclear physics there is always an uncertainty in our observations as expressed by Heisenberg's uncertainty relations. This uncertainty is caused by the influence of our observations on the nuclear particles. We have a similar uncertainty relation in ecology and environmental sciences caused by the complexity of the systems (Jørgensen & Fath, 2006). A further presentation of these ideas is given in Chapter 2, Section 2.6, where the complexity of ecosystems is discussed in more detail. In addition, many relatively simple physical systems such as the atmosphere show chaotic behavior, which makes long-term predictions impossible. The conclusion is unambiguous: We cannot and will not be able to know the world with complete accuracy and in complete detail. We have to acknowledge that these are the conditions for modern sciences.

4. Many systems in nature are irreducible systems (Wolfram 1984a,b); that is, it is impossible to reduce observations on system behavior to a law of nature, because the system has so many interacting elements that the reaction of the system cannot be surveyed without using models. For such systems other experimental methods must be applied. It is necessary to construct a model and compare the reactions of the model with our observations to test its reliability and get ideas for model improvements, construct an improved model, compare its reactions with the observations again to get new ideas for further improvements, and so forth. By such an iterative method we may be able to develop a satisfactory model that can describe our observation properly. These observations have not resulted in a new law of nature but in a new model of a piece of nature. As seen by the description of the details in the model development, the model should be constructed based on causalities, which inherit basic laws.

5. As a result of previous tendencies 1–4, modelling as a tool in science and research has developed and expanded. Ecological or environmental modelling has become a scientific discipline of its own — a discipline that has experienced rapid growth during the last decades. The core scientific journal in ecological modelling, *Ecological Modelling*, now publishes more than 4000 pages per year, while it published 320 pages in 1975. Developments in computer science and ecology have also favored this rapid growth in modelling, as they are the components on which modelling is founded.

6. The scientific analytical method has always been a very powerful tool in research. Yet, there has been an increasing need for scientific synthesis, that is, for combining the analytical results to form a holistic picture of natural systems. Due to the extremely high complexity of natural systems, it is impossible to obtain a complete and comprehensive picture of natural systems by analysis alone; it is necessary to synthesize important analytical results to get system properties. Synthesis and analysis must work hand-in-hand. The synthesis (e.g., in the form of a model) will show that further analytical results are needed to improve the synthesis and new analytical results may be used as components in better syntheses. The recent tendency in sciences is to give synthesis a higher priority than previously, but this does not imply that the analyses should be given a lower priority. Analytical results are needed to provide components for the synthesis, and the synthesis must be used to give priorities for the needed analytical results. No science exists without observations, but no science can be developed without the digestions of the observations to form a "picture" or "pattern" of nature either. Analyses and syntheses should be considered as two sides of the same coin.

7. A few decades ago, the sciences were more optimistic than they are today, because it was expected that a complete description of nature would soon be a reality. Einstein even talked about a "world equation" as the basis for all physics of nature. Today, we realize that nature is far more complex than a single world equation, and complex systems are nonlinear and sometimes chaotic. The sciences have a long way to go and it is not expected that the secret of nature can be revealed by a few equations. It may work in controlled laboratory conditions where the results usually can be described by using simple equations, but when we turn to natural systems, it will be necessary to apply many and complex models to describe our observations.

1.5. The Ecosystem as an Object for Research

Ecologists generally recognize ecosystems as a specific level of organization, but what is the appropriate selection of time and space scales? Any size area could be selected, but in the context of ecological modelling, the following definition presented by Morowitz (1968) will be used: "An ecosystem sustains life under present-day conditions, which is considered a property of ecosystems rather than a single organism or species." This means that a few square meters may seem adequate for microbiologists, while $100 \ km^2$ may be insufficient if large carnivores are considered (Hutchinson, 1970, 1978). Population-community ecologists tend to view ecosystems as networks of interacting organisms and populations. Tansley (1935) claimed that an ecosystem includes both organisms and chemical-physical components. It inspired Lindeman (1942) to use the following definition: "An ecosystem is composed of physical-chemical-biological processes active within a space-time unit." E.P. Odum (1953, 1959, 1969, 1971) followed these lines and is largely responsible for developing the process-functional approach, which has dominated ecosystem ecology for the last 50 years.

This does not mean that different views cannot be a point of entry. Hutchinson (1978) used a cyclic causal approach, which is often invisible in population-community problems. Measurement of inputs and outputs of total landscape units was the emphasis in the functional approaches by Bormann and Likens (1967). O' Neill (1976) emphasized energy capture, nutrient retention, and rate regulations. H.T. Odum (1957) underlined the importance of energy transfer rates. Quilin (1975) argued that cybernetic views of ecosystems are appropriate, and Prigogine (1947), Mauersberger (1983), and Jørgensen (1981, 1982, 1986) all emphasized the need for a thermodynamic approach for a proper holistic description of ecosystems.

For some ecologists ecosystems are either biotic assemblages or functional systems; the two views are separate. It is, however, important in the context of ecosystem theory to adopt both views and integrate them. Because an ecosystem cannot be described in detail, it cannot be defined according to Morowitz's (1968) definition before the objectives of our study are presented. Therefore, the definition of an ecosystem used in the context of system ecology and ecological modelling, becomes:

> An ecosystem is a biotic and functional system or unit, which is able to sustain life and includes all biotic and abiotic variables in that unit. Spatial and temporal scales are not specified *a priori*, but are entirely based upon the objectives of the ecosystem study.

Currently there are several approaches (Likens, 1985) used to study ecosystems:

1. Empirical studies — Bits of information are collected, and an attempt is made to integrate and assemble these into a complete picture.
2. Comparative studies — Structural and functional components are compared for a range of ecosystem types.
3. Experimental studies — Manipulation of a whole ecosystem is used to identify and elucidate mechanisms.
4. Modelling or computer simulation studies.

The motivation (Likens, 1985) in all of these approaches is to achieve an understanding of the entire ecosystem, giving more insight than the sum of knowledge about its parts relative to the structure, metabolism, and biogeochemistry of the landscape.

Likens (1985) presented an excellent ecosystem approach to Mirror Lake and its environment. The research contains all the previously mentioned studies, although the modelling part is less developed than the others. The study clearly demonstrates that it is necessary to use all four approaches simultaneously to achieve a good representation of the system properties of an ecosystem. An ecosystem is so complex that you cannot capture all the system properties by one approach.

Ecosystem studies widely use the notions of order, complexity, randomness, and organization. They are often interchangeably applied in the literature, which causes much confusion. As the terms are used in relation to ecosystems throughout the volume, it is necessary to give a clear definition of these concepts in this introductory chapter.

According to the Third Law of Thermodynamics about entropy at 0 K (Jørgensen, 2008a), randomness and order are the antithesis of each other and may be considered as relative terms. Randomness measures the amount of information required to describe a system. The more information required to describe the system, the more random it is.

Organized systems are to be carefully distinguished from ordered systems. Neither kind of system is random; whereas ordered systems are generated according to simple algorithms and may therefore lack complexity, an organized system must be assembled element by element according to an external wiring diagram with a high level of information. Organization is a functional complexity and carries functional information. It is nonrandom by design or by selection, rather than by *a priori* necessity. Complexity is a relative concept dependent on the observer (Jørgensen & Svirezhev, 2004). We may distinguish between structural complexity, which is defined as the number of interconnections between components in the system and functional complexity and defined as the number of distinct functions carried out by the system.

1.6. The Development of Ecological and Environmental Models

This section attempts to present briefly the history of ecological and environmental modelling. From the history we can learn why it is essential to draw upon the previously gained experience and what goes wrong when we do not follow the recommendations set up to avoid previous flaws.

Figure 1.4 gives an overview of the development in ecological modelling. The nonlinear time axis gives approximate information on the year

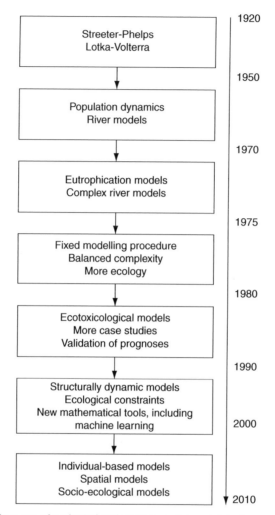

FIGURE 1.4 The development of ecological and environmental models is shown schematically.

when the various development steps took place. The first models of the oxygen balance in a stream (the Streeter-Phelps model, presented in Chapter 7) and of the prey–predator relationship (the Lotka-Volterra model, presented in Chapter 5) were developed back in the early 1920s. In the 1950s and 1960s, further development of population dynamic models took place. More complex river models were also developed in the 1960s. These developments could be named the second generation of models.

The wide use of ecological models in environmental management started around 1970, when the first eutrophication models emerged and very complex river models were developed. These models could be named the third generation of models. They are characterized by often being too complex, because it was so easy to write computer programs that could handle rather complex models. To a certain extent it was the revolution in computer technology that created this model generation. However, it became clear in the mid-1970s that the limitations in modelling were not from the computer and the mathematics, but from the available data and our knowledge about ecosystems and ecological processes. So, the modellers became more critical in their acceptance of models. They realized that a profound knowledge of the ecosystem — the problems and the ecological components — was the basis for development of sound ecological models. This period resulted in recommendations that are given in the Chapter 2:

- Strictly follow all steps of the procedure, such as conceptualization, selection of parameters, verification, calibration, examination of sensitivity, validation, and so forth.
- Find a balance between data, problem, ecosystem, and knowledge.
- A wide use of sensitivity analyses is recommended in the selection of model components and model complexity.
- Make parameter estimations by using all the methods, such as literature review, determination by measurement in laboratory or in situ, use of intensive measurements, calibration of submodels and the entire model, theoretical system ecological considerations, and various estimation methods based on allometric principles and chemical structure of the considered chemical compounds.

Parallel to this development, ecologists became more quantitative in their approach to environmental and ecological problems, probably because of the needs formulated by environmental management. The quantitative research results from the late 1960s onward have been of enormous importance for the quality of ecological models. They are probably just as important as the developments in computer technology.

The models from this period, going from the mid-1970s to the mid-1980s, could be called the fourth generation of models. The models from this period are characterized by a relatively sound ecological basis, along with an emphasis on realism and simplicity. Many models were validated in this period with an acceptable result and for some (few) it was even possible to validate the prognosis.

The conclusions from this period may be summarized in the following three points:

1. Provided that the previously listed recommendations are followed and the underlying database is of good quality, it is possible to develop models used as prognostic tools.
2. Models based upon a database of less than acceptable quality should not be used as a prognostic tool, but they could give an insight into the mechanisms behind the environmental management problem, which is valuable in most cases. Simple models are often of particular value in this context.
3. Ecologically sound models, that is, models based upon ecological knowledge, are powerful tools in understanding ecosystem behavior and as tools for setting up research priorities. The understanding may be qualitative or semiquantitative, but has in any case proved to be of importance for ecosystem theories and a better environmental management.

1.7. State of the Art in the Application of Models

The shortcomings of modelling have also been revealed. It became clear that the models were rigid in comparison with the enormous flexibility, which is characteristic of ecosystems. The hierarchy of feedback mechanisms that ecosystems possess was not accounted for in the models, which made them incapable of predicting adaptation and structural dynamic changes. Since the mid-1980s, modellers have proposed many new approaches such as (1) fuzzy modelling, (2) examinations of catastrophic and chaotic behavior of models, and (3) application of goal functions to account for adaptation and structural changes. Application of objective and individual modelling, expert knowledge, and artificial intelligence offers some new additional advantages in modelling. This will discussed in Chapter 3 of this volume as well as when it is advantageous to apply these approaches and what can be gained by their application.

Table 1.1   Biogeochemical Models of Ecosystems

Ecosystem	Modelling Effort
Rivers	5
Lakes, reservoirs, ponds	5
Estuaries	5
Coastal zone	4
Open sea	3
Wetlands	5
Grassland	4
Desert	1
Forests	5
Agriculture land	5
Savanna	2
Mountain lands (above timberline)	1
Arctic ecosystems	2
Coral reef	3
Waste water systems	5

All these recent developments could be named the fifth generation of modelling, which is covered in Chapters 3, 9, 10 and 11.

Table 1.1 reviews types of ecosystems, which have been modelled by biogeochemical models up to the year 2000. An attempt has been made to indicate the modelling effort by using a scale from 0–5 where 5 means very intense modelling effort, more than 50 different modelling approaches can be found in the literature; 4 means intense modelling effort with 20 to 50 different modelling approaches found in the literature; 4–5 may be translated to class 4 but on the edge of an upgrading to class 5; 3 means some modelling effort with 6 to 19 different modelling approaches published; 2, few (2 to 5) different models have been well studied and published; 1, one good study and/or a few insufficiently well-calibrated and validated models; and 0, almost no modelling efforts have been published with no well-studied models. Notice that the classification is based on the number of different models, not on the

Table 1.2   Models of Environmental Problems

Problem	Modelling Effort
Oxygen balance	5
Eutrophication	5
Heavy metal pollution, all types of ecosystems	4
Pesticide pollution of terrestrial ecosystems	4–5
Other toxic compounds include ecological risk assessment (ERA)	5
Regional distribution of toxic compounds	5
Protection of national parks	3
Management of populations in national parks	3
Endangered species (includes population dynamic models)	3
Groundwater pollution	5
Carbon dioxide/greenhouse effect	5
Acid rain	5
Total or regional distribution of air pollutants	5
Change in microclimate	3
As ecological indicator	4
Decomposition of the ozone layer	4
Relationships health-pollution	3
Consequences of climate changes	4

number of case studies where these models have been applied. In most cases, the same models have been used in several case studies.

Table 1.2 similarly reviews environmental problems that have been modelled through the years. The same scale is applied to show the modelling effort seen in Table 1.1. Table 1.2 covers biogeochemical models, as well as models used for management of population dynamics in national parks and steady-state models applied as ecological indicators. It is advantageous to apply goal functions in conjunction with a steady-state model to obtain good ecological indication, as proposed by Christensen (1991, 1992).

2

Concepts of Modelling

CHAPTER OUTLINE

2.1. Introduction

This chapter covers the topic of modelling theory and its application in the development of models. After the definitions of model components and modelling steps are presented, a tentative modelling procedure is given. The steps in the modelling procedure are discussed in detail and they include: model conceptualization, mathematical formulation, parameter estimation and calibration, sensitivity analysis, and validation. This chapter focuses on model selection or the selection of model components, processes, equations, and in particular, model complexity. Various methods to select "close to the right" complexity of the model are presented. Several model formulations are always available, and to choose among these will require that sound scientific constraints are imposed on the model. Many different model types with different advantages and disadvantages are available. The selection of the best model type for a well-defined ecological or environmental management problem will be discussed in Chapter 3, where an overview of the available model types will be presented. A mathematical model usually requires the use of a computer and a computer language. The selection

Fundamentals of Ecological Modelling. DOI: 10.1016/B978-0-444-53567-2.00002-8

19

of a computer language is not discussed here, because there are many possibilities and new languages emerge from time to time. In the models used as illustrative examples, STELLA (c) (High Performance Systems) software is applied.

2.2. Modelling Elements

In its mathematical formulation, an ecological model has five components:

1. **Forcing functions or external variables:** Functions or variables of an external nature that influence the state of the ecosystem. In a management context, the problem to be solved can often be reformulated. If certain forcing functions are varied, then how will this influence the state of the ecosystem? The model is used to predict what will change in the ecosystem when forcing functions are varied with time. The forcing functions, due to the human impact on ecosystems, are called *control functions*, because it is in our hands to change them. The control function in ecotoxicological models is, for instance, the discharge of toxic substances to the ecosystems; in eutrophication models it is discharge of nutrients. Other forcing functions of interest could be climatic and natural external variables, which influence the biotic and abiotic components and the process rates. In contrast to the control functions, they are not controllable by humans. By using models we will be able to address the crucial question: Which changes in the control functions are needed to obtain well-defined conditions for a considered ecosystem?

2. **State variables:** Describe, as the name indicates, the state or the conditions of the ecosystem. The selection of state variables is crucial to the model structure, but often the choice is obvious. If, for instance, we want to model the bioaccumulation of a toxic substance, then the state variables should be the organisms in the most important food chains and concentrations of the toxic substance in the organisms. In eutrophication models, the state variables are the concentrations of nutrients and phytoplankton. When the model is used in a management context, the values of the state variables simulated by changing the controllable forcing functions provide model results that contain the direct and indirect relations between the forcing functions and the state variables.

3. **Mathematical equations:** Used to represent the biological, chemical, and physical processes. They describe the relationship between the forcing functions and state variables. The same type of process may be found in many different environmental contexts, which implies that the same equations can be used in different models. However, this does not imply that the same process is always formulated using the same equation. First, the considered process may be better described by another equation because of the influence of other factors. Second, the number of details needed or desired to be included in the model may be different from case-to-case due to a difference in complexity of the system and/or the problem. Some modellers refer to the description and mathematical formulation of processes as submodels. The most applied process formulations are presented by a short overview in Section 2.3.

4. **Parameters:** Coefficients in the mathematical representation of processes. They may be considered constant for a specific ecosystem or part of an ecosystem for a certain time, but they may also be a function of time or vary spatially. In causal models, the parameter will have a scientific definition and a well-defined unit, for instance, the excretion rate of cadmium from a fish — the unit could be mgCd/(24h * kg of fish). Many parameters are indicated in the literature as ranges not constants, but even that is of great value in the parameter estimation as will be discussed further in the following text. In Jørgensen et al. (2000), a comprehensive collection of parameters in environmental sciences and ecology can be found. Our limited knowledge of parameters is one of the weakest points in modelling, a point that will be touched on often throughout this book. Furthermore, the applications of parameters as constants in our models are unrealistic due to the many feedback systems in real ecosystems. The flexibility and adaptability of ecosystems is inconsistent with the application of constant parameters in the models. A new generation of models that attempts to use varying parameters according to ecological principles seems a possible solution to the problem, but further development in this direction is absolutely necessary before we can achieve an improved modelling procedure that reflects the processes in real ecosystems. This topic will be further discussed in Chapter 10.

5. **Universal constants:** Such as the gas constant and atomic weights are also used in most models.

Models can be defined as formal expressions of the essential elements of a problem in mathematical terms. The first recognition of the problem is often verbal. This may be recognized as an essential preliminary step in the modelling procedure, which will be treated in more detail in the next section. The verbal model is, however, difficult to visualize so it is translated into a more convenient *conceptual diagram,* which contains the state variables, the forcing functions, and how these components are interrelated by mathematical formulations of processes. The conceptual diagram shows how the previous modelling elements 1 through 3 are related and connected.

Figure 2.1 illustrates a conceptual diagram of the nitrogen cycle in a lake. The state variables are nitrate, ammonium (which is toxic to fish in the un-ionized form of ammonia), nitrogen in phytoplankton, nitrogen in zooplankton, nitrogen in fish, nitrogen in sediment, and nitrogen in detritus. The state variables in this conceptual diagram are indicated as boxes connected by processes (indicated as arrows).

The forcing functions are outflows and inflows, concentrations of nitrogen components in the inflows and outflows, solar radiation, and the temperature (not shown in the diagram), which influence all of the process rates. The processes are formulated using quantitative expressions in the mathematical part of the model. Three significant steps in the modelling procedure need to be defined in this section before we go into the modelling procedure in detail. These are verification, calibration, and validation.

1. *Verification* is a test of the *internal logic* of the model. Typical questions in the verification phase include: Does the model behave as expected and intended? Is the model long-term stable, as one should expect in an ecosystem? Does the model follow the law of mass conservation, which is often used as the basis for the differential equations of the model (as discussed in the next section)? Is the use of units consistent? Verification is, to some extent, a subjective assessment of the model behavior and will continue during the model use before the calibration phase.

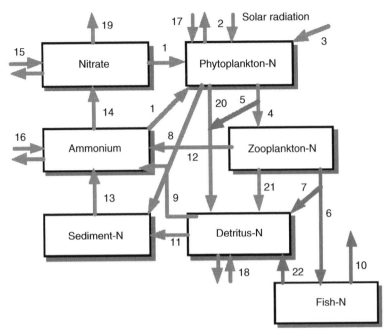

FIGURE 2.1 The conceptual diagram of a nitrogen cycle in an aquatic ecosystem. The processes are (1) uptake of nitrate and ammonium by algae; (2) photosynthesis; (3) nitrogen fixation; (4) grazing with loss of undigested matter; (5), (6), and (7) predation and loss of undigested matter; (8) settling of algae; (9) mineralization; (10) fishery; (11) settling of detritus; (12) excretion of ammonium from zooplankton; (13) release of nitrogen from the sediment; (14) nitrification; (15), (16), (17), and (18) inputs/outputs; (19) denitrification; and (20), (21), and (22) mortality of phytoplankton, zooplankton, and fish.

2. *Calibration* is an attempt to find the best agreement between the computed and observed data by variation of some selected parameters. It may be carried out by trial and error or by using software developed to find the parameters that best fit between observed and computed values. In some static and simple models, which contain only a few well-defined or directly measured parameters, calibration may not be required, but it is generally recommended to calibrate the model if observations of a proper quality and quantity are available.

3. *Validation* must be distinguished from verification. Validation consists of an objective test to show how well the model output fits the data. We distinguish between a structural (qualitative) validity and a predictive (quantitative) validity. A model is said to be structurally valid if the model structure reasonably and accurately

represents the cause-effect relationship of the real system. The model exhibits predictive validity if its predictions of the system behavior reasonably align with observations of the real system. The selection of possible objective tests will be dependent on the purposes of the model, but the standard deviations between model predictions and observations and a comparison of observed and predicted minimum or maximum values of a particularly important state variable are frequently used. If several state variables are included in the validation, then they may be given different weights.

Further detail on these three important steps in modelling will be given in the next section where the entire modelling procedure is presented as well as additional information given in Sections 2.4–2.7.

2.3. The Modelling Procedure

A tentative modelling procedure is presented in this section. The authors have successfully used this procedure numerous times and strongly recommend that all steps of the procedure are used very carefully. To make shortcuts in modelling is not recommended. Other scientists in the field have published other slightly different procedures, but detailed examination reveals that the differences are only minor. The most important steps of modelling are included in all the recommended modelling procedures.

Always, the initial focus of research is the definition of the problem. This is the only way in which the limited research resources can be correctly allocated.

The first modelling step is therefore a *definition of the problem.* This will need to be bound by the constituents of *space, time,* and *subsystems.* The bounding of the problem in space and time is usually easy, and consequently more explicit, than the identification of the subsystems to be incorporated in the model.

Systems thinking is important in this phase. You must try to grasp the big picture. The focal system behavior must be interpreted as a product of dynamic processes, preferably described by causal relationships.

Figure 2.2 shows the procedure proposed by the authors, but it is important to emphasize that this procedure is unlikely to be correct in the first attempt, so there is no need to aim for perfection in one step. The procedure should be considered as an iterative process and the main requirement is to get started (Jeffers, 1978).

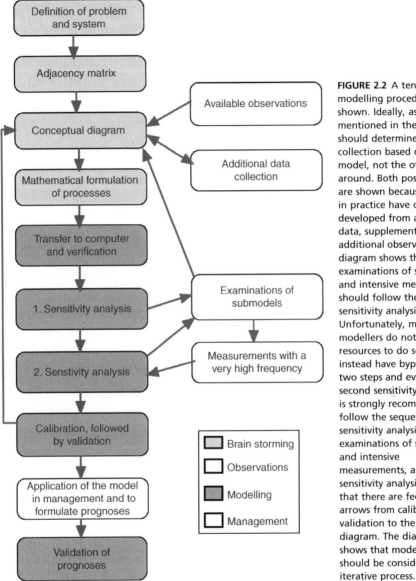

FIGURE 2.2 A tentative modelling procedure is shown. Ideally, as mentioned in the text, one should determine the data collection based on the model, not the other way around. Both possibilities are shown because models in practice have often been developed from available data, supplemented by additional observations. This diagram shows that examinations of submodels and intensive measurements should follow the first sensitivity analysis. Unfortunately, many modellers do not have the resources to do so and instead have bypassed these two steps and even the second sensitivity analysis. It is strongly recommended to follow the sequence of first sensitivity analysis, examinations of submodels and intensive measurements, and second sensitivity analysis. Notice that there are feedback arrows from calibration and validation to the conceptual diagram. The diagram shows that modelling should be considered an iterative process.

It is difficult, at least in the first instance, to determine the optimum number of subsystems to be included in the model for an acceptable level of accuracy defined by the scope of the model. Due to lack of data, it will often become necessary at a later stage to accept a lower number than intended at the start or to provide additional data for improvement of the model. It has often been argued that a more complex model should account more accurately for the behavior of a real system, but this is not necessarily true. Additional factors are involved, but a more complex model has more parameters and increases the level of uncertainty because parameters have to be estimated either by field observations, laboratory experiments, or calibrations, which again are based on field measurements. Parameter estimations are never completely without errors, and the errors are carried through into the model contributing to its uncertainty. The problem of selecting the *right model complexity* will be further discussed in Section 2.8. This is a problem of particular interest for modelling in ecology because ecosystems are very complex, but it does not imply that an ecological model to be used in research or environmental management should be very complex. It depends on the ecosystem and the problem.

A first approach to the data requirement can be made at this stage, but it is most likely to be changed later once experience with the verification, calibration, sensitivity analysis, and validation has been gained. Development of an ecological model should be considered an iterative process.

In principle, data for all the selected state variables should be available; in only a few cases would it be acceptable to omit measurements of selected state variables, as the success of the calibration and validation is closely linked to the *data quality and quantity.*

It is helpful at this stage to list the state variables and attempt to get an overview of the most relevant processes by setting up an *adjacency matrix.* The state variables are listed vertically and horizontally. A 1 is used to indicate that a direct link exists between the two state variables, while 0 indicates that there is no link between the two components. The conceptual diagram in Figure 2.1 can be used to illustrate the application of an adjacency matrix in modelling:

Adjacency matrix for the model in Figure 2.1

From	Nitrate	Ammonium	Phyt-N	Zoopl-N	Fish N	Detritus-N	Sediment-N
To							
Nitrate	–	1	0	0	0	0	0
Ammonium	0	–	0	1	0	1	1
Phyt-N	1	1	–	0	0	0	0
Zoopl-N	0	0	1	–	0	0	0
Fish N	0	0	0	1	–	0	0
Detritus-N	0	0	1	1	1	–	0
Sediment-N	0	0	1	0	0	1	–

In this example, the adjacency matrix is made from the conceptual diagram for illustrative purposes, but in practice it is recommended to set up the adjacency matrix before the conceptual diagram. The modeller should ask for each of the possible links: Is this link possible? If yes, is it sufficiently significant to be included in the model? If yes write 1, if no write 0. The adjacency matrix shown above may not be correct for all lakes. If resuspension is important, then there should be a link between sediment-N and detritus-N. If the lake is shallow, then resuspension may be significant, while the process is without any effect in deep lakes. This example clearly illustrates the idea behind the application of an adjacency matrix, which is to get the very first overview of the state variables and their interactions. The adjacency matrix can be considered as a checklist to assess which processes of all the possible linkages actually realized should be included in the model.

Once the *model complexity*, at least at the first attempt, has been selected, it is possible to *conceptualize the model*, for instance, in the form of a diagram as shown from Figure 2.1. This diagram will provide information on which state variables, forcing functions, and processes are required in the model.

Ideally, one should determine which data are needed to develop a model according to a conceptual diagram; that is, to let the conceptual model or even some first more primitive mathematical models determine the data at least within some given economic limitation. In real life, most models have been developed *after* the data collection as a compromise between model scope and available data. There are developed methods to determine the ideal data set needed for a given model to minimize the uncertainty of the model, but unfortunately the application of these methods is limited.

The conceptual diagram in Figure 2.1 indicates the state variables as boxes; for instance, nitrate, and the processes as arrows between boxes. The forcing functions are symbolized by arrows to or from a state variable like 15 and 16. It is possible to use other symbols for the modelling components.

The STELLA software will be used to illustrate the development of models throughout this book. It uses boxes for state variables (compartments), thick arrows with a symbol of a valve for the processes (connections), thick arrows coming or going to a cloud for the forcing functions (which require a constant, an equation, a table, or a graph), and a thin arrow to indicate the transfer of information or variables (controls such as forcing function, parameter, and/or a state variable calculated by an algebraic expression from another state variable and so on). See Figure 2.3.

There are other symbolic languages for development of conceptual diagrams, for instance, Odum's energy circuit language. It has more symbols than STELLA, so it is more informative but also more time-consuming to develop. For an overview of the most used symbolic languages including Odum's energy circuit language, see Jørgensen and Bendoricchio (2001).

For each state variable, a differential equation is constructed: accumulation = inputs − outputs. For detritus-N in Figure 2.1, the inputs are the processes 20 + 5 +21 + 7 + 22 +18 (in) and the outputs are the processes 11 + 9 + 18 (out). The differential equations are solved analytically in mathematics, but it is rarely possible with most ecological models because they are too complex. The differential equations are therefore solved

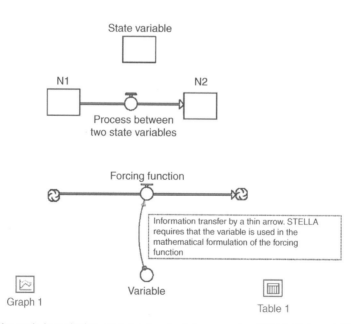

FIGURE 2.3 The symbols applied to erect a conceptual diagram using STELLA. State variables are boxes for which differential equations are erected as accumulation = inputs − outputs. Processes are thick arrows with the valve symbol. Forcing functions are thick arrows starting or ending as a cloud. Circles are variables in general. Graph 1 and Table 1 indicate that the results can be presented as graphs or as tables.

numerically within the computer software. A time step is selected for the model calculations. The shorter the time step, the closer the computer calculations come to the real-time variations of inputs and outputs, but the shorter the time step, the longer the simulation takes to run. It is recommended to test different time steps and use the longest time step that does not significantly change the model results by decreasing the time step further. The term "significant changes" is evaluated relative to the accuracy of the observations used as basis for the development of the model.

The STELLA software develops the differential equations directly from the conceptual diagram, which is input on the main user interface. The time derivative of the state variables will be equal to all the inputs = all process arrows going into the state variables minus all outputs = all process arrows going out from the state variables. The processes must, however, be formulated as an algebraic equation.

The next step is formulating the processes as *mathematical equations*. Many processes may be described by more than one equation, and it may be of great importance for the results of the final model that the right one is selected for the case under consideration. The ecological literature contains mathematical formulations of most ecological processes, but a short overview of the most applied mathematical equations is presented here. More than 95% of all ecologically relevant processes can be formulated mathematically by one of the following six equations:

1. A constant flow rate, also denoted zero order expression:

$$\frac{dC}{dt} = k_1 \tag{2.1}$$

2. A first-order rate expression, where the rate is proportional to a variable such as a concentration of a state variable:

$$\frac{dC}{dt} = k_1 C \tag{2.2a}$$

This expression corresponds to exponential growth and the following solution can be obtained by integration:

$$C(t) = C_0 e^{k_1 t} \tag{2.2b}$$

This is often used to for modelling population growth (see Chapter 5). Decomposition processes and radioactive decay can also be approximated as first order reactions, in which the rate is negative.

3. A second-order rate expression occurs when the rate is proportional to two state variables simultaneously, for instance:

$$\frac{dC_1}{dt} = k_2 C_1 C_2 \qquad (2.3)$$

4. This is a first-order rate expression with a regulation due to environmental constraints, for instance, space or resources.

$$\frac{dC}{dt} = k_4 C \left(1 - \frac{C}{K}\right) \qquad (2.4)$$

where K is the carrying capacity. When the concentration reaches the carrying capacity the factor becomes zero and the growth stops. This process rate expression is denoted logistic growth and it is illustrated in more detail in Chapter 5. These two growth expressions are both extensively applied in population dynamic models.

5. A Michaelis-Menten expression or Monod kinetics known from enzymatic processes in biochemistry is given by:

$$\frac{dC}{dt} = \frac{k_3 C}{(C + k_m)} \qquad (2.5)$$

Where k_3 is the maximum reaction rate and k_m is the Michaelis constant. At small concentrations of the substrate, this process rate is proportional to the substrate concentration, while the process rate is at maximum and constant at high substrate concentrations where the enzymes are fully utilized. The same expression is used when the growth rate of plants is determined by a limiting nutrient according to Liebig's minimum law. The Michaelis-Menten's constant, k_m, or the half saturation constant, corresponds to the concentration that gives half the maximum rate. At small concentrations of substrate or nutrients, the rate is very close to a first-order rate expression, whereas it is close to a zero order rate expression at high concentrations. Notice, that the rate is regulated from a first-order to a zero order expression more and more as the concentration increases.

6. A rate governed by diffusion often uses a concentration gradient to determine the rate as it is expressed in Fick's First Law:

$$\frac{dC}{dt} = k_5 \frac{dC}{dx} \qquad (2.6)$$

There are several modifications of these six expressions. For instance, a threshold concentration tr, is often used in the Michaelis-Menten expression. The concentration (state variable) is replaced by the concentration –tr. The concentration therefore has to exceed tr to generate any rate. For grazing and predation processes, the Michaelis-Menten's expression is often multiplied by (1 – concentration/carrying capacity) similar to what is used in the logistic growth expression. It implies that when food is abundant (concentration is high) another factor determines and limits the growth such as space or nesting area. These modifications will be used in Chapter 7 for development of a eutrophication model.

Once the system of mathematical equations is available, model *verification* can be carried out. As pointed out in Section 2.2, this is an important step, which unfortunately is omitted by some modellers. The next section presents the details of this modelling step.

2.4. Verification

The next step of the modelling procedure includes verification, which is a test of the internal model logic. Crucial questions about the model are asked and answered by the modeller. Verification is to some extent a subjective assessment of the behavior of the model.

Findeisen et al. (1978) gave the following definition of verification: "A model is said to be verified if it behaves in the way the model builder wanted it to behave." This definition implies that there is a model to be verified, which means that not only the model equations have been set up, but also that the parameters have been given reasonable realistic values. Consequently, the sequence verification, sensitivity analysis, and calibration must not be considered a rigid step-by-step procedure, but rather as an iterative operation, which must be repeated a few times. The model is first given realistic parameters from the literature, then it is calibrated coarsely, and finally the model can be verified followed by a sensitivity analysis and a finer calibration. The model builder will have to go through this procedure several times before the verification and the model output in the calibration phase will be satisfactory.

It is recommended at this step that answers to the following questions are provided:

1. **Is the model stable in the long term?** The model is run for a long period with the same annual variations in the forcing functions to observe whether the state variable values remain at approximately the same levels. During the first period, state variables are dependent on the initial values for these, and it is recommended that the model is also run with initial values corresponding to the long-term values of the state variables. The procedure also can be recommended for finding the initial values if they are not measured or known by other means. This question presumes that real ecosystems are long-term stable, which is not necessarily the case.The model is run for a long period using a certain pattern in the fluctuations of the forcing functions. It should then be expected that the state variables, too, show a certain pattern in their fluctuations. The simulation period should be long enough to allow the model to demonstrate any possible instability.

2. **Does the model react as expected?** For example, if the input of toxic substances is increased, then we should expect a higher concentration of the toxic substance in the top carnivores. If this is not so, then it shows that some formulations may be wrong and these should be corrected. This question assumes that we actually know at least some behavior of the ecosystem, which is not always the case. In general, playing with the model is recommended at this phase. Through such exercises the modeller gets acquainted with the model and its reactions to perturbations. Models should generally be considered an experimental tool. The experiments are carried out to compare model results with observations, and changes of the model are made according to the modeller's intuition and knowledge of the model's behavior. If the modeller is satisfied with the accordance between model and observations, then the model is accepted as a useful description of the real ecosystem — at least within the framework of the observations. This part of the verification is based upon more subjective criteria. Typically, the model builder formulates several questions about the model behavior and tests the model response by provoking changes in forcing functions or initial conditions. If the responses are not as expected, then the model structure or equations will have to be changed, provided that the parameter space is approved. Examples

of typical questions will illustrate this operation: Will increased BOD_5-loading in a stream model imply decreased oxygen concentration? Will increased temperature in the same model imply decreased oxygen concentration? Will the oxygen concentration be at a minimum at sunrise when photosynthesis is included in the model? Will decreased predator concentration in a prey-predator model imply increased prey concentration? Will increased nutrient loadings in a eutrophication model give increased concentration of phytoplankton? Numerous other questions can be asked.

3. **It is also recommended to check all the units at this phase of model development.** Check all equations for consistency of units. Are the units the same on both sides of the equation sign? Are the parameters used in the model consistent for the type of equations used and do the units match with the available data?

4. **Investigate the statistical properties of the noise in the model.** To conform to the properties of white noise, any error sequence should broadly satisfy the following constraints: that its mean value is zero, that it is not correlated with any other error sequence, and that it is not correlated with the sequences of measured input forcing functions. Evaluation of the error sequences in this fashion can therefore essentially provide a check on whether the final model invalidates some of the assumptions inherent in the model. If the error sequences do not conform to their desired properties, then this suggests that the model does not adequately characterize all of the more deterministic features of the observed dynamic behavior. Consequently, the model structure should be modified to accommodate additional relationships. To summarize this part of the verification the errors:

   1. (Comparison model output/observations) must have mean values of approximately zero
   2. Are not mutually cross-related
   3. Are not correlated with the measured input forcing functions

Results of this kind of analysis are illustrated in detail in Beck (1987). Notice that this analysis requires good estimates of standard deviations in sampling and analysis (observations).

Notice finally that during verification it is possible to perform multiple scenario analyses or "Gedanken Experiments." For example, we can test a eutrophication model by its response to the following test. We rent a helicopter and buy 100,000 kg of phosphorus fertilizer and drop it instantly to the lake. The experiment could be made at no cost using the model, while it would be very expensive to rent a helicopter and buy 100,000 kg of fertilizer. A major advantage of models is how easy it is to assess the system behavior under a wide array of scenarios.

Model verification may seem very cumbersome, but it is a very necessary step for the model development process. Through the verification one learns the model through its behavior, and the verification becomes an important checkpoint in the construction of a workable model. This also emphasizes the importance of good ecological knowledge of the ecosystem without which the right questions as to the internal logic of the model cannot be posed.

Unfortunately, many models have not been verified properly due to lack of time, but the experience shows that what might seem to be a shortcut will lead to an unreliable model, which at a later stage might require more time to compensate for the lack of verification. It must therefore be strongly recommended to invest enough time in the verification and to plan for the necessary allocation of resources in this important phase of the modelling procedure.

2.5. Sensitivity Analysis

Sensitivity analysis follows verification. Through this analysis the modeller gets a good overview of the most *sensitive components of the model.* Thus, sensitivity analysis attempts to provide a measure of the sensitivity of parameters, forcing functions, or submodels to the state variables of greatest interest in the model. If a modeller wants to simulate a toxic substance concentration in carnivorous insects as a result of the use of insecticides, then one will choose this state variable as the most important one for a sensitivity analysis along with the concentration of the toxic substance concentration in plants and herbivorous insects.

In practical modelling, the sensitivity analysis is carried out by changing the parameters, the forcing functions, or the submodels.

The corresponding response on the selected state variables is observed. Thus, the sensitivity, S, of a parameter, P, is defined as follows:

$$S = [\partial x/x]/[\partial P/P] \tag{2.7}$$

where x is the state variable under consideration.

The relative change in the parameter value is chosen based on our knowledge of the certainty of the parameters. If the modeller estimates the uncertainty to be about 50%, then a change in the parameters at $\pm 10\%$ and $\pm 50\%$ is chosen and the corresponding change in the state variable(s) recorded. It is often necessary to find the sensitivity at two or more levels of parameter changes as the relationship between a parameter and a state variable is rarely linear.

A sensitivity analysis makes it possible to distinguish between high-leverage variables, whose values have a significant impact on the system behavior and low-leverage variables, whose values have minimal impact on the system. Obviously, the modeller must concentrate the effort on improvements of the parameters and the submodels associated with the high-leverage variables. The result of a sensitivity analysis of a eutrophication model with 18 state variables, presented in Chapter 7, is shown in Table 2.1. The sensitivity of the examined parameters by a 10% increase to phytoplankton, s-phyt; to zooplankton, s-zoo; to soluble nitrogen, s-nit; and to soluble phosphorus, s-phos, is shown. These results clearly indicate that the parameters "maximum growth rate of phytoplankton and zooplankton," "mortality of zooplankton," and the "settling rate of

Table 2.1   Results of a $\pm 10\%$ Sensitivity Analysis of the 18 State Variable Model in Chapter 7

Parameter	s-phyt	s-zoo	s-nit	s-phos
Maximum growth rate of phytoplankton	0.488	0.620	-0.356	-0.392
Maximum growth rate of zooplankton	-2.088	-4.002	2.749	4.052
Denitrification rate	-0.19	-0.010	-0.579	0.013
Fish concentration	0.008	0.012	-0.011	-0.014
Rate of mineralization	0.003	0.010	0.038	0.001
Mortality zooplankton	2.063	1.949	-3.479	-3.350
Settling rate	-1.042	-0.0823	0.321	0.388

phytoplankton," are very important parameters to determine accurately because they all have a sensitivity to the most important state variable, the phytoplankton, which is more than 0.5 or 50%, meaning that a change of the parameters by 10% would make a change *of the phytoplankton concentration of more than 50%.* On the other hand, the parameters "maximum denitrification rate," the "mortality of fish," and the "rate of mineralization" are significantly less important parameters. They all have a sensitivity of less than 0.1 or 10%. Therefore, they would change the phytoplankton less than 1% if the parameters are changed 10%.

The interaction between the sensitivity analysis and the calibration could consequently work along the following lines:

1. A sensitivity analysis is carried out at two or more levels of parameter changes. Relatively large changes are applied at this stage.
2. The most sensitive parameters are determined more accurately either by a calibration or by other means (see Section 2.9).
3. Under all circumstances, great efforts are made to obtain a relatively well calibrated model.
4. A second sensitivity analysis is then carried out using more narrow intervals for the parameter changes.
5. Still further improvements of the parameter certainty are attempted.
6. A second or third calibration is then carried out focusing mainly on the most sensitive parameters.

A sensitivity analysis on submodels (process equations) can also be carried out. Then the change in a state variable is recorded when the equation of a submodel is deleted from the model or changed to an alternative expression, for instance, with more details built into the submodel. Such results may be used to make structural changes in the model. For example, if the sensitivity shows that it is crucial for the model results to use a more detailed submodel, then this result should be used to change the model correspondingly.

If it is found that the state variable in focus is very sensitive to a certain submodel, then it should be considered which alternative submodels could be used and they should be tested and/or examined in further detail either in vitro or in the laboratory.

It can generally be stated that those submodels, which contain sensitive parameters, are also submodels that are sensitive to the important state variable. On the other hand, it is not necessary to have a sensitive parameter included in a submodel to obtain a sensitive submodel. A modeller with a certain experience will find that these statements agree with intuition, but it is also possible to show that they are correct by analytical methods.

A sensitivity analysis of forcing functions gives an impression of the importance of the various forcing functions and tells us what accuracy is required of the forcing functions.

2.6. Calibration

The goal of *calibration* is to improve the parameter estimation. Some parameters in causal ecological models can be found in the literature, not necessarily as constants but as approximate values or intervals. To cover all possible parameters for all possible ecological models including ecotoxicological models, we need to know more than one billion parameters. Therefore, in modelling there is a particular need for *parameter estimation methods*. This will be discussed later in this chapter and further in Chapter 8, where methods to estimate ecotoxicological parameters based upon the chemical structure of the toxic compound are presented. In all circumstances, it is a great advantage to give even approximate values of the parameters before the calibration gets started as previously mentioned. It is, of course, much easier to search for a value between 1 and 10 than to search between 0 and $+\infty$.

Even where all parameters are known within intervals either from the literature or from estimation methods, it is usually necessary to calibrate the model. Several sets of parameters are tested by the calibration and the various model outputs of state variables are compared with measured values of the same state variables. The parameter set that gives the best agreement between model output and measured values is chosen.

The need for the calibration can be explained by using the following characteristics of ecological models and their parameters:

1. Most parameters in environmental science and ecology are not known as exact values. Therefore, all literature values for parameters (Jørgensen et al., 1991, 2000). Parameter estimation methods must be used when no literature value can be found, particularly ecotoxicological models. See, Jørgensen (1991, 1992a) and Chapter 8. In addition, we must accept that unlike many physical parameters, ecological ones are not constant but change in time or situation (Jørgensen, 1986, 1992b, 2002). This point will be discussed further in Chapter 10.

2. All models in ecology and environmental sciences are simplifications of nature. The most important components and processes may be included, but the model structure does not account for every detail. To a certain extent the influence of some unimportant components and processes can be taken into account by the calibration. This will give slightly different values for the parameters from the real, but unknown, values in nature, but the difference may partly account for the influence from the omitted details.

3. Most models in environmental sciences and ecology are "lumped models," which means that one parameter represents the average values of several species. As each species has its own characteristic parameter value, the variation in the species composition with time will inevitably give a corresponding variation in the average parameter used in the model. Adaptation and shifts in species composition will require other approaches. This will be discussed in more detail in Chapter 10.

A calibration cannot be carried out randomly if more than a couple of parameters have been selected for calibration. If, for instance, 10 parameters have to be calibrated and the uncertainties justify the testing of 10 values for each parameter, the model has to be run 10^{10} times, which is an impossible task. Therefore, the modeller must learn the behavior of the model by varying one or two parameters at a time and observing the response of the most crucial state variables. In some (few) cases it is possible to separate the model into several submodels, which can be calibrated approximately independently. Although the calibration described is based to some extent on a systematic approach, it is still a trial-and-error procedure.

However, procedures for automatic calibration are available. This does not mean that the trial-and-error calibration described earlier is redundant. If the automatic calibration should give satisfactory results within a certain frame of time, then it is necessary to calibrate only 6–9 parameters simultaneously. In any circumstances, the narrower the ranges of the parameters before the calibration gets started, the easier it is to find the optimum parameter set.

In the trial-and-error calibration, the modeller has to set up, somewhat intuitively, some calibration criteria. For instance, you may want to simulate accurately the minimum oxygen concentration for a stream model and/or the time at which the minimum occurs. When you are satisfied with these model results, you may then want to simulate the shape of the oxygen concentration versus time curve properly, and so on. The model must be calibrated step-by-step to achieve these objectives step-by-step.

If an automatic calibration procedure is applied, then it is necessary to formulate objective criteria for the calibration. A possible function could be based on an equation similar to the calculation of the standard deviation:

$$Y = [(\Sigma((X_c - X_m)^2 / X_{m,a})/n]^{1/2} \tag{2.8}$$

where X_c is the computed value of a state variable, X_m is the corresponding measured value, $X_{m,a}$ is the average measured value of a state variable, and n is the number of measured or computed values. Y is computed during an automatic calibration with the goal to obtain the lowest Y value possible.

Often, the modeller is more interested in a good agreement between model output and observations for one or two state variables and less interested in a good agreement with other state variables. Therefore, weights are chosen for the various state variables to account for the emphasis put on each state variable in the model. For a model of the fate and effect of an insecticide, emphasis may be put on the toxic substance concentration of the carnivorous insects while considering the toxic substance concentrations in plants, herbivorous insects, and soil to be of less importance. Therefore, a weight of ten is applied for the first state variable and only one for the subsequent three.

If it is impossible to calibrate a model properly, then it is not necessarily due to an incorrect model. Instead, it may be due to the poor data quality, which is crucial for calibration. It is also of great importance that the *observations reflect the system dynamics*. If the objective of the model is to give a good description of one or a few state variables, then it is essential that the data show the dynamics of just these internal variables. The frequency of the data collection should therefore reflect the dynamics of the state variables in focus. This rule has unfortunately often been violated in modelling.

It is strongly recommended that the dynamics of all state variables are considered before the data collection program is determined in detail. Frequently, some state variables have particularly pronounced dynamics in specific periods — often in spring — and it may be of great advantage to have a dense data collection in this period in particular. Jørgensen et al. (1981) showed how a dense data collection program in a certain period can be applied to provide additional certainty for the determination of some important parameters. This question will be further discussed in Section 2.9.

From these considerations, recommendations can now be drawn about the feasibility of carrying out a calibration of a model in ecology:

1. Find as many parameters as possible from the literature (see Jørgensen et al., 1991, 2000). Even a *wide* range for the parameters should be considered very valuable, as approximate initial guesses for all parameters are urgently needed.
2. If some parameters cannot be found in the literature, which is often the case, then the *estimation methods* mentioned later in this Section 2.9 and in Chapter 8 may be used. For some crucial parameters it may be recommended to determine them by experiments *in situ* or in the laboratory.
3. A *sensitivity analysis* should be carried out to determine which parameters are most important to be known with high certainty. The estimation methods and the determination of the parameters by experiments should focus mainly on the most sensitive parameters.
4. An *intensive data collection program* for the most important state variables should be used to provide a better estimation for the most crucial parameters. For further details see Section 2.9.
5. First, at this stage, the *calibration* should be carried out using the data not yet applied. The most important parameters are selected and the calibration is limited to these, or, at the most, to eight to ten parameters. In the first instance, the calibration is carried out by using the trial-and-error method to get acquainted with the model

reaction to changes in the parameters. An automatic calibration procedure is used subsequently to polish the parameter estimation.

6. These results are used in a *second sensitivity analysis*, which may give results different from the first sensitivity analysis.

7. A *second calibration* is now used on the parameters that are most important according to the second sensitivity analysis. In this case, too, both the previous calibration methods may be used. In some cases, the modeller would repeat steps 6 and 7 one time more and make a third calibration. After this final calibration the model can be considered calibrated and we can go to the next step — validation.

2.7. Validation and Assessment of the Model Uncertainty

The calibration should *always* be followed by a *validation*. During this step the modeller tests the model against an *independent* data set to observe how well the model simulations fit these data. It may be possible, even in a data-rich situation, to force a wrong model by the parameter selection to give outputs that fit well with the data. It must, however, be emphasized that the validation only confirms the model behavior under the range of conditions represented by the available data. So, it is preferable to validate the model using data obtained from a period in which conditions other than those of the period of data collection for the calibration prevail. For instance, when a eutrophication model is tested, it should preferably have data sets for the calibration and the validation that differ by the level of eutrophication. This is often impossible or at least very difficult as it may correspond to a complete validation of the model predictions, which at best takes place at a later stage of the model development. However, it may be possible and useful to obtain data from a *certain* range of nutrient loadings, for instance, from a humid and a dry summer. Alternatively, it may be possible to get data from a similar ecosystem with approximately the same morphology, geology, and water chemistry as the modelled ecosystem. Similarly, a BOD/DO model should be validated under a wide range of BOD-loadings, a toxic substance model under a wide range of concentrations of the considered toxic substances, and a population model by different levels of the populations, and so forth.

If an ideal validation cannot be obtained, then it is still important to validate the model as best as possible. The method of validation is dependent on the model objectives. A comparison between measured and computed data using an objective function Eq. (2) is an obvious test. This is, however, often not sufficient, as it may not focus on *all* the main objectives of the model, but only on the general ability of the model to describe correctly the state variables of the ecosystem. It is necessary, therefore, to translate the main objectives of the model into a few validation criteria. They cannot be formulated generally, but are individual for the model and the modeller. For instance, if we are concerned with the eutrophication in an aquatic ecosystem, it would be useful to compare the measured and computed maximum concentrations of phytoplankton. The validation discussion can be summarized by the following issues:

1. Validation is always required to get a picture of the model reliability.
2. Attempts should be made to get data for the validation that are entirely different from those used in the calibration. It is important to have data from a wide range of forcing functions that are defined by the model objectives.
3. The validation criteria are formulated based on the model objectives and the quality of the available data. The main purpose of the model may, however, be an exploratory analysis to understand how the system responds to the dominating forcing functions. In this case, a structural validation is probably sufficient.

Validation is a very important modelling step because it gives the uncertainty of the model results. It attempts to answer the question: Which model uncertainty should we consider when using the model to develop strategies for environmental management? If we use the model as research tool, then the validation will tell us whether the model results can be used to support or reject a hypothesis. The uncertainty determined by the validation relative to the difference between the hypothesis and the model results will be decisive. In Chapter 7, a eutrophication model with 18 state variables will be applied as a case study to demonstrate how the validation results can be used to assess the expected uncertainty of the prognoses developed by the model.

The validation result can also be used to consider the model revisions that would be needed to reduce the uncertainty. In our effort to improve the model, we should ask the following pertinent questions:

1. What is the uncertainty of the observations (measurements)? If the uncertainty of the model is not very different from the uncertainty of the observations, then it will probably be beneficial to get more reliable observations with less uncertainty.

2. Do the observations represent the system dynamics? If not, then more frequent monitoring should be considered for some period to capture the system dynamics. See the discussion of this question in Section 2.9.

3. Are some important processes or components missing or described wrongly in the model? In this context, as previously mentioned, it is important to set up a mass and/or energy balance to reveal the most important processes and sources.

It is recommended to give a sufficiently comprehensive answer to question 3 and eventually use the model experimentally to find the best answer. It is quite easy in most cases to replace important equations by other expressions or add new components or processes and so on. Such experiments are very elucidating for the importance of formulations and inclusion of processes. Small changes in process equations that make big changes in the model results uncover the soft points of the model and may inspire additional experiments or observations in situ or in the laboratory, and eventually to further changes of the model.

It should be emphasized that the "ideal" model can never be achieved, but step-by-step by steadily questioning the model and using these three points again and again, we can improve the model quality moving asymptotically toward the ideal model. An ideal model is, however, not necessary to have a useful and powerful tool in environmental management and ecosystem research. A satisfactory calibration and validation with sufficiently low uncertainties to allow application in a defined context would be the general requirement for the pragmatic modeller.

2.8. Model Classes

It is useful to distinguish between various model classes and briefly discuss the selection of model classes.

Pairs of models are shown in Table 2.2. The first division of models is based on the application *scientific and management models.* This initial

Table 2.2 Classification of Models (Pairs of Model Types)

Type of Models	Characterization
Research models	Used as a research tool
Management models	Used as a management tool
Deterministic models	The predicted values are computed exactly
Stochastic models	The predicted values depend on probability distribution
Compartment models	The variables defining the system are quantified by means of time-dependent differential equations
Matrix models	Uses matrices in the mathematical formulation
Reductionistic models	Include as many relevant details as possible
Holistic models	Uses general principles
Static models	The variables defining the system are not dependent on time
Dynamic models	The variables defining the system are a function of time (or perhaps of space)
Distributed models	The parameters are considered functions of time and space
Lumped models	The parameters are within certain prescribed spatial locations and time, considered as constants
Linear models	First-degree equations are used consecutively
Nonlinear models	One or more of the equations are not first degree
Causal models	The inputs, states, and the outputs are interrelated by using causal relations
Black-box models	The input disturbances effect only the output responses, no causality is required
Autonomous models	The derivatives are not explicitly dependent on the independent variable (time)
Non-autonomous models	The derivatives are explicitly dependent on the independent variable (time)

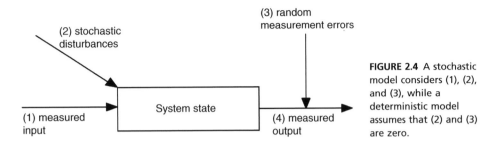

FIGURE 2.4 A stochastic model considers (1), (2), and (3), while a deterministic model assumes that (2) and (3) are zero.

distinction guides the objectives of the model development toward either research or application orientation and influences the choice of which processes and state variables to emphasize.

The next pair is *stochastic and deterministic models*. A stochastic model contains stochastic input disturbances and random measurement errors, as shown in Figure 2.4. If they are both assumed to be zero, then the stochastic model will reduce to a deterministic model provided the parameters are not estimated in terms of statistical distributions. A deterministic model assumes that the future response of the system is completely determined by knowledge of the present state and future measured inputs. Stochastic models are not frequently applied in ecology.

The third pair in Table 2.2 is *compartment and matrix models*. Some modellers refer to compartment models as models based on the use of compartments in the conceptual diagram, while other modellers distinguish between the two model classes entirely by the mathematical formulation as indicated in Table 2.2. Both model types are applied in ecological modelling, although the use of compartment models is far more pronounced.

The classification of *reductionistic and holistic models* is based upon a difference in the scientific ideas behind the model. The reductionistic modeller will attempt to incorporate as many details of the system as possible to capture its behavior, believing that the properties of the system are the sum of the details. A holistic modeller will abstract some detail to capture broader scale patterns. The bridge between these bottom-up and top-down approaches is spanned by the use of hierarchical models that include lower level micro-scale interactions constrained by higher level macro-scale processes.

Most problems in environmental sciences and ecology may be described by dynamic models, which use differential or difference equations to describe the system response to external factors. Differential equations are used to represent continuous changes of state with time, while difference equations use discrete time steps. The steady state corresponds to the situation when all derivatives equal zero. The oscillations around the steady state are described by use of a dynamic model, while the steady state can be described by use of a static model (see Figure 2.5), which can be reduced to algebraic equations.

Some dynamic systems have no steady state; for instance, systems that show limit cycles. This situation obviously requires a dynamic model to describe the system behavior. In this case, the system is always nonlinear, although there are nonlinear systems that have steady states.

A static model assumes, consequently, that all variables and parameters are time independent. The advantage of the static model is its potential for simplifying subsequent computational effort through the elimination of one of the independent variables in the model relationship, but static models may give unrealistic results because oscillations caused by seasonal and diurnal variations may be utilized by the state variables to obtain higher average values.

A distributed model accounts for variations of variables in time and space. A typical example would be an advection-diffusion model for transport of a dissolved substance along a stream. It might include

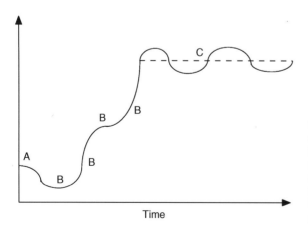

Time

FIGURE 2.5 Y is a state variable expressed as a function of time. A is the initial state and B the transient states. C oscillates around a steady state. The dotted line corresponds to the steady state that can be described by a static model. The transient state requires the use of a dynamic model.

variations in the three orthogonal directions. The analyst might decide, based on prior observations, that gradients of dissolved material along one or two directions are not sufficiently large to merit inclusion in the model. The model would then be reduced by that assumption to a lumped parameter model. Whereas the lumped model is frequently based upon ordinary differential equations, the distributed model is usually defined by partial differential equations.

The causal, or internally descriptive, model characterizes the manner in which inputs are connected to states and how the states are connected to each other and to the outputs of the system, whereas the black-box model reflects only what changes in the input will affect the output response. In other words, the causal model describes the internal mechanisms of process behavior. The black-box model deals only with what is measurable at the boundary: the input and the output. The relationship may be found by a statistical analysis. If, on the other hand, the processes are described by model equations that represent the relationships, then the model will be causal.

The modeller may prefer to use black-box descriptions in the cases where knowledge about the processes is limited. The disadvantage of the black-box model is that it has limited application to the ecosystem under consideration or at least to a similar ecosystem, and that it cannot consider changes of the system.

If general applicability is needed, then it is necessary to set up a causal model. The latter type is more widely used in environmental sciences than the black-box model, mainly because the causal model gives the user deeper understanding about the function of the system, including the many chemical, physical, and biological reactions.

Autonomous models are not explicitly dependent on time (the independent variable):

$$dy/dt = a^*y^b + c^*y^d + e \tag{2.9}$$

Non-autonomous models contain terms, $g(t)$, that make the derivatives dependent on time, exemplified by the following equation:

$$dy/dt = a^*y^b + c^*y^d + e + g(t) \tag{2.10}$$

The pairs in Table 2.2 may be used to define the type of model that is most applicable to solve a given problem. It will be further discussed

Table 2.3 Model Identification

Model Types	Organization	Pattern	Measurements
Biodemographic	Conservation of genetic information	Life cycles of species	Number of species or individual
Bioenergetic	Conservation of energy	Energy flow	Energy
Biogeochemical	Conservation of mass	Element cycles	Mass of concentrations

in the next section, where a practical model classification will also be presented.

Table 2.3 shows another way to classify models. The differences among the three model types are the choice of components used as state variables. If the model describes a number of individuals, species, or classes of species, then it is called *biodemographic*. A model that describes the energy flows is *bioenergetic* and the state variables will typically be expressed in kJ or kJ per unit of volume or area. *Biogeochemical models* consider the flow of material and the state variables are indicated as kg or kg per unit of volume or area. This model type is mainly used in ecology.

The problem, the ecosystem characteristics, and the available database should be reflected in the choice of model class. The two model classifications presented earlier are useful for defining the modelling problem. Is the problem related to a description of populations, energy flows, or mass flows? The answer determines whether we should develop a biodemographic, bioenergetic, or biogeochemical model. Biodemographic models that include a description of age structure can be elegantly developed by a matrix model, provided that first-order processes can be assumed. This will be demonstrated in Chapter 5, Section 5.4.

If the model is developed on the basis of a database that has limited quality and/or quantity, then the model should have relatively low complexity. A dynamic model is generally more demanding to calibrate and validate than a static model. Therefore, the latter type would often be selected in a data-poor situation, provided that a description of the steady state is sufficient to solve the problem. Steady-state descriptions imply that an equation input = output for each state variable can be applied to find (estimate) one (otherwise unknown) parameter. Chapter 6 shows how a steady-state model can be developed and used to get a

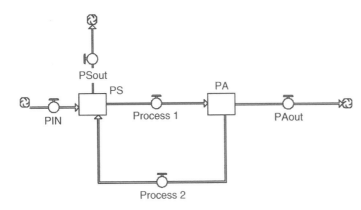

FIGURE 2.6 A conceptual diagram of a simple model with two state variables, PS and PA, is shown. PIN is a forcing function. (1) and (2) are processes.

good overview of an ecological situation, even in a relatively data-poor situation.

Dynamic models are able to make predictions about the variations of state variables in time and/or space. Differential equations are used to express the variation. With reference to Figure 2.6, the following differential equations are valid:

$$dPS/dt = PIN + Process(2) - Process(1) - PS * Q/V \qquad (2.11)$$

$$dPA/dt = Process(1) - PA * Q/V \qquad (2.12)$$

where PIN represents the input (a forcing function), Q the flow rate out of the system, V the volume of the system and (1) and (2) processes that can be formulated as mathematical equations with PS and PA as variables; for instance $(1) = kPS/(0.5 + PS)$ (a Michaelis-Menten expression) and $(2) = k'*PA$, where k and k' are two parameters.

The corresponding steady-state model gives us two equations:

$$PIN + k'PA = PS(Q/V + k/(0.5 + PS)) \text{ and } PA * Q/V = kPS/(0.5 + PS)$$

that can be used to find k and k', presuming that we know the two state variables at steady state and the forcing functions.

Many population dynamic, biogeochemical, and ecotoxicological models apply differential equations because the time variations are important.

It is known that ecosystems are adaptable. Over time, species can change their properties to meet changing conditions (i.e., change of forcing functions or disturbances). If the changes are major, then there may even be a shift to other species with properties better fitted to the emerging conditions. Models that account for the change of properties of the biological components have variable parameters and are described by nonstationary, time-varying differential equations. They are often called structurally dynamic models (SDMs; see Jørgensen, 1986, 1997, 2002), because they are able to predict the changes in properties of the biological components. Chapter 10 covers this model type and its application. Structurally dynamic models are distributed models, because the parameters are considered functions of time and space. While distributed models in most cases are based on mathematical formulations of these functions when the model is developed, we will only use the term structurally dynamic models for models that can simulate change in the structure (shifts in parameter values). Structurally dynamic models are an important recent development in ecological modelling because the parameters found on the basis of the observations in the ecosystem under the present prevailing conditions cannot be valid when the conditions are changed due to adaptation. Therefore, models without dynamic structure often give unreliable results, particularly if the forcing functions are significantly changed.

In Chapter 3, an overview of the model types that are available for the development of ecological models is presented. The choice of model type for development in a particular situation depends on the different mathematical methods, different goals, and different applications and may also use different types of databases. While the model classes are characterized by a difference in one property only (e.g., steady state vs. dynamic state and mass flows vs. energy flows), the different model types are significantly different. They have been developed to solve some fundamental modelling problems in ecology during the last couple of decades, including: (1) How do we account for the individuality of organisms? (2) How do we account for adaptation and shifts in species composition? (3) What model approach is best when our data set is uncertain (i.e., fuzzy)? (4) How can we make an effective model from a very heterogeneous database? (5) How can we improve model parameter estimation? We have solved these problems by development of several different model types that have expanded the range and application of ecological models in many different directions.

2.9. Selection of Model Complexity and Structure

The literature of environmental modelling contains several methods that are applicable to the selection of model complexity. References can be given to the following papers devoted to this question: Halfon (1983, 1984, 1986), Halfon, Unbehauen, and Schmid (1979), Costanza and Sklar (1985), Bosserman (1980, 1982) and Jørgensen and Mejer (1977).

It is clear from the previous discussions in this chapter that selection of the model complexity is a matter of balance. On one hand, it is necessary to include the state variables and the processes essential for the problem in focus. On the other hand, it is important not to make the model more complex than appropriate for the available data set. As Einstein once quipped, "A scientific theory should be as simple as possible, but no simpler." The same applies to models. Our knowledge of processes and state variables together with our data set determine the selection of model complexity. If our knowledge is poor, then the model will include few details and will have a relatively high uncertainty. If we have a profound knowledge of the problem we want to model, then we can construct a more detailed model with a relatively low uncertainty. Many researchers claim that a model cannot be developed before one has a certain level of knowledge, and that it is a flaw to attempt to construct a model in a data poor situation. This is wrong because a model can always assist the researcher by synthesizing the present knowledge and by visualizing the system. But the researcher must always present the shortcomings and the uncertainties of the model and not try to pretend that the model is a complete and detailed picture of reality. A model will often be a fruitful instrument to test hypotheses in the hands of the researcher, but only if the incompleteness of the model is fully acknowledged.

It should not be forgotten in this context that models have always been applied in science. The difference between present and previous models is only that today, with modern computer technology, we are able to work with very complex models. However, it has been a temptation to construct models that are too complex — it is easy to add more equations and more state variables to the computer program, but much harder to get the data needed for calibration and validation of the model.

Even if we have very detailed knowledge about a problem, we will never be able to develop a model capable of accounting for the complete input-output behavior of a real ecosystem and valid for all frames

(Zeigler, 1976). This ideal model is named "the base model" by Zeigler, and it would be very complex and require such a great number of computational resources that it would be almost impossible to simulate. The base model of a problem in ecology will never be fully known because of the complexity of the system and the impossibility to observe all states. However, given an experimental frame of current interest, a modeller is likely to find it possible to construct a relatively simple model that is workable in that frame.

According to this discussion, a model may be made more realistic by adding more connections. Additions of new parameters up to a point do not contribute further to improve the simulation; on the contrary, more parameters imply more uncertainty because of the possible lack of information about the flows the parameters can quantify. Given a certain amount of data, the addition of new state variables or parameters beyond a certain model complexity does not add to our ability to model the ecosystem; it only adds to unaccountable uncertainty. These ideas are visualized in Figure 2.7. The relationship between knowledge gained through a model and its complexity is shown for two levels of data quality and quantity. The question under discussion can be formulated with

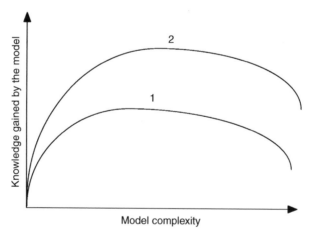

FIGURE 2.7 Knowledge plotted versus model complexity measured by the number of state variables. The knowledge increases up to a certain level. Increased complexity beyond this level will not add to the knowledge gained about the modelled system. At a certain level, the knowledge might even be decreased due to uncertainty caused by too high a number of unknown parameters. (2) corresponds to an available data set, which is more comprehensive or has a better quality than (1). Therefore the knowledge gained and the optimum complexity is higher for data set (2) than for (1). *(Reproduced from Jørgensen, 1988.)*

relation to this figure: How can we select the optimum model complexity and structure for the given understanding for the question at hand?

We will discuss in the following section the methods available to select a good model structure. If a rather complex model is developed, then the use of one of the methods presented in the previous references is recommended, but for simpler models it is often sufficient to select a model of balanced complexity, as discussed earlier.

Costanza and Sklar (1985) have examined 88 different models, and showed that more theoretical discussion behind Figure 2.7 is valid in practice. Their results are summarized in Figure 2.8, where effectiveness is plotted versus articulation (= expression for model complexity). Effectiveness is understood as a product of model results and confidence (i.e., certainty), while articulation is a measure of the complexity of the model with respect to number of components, time, and space. The measures of articulation or complexity and effectiveness are relative. Some other authors may have applied other measures, but it is

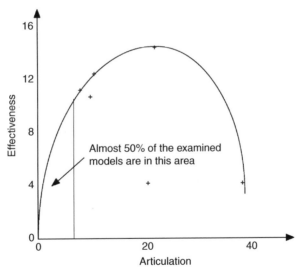

Almost 50% of the examined models are in this area

FIGURE 2.8 Plot of articulation index versus effectiveness = articulation*certainty for the 88 models reviewed by Costanza and Sklar (1985). As almost 50% of the models were not validated, they had an effectiveness of 0. These models are not included in the figure, but are represented by the line effectiveness = 0. Notice that nearly 50% of the models have a relatively low effectiveness due to too little articulation, and that only one model had an articulation that was too high, which implies that the uncertainty by drawing the effectiveness frontier as shown in the figure is high at articulations above 25. *(This figure is partly reproduced from Costanza and Sklar, 1985.)*

clearly seen by comparison of Figures 2.7 and 2.8 that they show the same type of relationship.

Selection of the right complexity is of great importance in environmental and ecological models as already stated. The methods presented and discussed in the following section provide an objective procedure to select the correct level of model complexity. However, the model selection always requires that the application of these methods is combined with a good knowledge of the system being modelled. The methods must work hand-in-hand with an intelligent answer to the question: Which components and processes are most important for the problem in focus? The conclusion is therefore: Know your system and your problem before you select your model, including the complexity of the model. It should not be forgotten that the model will always be an extreme simplification of nature. This implies that we cannot make a model of an ecosystem, but we can develop a model of some aspects of that ecosystem.

A parallel to the application of geographical maps (see Section 1.1) can be made again: We cannot make a map (model) of a state with all its details, instead we show some geographic aspects on a certain scale. Therein lays our limitations, which are due to the immense complexity of nature. We have to accept these limitations since we cannot produce a complete model or get a total picture of a natural system. Some kind of map is always more useful than no map, so some kind of model of an ecosystem is better than no model at all. As the map quality improves due to better techniques and knowledge, so does the model of an ecosystem as we gain more experience in modelling and improve our ecological knowledge. We do not need a complete set of details to get a proper overview and a holistic picture; we need some details and we need to understand how the system works on the system level.

Therefore, the conclusion is that although we can never know all of the details needed to make a complete model, we can produce good workable models that expand our knowledge of ecosystems, particularly of their properties as systems. This is completely consistent with Ulanowicz (1979) who points out that the biological world is a sloppy place. Very precise predictive models will inevitably be wrong. It would be more fruitful to build a model that indicates the general trends and take into account the probabilistic nature of the environment.

Furthermore, it seems possible in most situations to apply models as a management tool (Jørgensen et al., 1995). Models should be considered as tools — tools to overview complex systems, and tools to obtain a picture of the systems properties on the system level. Already, a few interactive state variables make it impossible to overview how the system reacts to perturbations or other changes. There are only two possibilities to get around this dilemma: Either limit the number of state variables in the model, or describe the system by use of holistic methods and models, preferably by using higher level scientific laws. See also the discussion about holistic and reductionistic approaches in Sections 2.3 and 2.5. The trade-off for the modeller is between knowing a lot about a little or a little about a lot.

Through a good knowledge of the system, it is possible to set up mass or energy flow diagrams. This might be considered a conceptual model of its own, but the idea is to use the diagram to recognize the most important flows for the model in question. Let us use an energy flow diagram for Silver Springs (Figure 2.9). If the goal of the model is to predict the net primary production for various conditions of temperature and input

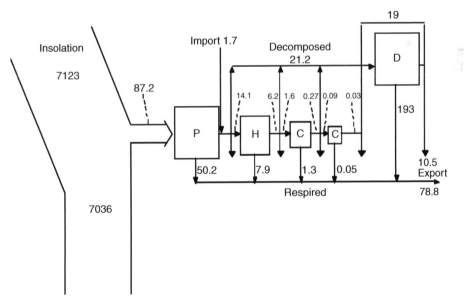

FIGURE 2.9 Energy flow diagram for Silver Springs, Florida. Figures in cal/m²/year. *(Adapted from H. T. Odum, 1957.)*

of fertilizers, then it is important to include plants, herbivores, carnivores, and decomposers (as they mineralize the organic matter). A model consisting of these four state variables might be sufficient and the top carnivores, import, and export can be excluded.

As energy flows are different from ecosystem to ecosystem, the selected model should also be different. A general model for one type of ecosystem, for example, a lake, does not exist; on the contrary, it is necessary to adopt the model to the characteristic features of the ecosystem. Figures 2.10 and 2.11 show the phosphorus flows of two eutrophication models for two different lakes: a shallow lake in Denmark and Lake Victoria in East Africa. From time to time the latter has a thermocline, which implies that the lake should be divided into at least two horizontal layers, (Jørgensen et al. 1982). The food web is also different

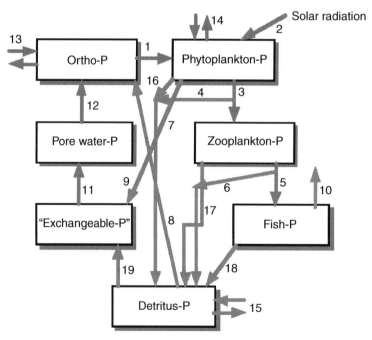

FIGURE 2.10 The phosphorus cycle in an aquatic ecosystem. The processes are (1) uptake of phosphorus by algae; (2) photosynthesis; (3) grazing with loss of undigested matter; (4), (5) predation with loss of undigested material; (6), (7), and (9) settling of phytoplankton; (8) mineralization; (10) fishery; (11) mineralization of phosphorous organic compounds in the sediment; (12) diffusion of pore water P; (13), (14), and (15) inputs/outputs; (16), (17), and (18) represent mortalities; and (19) is settling of detritus.

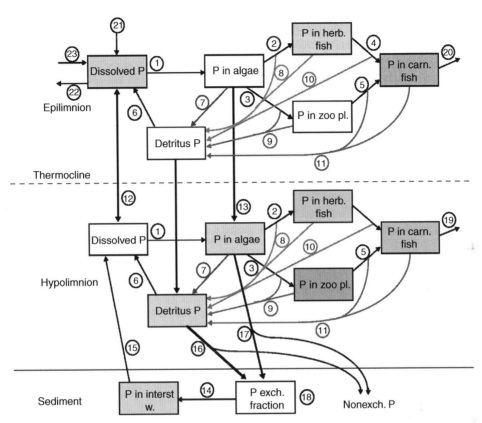

FIGURE 2.11 Eutrophication model of an aquatic ecosystem illustrated by use of P-cycling. Arrows indicate processes. A thermocline is considered. Explanation of numbers are as follows: (1) uptake of phosphorus by algae; (2) grazing by herbivorous fish; (3) grazing by zooplankton; (4), (5) predation on fish and zooplankton, respectively, by carnivorous fish; (6) mineralization; (7) mortality of algae; (8), (9), (10), (11) grazing and predation loss; (12) exchange of P between epilimnion and hypolimnion; (13) settling of algae (epilimnion-hypolimnion); (14) settling of detritus (epilimnion-hypolimnion); (15) diffusion of P from interstitial to lake water; settling of detritus (16) and algae (17) (hypolimnion-sediment, a part goes to the non-exchangeable fraction); (18) mineralization of P in exchangeable fraction; (19), (20) fishery; (21) precipitation; (22) outflows; and (23) inflows (tributaries).

in the two lakes: Lake Victoria herbivorous fish graze on phytoplankton, while in the Danish lake the grazing is entirely by zooplankton. These differences were also reflected in the models set up for the two ecosystems.

In many shallow lakes, the physical processes caused by wind play an important role. In Lake Balaton, the wind stirs up the sediment, which consists almost entirely of calcium compounds with a high adsorption capacity for phosphorous compounds. Consequently, studies on Lake Balaton have shown that the mass flows of phosphorous compounds from the water column to the sediment due to this effect are significant. Therefore, an adequate description of the sediment stirring, the adsorption of phosphorous compounds on the suspended matter, and sedimentation must be included in a eutrophication model for this lake.

Jørgensen and Mejer (1977, 1979) examined the inverse sensitivity, called the ecological buffer capacity, to select the number of state variables. The concept of ecological buffer capacity is illustrated in Figure 2.12 and is defined as:

$$\beta = \frac{1}{(\partial(St)/\partial F)} \tag{2.13}$$

where St is a state variable and F a forcing function. It is possible to define many different buffer capacities corresponding to all possible combinations of state variables and forcing functions. However, the model scope will often point out which buffer capacity should be in focus. For a eutrophication model, the most sensitive factor would be the change in input of phosphorus (or nitrogen) to the concentration of phytoplankton. Now the modeller examines the relationship between the buffer capacity in focus and the number of state variables.

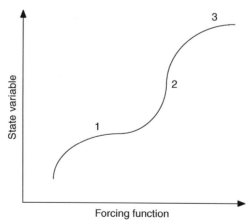

FIGURE 2.12 A relation between a state variable and a forcing function is shown. At points 1 and 3 the buffer capacity is high; at point 2 it is low.

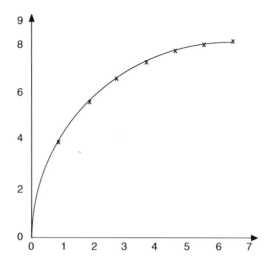

FIGURE 2.13 Illustrates the buffer capacity for a eutrophication model of a shallow Danish lake. In this case, a model with six state variables for each of the important nutrients (C, P, and N) was selected. Adding a seventh state variable representing an additional zooplankton species and an additional phytoplankton species produced only minor changes to the buffer capacity. Other possibilities could also have been tested. In this context it must be pointed out that the buffer capacity does not necessarily increase with the number of state variables as in Figure 2.12. The change in buffer capacity only decreases with the number of state variables when their sequence is selected according to decreasing importance.

As long as the buffer capacity is changed significantly by adding an extra state variable, the model complexity should be increased. But if additional state variables only change the buffer capacity insignificantly, an increased model complexity will only augment the number of parameters, adding to the uncertainty without contributing to a more accurate model.

Figure 2.13 illustrates the buffer capacity for a eutrophication model of a shallow Danish lake. In this case, a model with six state variables for each of the important nutrients — carbon, nitrogen, and phosphorus — was selected. Inclusion of a seventh state variable created only a minor change to the buffer capacity.

Flather (1992, 1996) recommended using Akaike's Information Criterion (AIC), to select a best model from the *a priori* best candidate models:

$$\text{AIC} = n \log(\text{RSS}/n)^2 + 2K, \tag{2.14}$$

where n is the number of observations, RSS is the residual sum of squares (model outputs-observations), and K is the number of parameters +1.

The model with the lowest AIC is preferable. The application of this equation is recommended to select submodels. This equation can also be applied in principle to large models, but not in practice where a comparison of several large models would be too time-consuming.

For other applicable methods used to select the model complexity, see Halfon (1983) and Bosserman (1980, 1982) where the use of the connectivity is presented. Experience shows some model corrections at a later stage will be unnecessary if the model has been calibrated and the validation phase indicates that improvements might be needed. This does not, however, imply that corrections of the model structure at a later stage can be omitted. The methods presented for the selection of model structure are not so rigorous that the very best model is always selected at the first instance. The methods presented earlier assist the modeller to exclude some unworkable models, but not necessarily to choose the very best model. Remember, there is no one right model.

2.10. Parameter Estimation

Many parameters in causal ecological models can be found in the literature, not necessarily as constants but as approximate values or

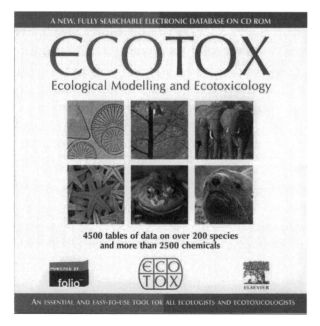

FIGURE 2.14 Jørgensen et al., (2000) contains about 120,000 parameters of interest for ecological modellers.

intervals. Jørgensen et al. (2000) contains about 120,000 parameters of interest for ecological modellers (see Figure 2.14).

However, even if all parameters are known in a model from the literature, calibrating the model is usually required because the biological parameters are only known within ranges. Several sets of parameters are tested by calibration and the various model outputs of state variables are compared with measured or observed values of the same state variables. The parameter set that gives the best agreement between model outputs and measured state variables is chosen.

A eutrophication model is generally calibrated based on an annual measurement series with a sampling frequency of once or twice per month. This sampling frequency is not sufficient to describe the lake dynamics. If it is the scope of the model to predict maximum values and related data for phytoplankton concentrations and primary production, then it is necessary to have a sampling frequency that gives an estimate of the maximum value in phytoplankton concentration and the primary production.

Figure. 2.15 shows characteristic algae concentrations plotted versus time (April 1–May 15) in a hypertrophic lake with a sampling frequency of (1) twice per month and (2) three times per week (denoted as the "intensive" measuring program). The two plots are significantly different and an attempt to get a realistic calibration based on (1) will fail, provided it is the aim to model the day-to-day variation in phytoplankton concentration according to (2). This example illustrates that it is

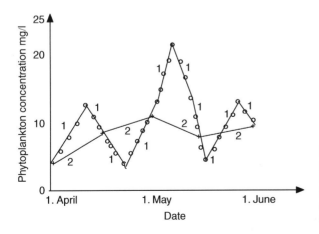

FIGURE 2.15 Algae concentration plotted versus time: (1) = sampling frequency twice a month (+) and (2) = sampling frequency three times a week (*). Note the difference of d(PHYT)/dt between the two curves.

important not only to have data with low uncertainty, but also data sampled with a frequency corresponding to the dynamics of the system.

The rule to match appropriate sampling frequency has often been neglected in modelling eutrophication, probably because limnological lake data, which are not sampled for modelling purposes, are often collected with a relatively low frequency. On the other hand, the model then attempts to simulate the annual cycle, and an annual sampling program with a frequency of three samples per week requires too many resources. A combination of an annual sampling program with a frequency of one to three samples per month and an intensive measuring program placed in periods, where different subsystems show maximum changes, is a good basis for parameter estimations.

The intensive measuring program can, as presented next, be used to estimate state variables' derivatives. For comparison of these estimations by low and high sampling frequency, see the slopes of curves (1) and (2) in Figure 2.15. These estimates can be used to set up an overdetermined set of algebraic equations, making the model parameters the sole unknown. An outline of the method runs as follows (see Figure 2.16; for further details, see Jørgensen et al., 1981):

Step 1. Find cubic spline coefficients, $S_i(t_j)$, that is, second-order time derivatives at time of observation t_j, of the spline function $s_i(t_j)$ approximating the observed variable $\psi_i(t)$, according to the cubic spline method. Alternatively, it is possible to find an nth order polynomium (4th–8th order is most often used) approximating the observations by an nth order regression analysis. Several statistical software packages are available to perform such regression analyses very rapidly.

Step 2. Find $\partial\psi_i(t_j)/\partial t = f(t)$ by differentiation of the function found in step 1: $\psi = (\psi, t, a)$, where a is a parameter.

Step 3. Solve the model equation of the form:

$$\partial\psi(t_j)/dt = f(\psi, \partial\psi/\partial r, \partial^2\psi/\partial r^2, t, a) \qquad (2.15)$$

with the average value of a, regarded as unknown.

Step 4. Evaluate the feasibility of the solution a_0 found in step 3. If not feasible, then modify the part of the model influenced by a_0 and go to step 1.

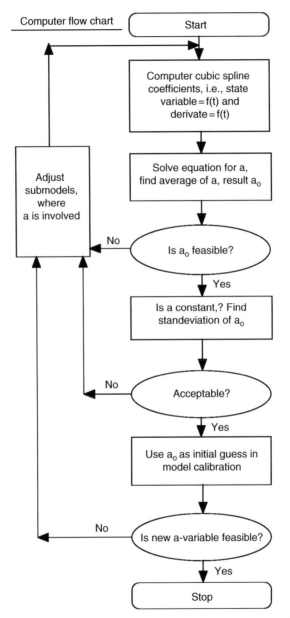

FIGURE 2.16 Computer flow chart of the method applied to estimate parameters by using "intensive measurements."

Step 5. Choose a significance level, and perform a statistical test on constancy of a_0. If the test fails, then modify appropriate submodels and go to step 1.

Step 6. Use a_0 as an initial guess in a computerized parameter search algorithm, such as Marquardt, Powell, or steepest descent algorithms, to minimize a performance index such as the one proposed in Eq. (2.2).

Although the model in hand may be highly nonlinear regarding the state variables, it usually turns out that this is not the case regarding the parameter set a, or the subset of a that is tuned by calibration. Since the number of differential equations is greater than the number of estimable parameters, Eq. (2.15) is overdetermined. It is easy to smooth the solution, but it is more important to evaluate the constancy of a_0, for example, by variance analysis, test of normality of white noise, and so forth. Information on standard deviation of a_0 around its average value may eventually be used as a point of departure for introducing stochasticity into the model, admitting the fact that parameters in real life may not be as constant as the modellers assume.

As a certain parameter, a_k, seldom appears at more than one or two places in the model equations, an unacceptable value of a_k found as solution to Eq. (2.15) quite accurately locates the inappropriate terms and constructs in the model. Experience with this method shows it to be a valuable diagnostic tool to single out unfitted model terms.

Since the method is based on cubic spline approximation, it is essential that observations are dense, for example, $t_{j+1} - t_j$ should be small in the sense that local third-degree polynomials should approximate observed values well. It is difficult, in general, to test whether this is fulfilled as the "true" $\psi_i(t)$ function might have microscopic curls that generate oscillating derivates (ψ_i/dt). However, if the method yields basically the same result on a random subset of observations, then it may be safe to assume that $\{s_i(t_j)/dt\}$ represents the true rates on a daily basis. After appropriate adjustment of model equations, an acceptable parameter set a_o may eventually be obtained.

With a_0 as an initial guess, a better parameter set may be found by systematic perturbation of the set until some norm (performance index) has reached a (local) minimum. At each perturbation, the model equations are solved. Gradients $\{\delta\psi_i/\delta a_k\}$ are hardly ever known analytically. All numerical methods currently in use to solve this kind of problem fail when the number of parameters surpasses four or five unless the initial guess is very close to a value that minimizes the performance index. This is why steps 1 and 2 mentioned previously are so important. The result of the application of intensive measurements to calibrate the eutrophication model is summarized in Table 2.4, where the difference in parameter estimation is pronounced. It is important to use the parameters determined by intensive measurements before the final calibration.

The illustrated use of intensive measurements for parameter estimation prior to the calibration was based upon determinations of the actual growth of phytoplankton. By determination of the derivatives, it was possible to fit the parameters to the unknown in the model equations.

Intensive measurements were used for the 18 state variable eutrophication model presented in Chapter 7. It was possible to determine the maximum phytoplankton growth rate by the previous method to be 1.6 day^{-1} ±10% relative. It was also possible to choose between possible expressions for the temperature influence on the phytoplankton growth.

Measurements and observations *in vitro* were used in the referred case to find the derivates. In principle, the same basic idea can be used either in the laboratory or by construction of a microcosm. In both cases, the measurements are facilitated by a smaller unit, where disturbing factors or processes might be kept constant. Current record of important state variables is often possible and provides a high number of data, which decreases the standard deviation.

As an example fish growth can be described by use of the following equation:

$$dW/dt = a \times W^b \qquad (2.16)$$

where W is the weight, and a and b are constants. In an aquarium or an aquaculture farm it is possible to measure the fish weight over time. If enough data are available, then it is easy by statistical methods to determine a and b in Eq. (2.16). In this case, the feeding is known to be at the optimum level, no predator is present, and the water quality, which influences growth, is maintained constant to assure the very best growth conditions for the fish. By varying these factors, it is even

Table 2.4   Comparison of Parameter Values

Parameter	Parameter (Symbol)	Unit	Application of Intensive Measurements	Glumsø Lake*	Lyngby Lake*	Literature ranges
Settling rate	SVS = D × SA	m d^{-1}	0.30 ± 0.05	0.2	0.05	0.1–0.6
Max. growth rate**	CDRmax (reduced)	d^{-1}	1.33 ± 0.51	2.3	1.8	1–3
Max. growth rate**	CDRmax (model)	d^{-1}	4.71 ± 1.8	4.11	3.21	2–6
Max. uptake rate P**	UPmax	d^{-1}	0.0072 ± 0.0007	0.003	0.008	0.003–0.01
Min. C:biomass Ratio**	FCAmin		0.4	0.15	0.15	0.3–0.7
Min. P: biomass Ratio**	FPAmin	0.03	0.013	0.013	0.013	0.013l–0.035
Min. N: biomass ratio	FNAmin		0.120.10	0.10	0.10	0.08–0.12
Max.uptake rate N**	UNmax	d^{-1}	0.023 ± 0.005	0.015	0.012	0.0l–0.035
Michaelis-Menten** constant N	KN	mg l^{-1}	0.34 ± 0.07	0.2	0.2	0.1–0.5
Denitrification rate	DENITX	g m^{-3} d^{-1}	0.83 ± l.05			
Respiration rate**	RC	d^{-1}	0.088	0.13	0.2	0.05–0.25
Mineralization rate P	KDPl0	d^{-1}	0.80 ± 0.47	0.40	0.25	0.2–0.8
Mineralization rate N	KDNl0	d^{-1}	0.21 ± 0.11	0.05	0.15	O.OS–0.3
Max. uptake rate c**	UCmax	d^{-1}	1.21 ± 0.97	0.65	0.40	0.2–1.4

Notes:

*Lyngby and Glumsø lakes have approximately the same biogeochemical characteristics and morphology;

**all parameters related to phytoplankton.

possible to find the influence of the water quality, and the available food on the growth parameters. The results of such experiments can often be found in the literature. Still, the modeller might not find the parameter for the species of interest, nor find the parameters in the literature under the specific conditions in the ecosystem being modelled. In such cases, it may be necessary to use experiments to determine important model parameters. The use of laboratory experiments is advisable also when the literature values for the crucial parameters are too wide for the most sensitive parameters.

However, parameters taken from the literature or resulting from such experiments should be applied with precaution because the discrepancy between the values in the laboratory or even the microcosms and those in nature is much greater for biological parameters than for chemical or physical parameters. The reasons for this can be summarized in the following points:

1. Biological parameters are generally *more sensitive to environmental factors than chemical or physical parameters*. An illustrative example would be: A small concentration of a toxic substance could change growth rates significantly.
2. Biological parameters are *influenced by many environmental factors*, of which some are quite variable. For instance, phytoplankton growth rate is dependent on the nutrient concentration, but the local nutrient concentration is again very dependent on the water turbulence, which is dependent on the wind stress, and so forth.
3. The example in point 2 shows that the *environmental factors influencing biological parameters are interactive*, which makes it almost impossible to predict an exact value for a parameter in nature from measurements in the laboratory where the environmental factors are all kept constant. On the other hand, if the measurements are carried out *in situ*, then it is not possible to interpret under which circumstances the measurement is valid, because that would require the simultaneous determination of too many interactive environmental factors.
4. Often, determinations of biological parameters or variables *cannot be carried out* directly, but it is necessary to measure another quantity that cannot be exactly related to the biological quantity in

focus. For instance, the phytoplankton biomass cannot be determined by any direct measurement, but it is possible to obtain an indirect measurement by using the chlorophyll concentration, the ATP concentration, the dry matter 1–70μ, and so forth. Still, none of these indirect measurements give an exact value of the phytoplankton concentration, as the ratio of chlorophyll or ATP to the biomass is not constant, and the dry matter 1–70μ might include other particles (e.g., clay particles). So, it is recommended in practice to apply several of these indirect determinations simultaneously to assure a reasonable estimate. Correspondingly, the phytoplankton growth rate might be determined by the oxygen method or the C-14-method. Neither method determines the photosynthesis; instead they determine the net production of oxygen and the net uptake of carbon, respectively; that is, the result of the photosynthesis and the respiration. The results of the two methods are therefore corrected to account for the respiration, but obviously the correction should be different in each individual case, which is difficult to do accurately.

5. Biological parameters are *influenced by several feedback mechanisms of a biochemical nature*. The past will determine the parameters in the future. For instance, the phytoplankton growth rate is dependent on the temperature — a relationship that can easily be included in ecological models. The maximum growth rate is obtained by the optimum temperature, but the past temperature pattern determines the optimum temperature. A cold period will decrease the optimum temperature. To a certain extent, this can be taken into account by the introduction of variable parameters (Straskraba, 1980). In other words, it is an approximation to consider parameters as constants. An ecosystem is a soft, flexible system, described with approximations as a rigid system with constant parameters (Jørgensen, 1981, 1992a,b).

The estimation of the settling velocity as a parameter in ecological models may be crucial whether the component is suspended matter or phytoplankton, as it determines the removal rate for a considered component. The sensitivity of this parameter to the phytoplankton concentration in a eutrophication model has been determined to be about -1.0 (see Table 2.3). It means that if the parameter is increased

1%, the phytoplankton concentration will decrease 1% (Jørgensen et al., 1978). Let us therefore use the estimation of the settling rate as another illustration of the needed considerations in our effort to obtain a proper determination of parameters.

Settling velocity may be determined in three ways:

1. Values from previous models found in the literature can be used to give a first estimation of the parameter. Tables 2.5 and 2.6 summarize values found in the literature. As can be seen, these values are indicated as ranges, therefore, it is necessary to calibrate the parameters using measured values for the stated variables.

Table 2.5 Phytoplankton Settling Velocities

Algal Type	Settling Velocity (m/day)	References
Total phytoplankton	0.05–0.5	Jørgensen et al. (1991, 2000);Tetra Tech (1980)
	0.05– 0.2	Di Toro & Connolly (1980); O'Connor et al. (1981); Thomann et al. (1974); Thomann & Fitzpatrick (1982)
	0.02– 0.05	Jørgensen et al. (1991, 2000)
	0.4	Lombardo (1972)
	0.03– 0.05	Scavia (1980)
	0.05	Bierman (1976)
	0.2–0.25	Youngberg (1977)
	0.04– 0.6 *	Jørgensen et al. (2000)
	0.01–4.0	*Jørgensen et al. (2000)
	0.1–2.0 *	Snape et al. (1995)
	0.15–2.0 *	Jørgensen et al. (2000)
	0.1–0.2 *	Brandes (1976)
Diatoms	0.05– 0.4	Bierman (1976); Brandes et al. (1974)
	0.1– 0.2	Jørgensen et al. (2000)
	0.1– 0.25	Tetra Tech (1980)
	0.03– 0.05	Snape et al. (1995)
Diatoms	0.3– 0.5	Jørgensen et al. (2000)
	2.5	Lehman et al. (1975)
	0.02–14.7 *	Jørgensen et al. (2000)
Green algae	0.05– 0.19	Jørgensen et al. (2000)
	0.05– 0.4	Bierman (1976)
Green algae	0.02	Snape et al. (1995)
	0.8	Lehman et al. (1975)
	0.1– 0.25	Tetra Tech (1980)
	0.08– 0.18 *	Jørgensen et al. (2000)
	0.27– 0.89 *	Jørgensen et al. (2000)

Continued

Table 2.5 Phytoplankton Settling Velocities—cont'd

Algal Type	Settling Velocity (m/day)	References
Blue-green algae	0.05– 0.15	Bierman (1976)
	0.08	Snape et al. (1995)
	0.2	Lehman et al. (1975)
	0.1	Jørgensen et al. (2000)
	0.08–0.2	Tetra Tech (1980)
Flagellates	0.5	Lehman et al. (1975)
	0.05	Bierman (1976)
	0.09– 0.2	Tetra Tech (1980)
	0.07–0.39 **	Jørgensen et al. (2000)
Dinoflagellates	2.8–6.0 **	Jøregensen et al. (2000)
Asterionella formosa	0.25– 0.76 **	Jørgensen et al. (2000)
Chaetoceros lauderi	0.46– 1.56 **	Jørgensen et al. (2000)
Chrysophytes	0.5	Lehman et al. (1975)
Coccolithophores	0.25– 13.6	Jørgensen et al. (2000)
	0.3– 1.5 **	Jørgensen et al. (2000)
Coscinodiscus lineatus	1.9– 6.8 **	Jørgensen et al. (2000)
Cyclotella meneghimiana	0.08– 0.31 **	Jørgensen et al. (2000)
Ditylum brightwellii	0.5– 3.1 **	Jørgensen et al. (2000)
Melosira agassizii	0.67– 1.87 **	Jørgensen et al. (2000)
Nitzschia seriata	0.26– 0.50 **	Jørgensen et al. (2000)
Rhizosolenia robusta	1.1– 4.7 **	Jørgensen et al. (2000)
R. setigera	0.22– 1.94 **	Jørgensen et al. (2000)
Scenedesmus quadracauda	0.04– 0.89 **	Jørgensen et al. (2000)
Skeletonema costatum	0.31– 1.35 **	Jørgensen et al. (2000)
Tabellaria flocculosa	0.22– 1.11 **	Jørgensen et al. (2000)
Thalassiosira nana	0.10– 0.28 **	Jørgensen et al. (2000)
T. pseudonana	0.15– 0.85 **	Jørgensen et al. (2000)
T. rotula	0.39– 17.1	Jørgensen et al. (2000)

Notes: Other values used in models.
*Model documentation values;
**literature values.

Table 2.6 Detritus, Settling Rate

Item	Settling Velocity (m/day)	References
Detritus	0.1–2.0	Jørgensen et al. (2000)
Nitrogen detritus	0.05– 0.1	Jørgensen et al. (2000)
Fecal pellets (fish)	23– 666	Jørgensen et al. (2000)

2. Values from calculations based upon knowledge of the size can be used as first estimations. Due to the influence of the many factors previously mentioned, a calibration is also required in this case. This method is hardly applicable for phytoplankton because of their ability to change the specific gravity, but it may be useful for other particles.

3. Measurements *in situ* by use of sedimentation traps. It is possible to determine the distribution of the material in inorganic and organic matter and partly also in phytoplankton and detritus by analysis of chlorophyll (fresh material), phosphorus, nitrogen, and ash. Measurements of phytoplankton settling velocities in the laboratory will unlikely give a reliable value, as they do not consider the various factors *in situ*.

It has been previously pointed out that the *calibration is facilitated significantly if we have good initial estimates of the parameters*. Some estimates might be found in the literature, but it is often only a few compared with the number of parameters needed if we want to model all interesting mass flows in all relevant ecosystems. For the nutrient flows, the parameters known from the literature are the most common species only. If we turn to flows of toxic substances in ecosystems, then the number of known parameters is even more limited. The Earth has millions of species and the number of substances of environmental interest is about 100,000. If we want to know 10 parameters for each interaction between substances and species, then the number of parameters needed is enormous. For example, if we need the interactions of only 10,000 species with the 100,000 substances of environmental interest, the number of needed parameters is $10 \times 10,000 \times 100,000 = 10^{10}$ parameters. In Jørgensen et al. (2000; see Figure 2.14) 120,000 parameters can be found, and if we estimate that this *Handbook* covers about 10% of the parameters, which can be found in the entire literature, then we know only about 0.012% of the needed parameters. Physics and chemistry have attempted to solve this problem by setting up some general relationships between the properties of the chemical compounds and their composition and structure. This approach is widely used in ecotoxicological modelling, and will be discussed in Chapter 8. If needed data cannot be found in the literature, then such relationships are widely used as the second-best approach to the problem.

If we draw a parallel to ecology, then we need some general relationships that give us good first estimations of the needed parameters. In many ecological models used in an environmental context, the required accuracy is not very high. In many toxic substance models, we need only to know whether we are far from or close to the toxic levels. Still, more experience with the application of general relationships is needed before a more general use can be recommended. It should be emphasized that in chemistry such general relationships are used very carefully.

Modern molecular theory provides a sound basis for predicting reliable quantitative data on the chemical, physical, and thermodynamic properties of pure substances and mixtures. The biological sciences are not based upon a similar comprehensive theory, but it is possible, to a certain extent, to apply basic biochemical mechanisms laws to ecology. Furthermore, the very basic biochemical mechanisms are the same for all plants and all animals. The spectrum of biochemical compounds is wide, but considering the number of species and the number of possible chemical compounds it is very limited. The number of different protein molecules is significant, but they are all constructed from only 24 different amino acids.

This explains why the elementary composition of all species is quite similar. All species need, for their fundamental biochemical function, a certain amount of carbohydrates, proteins, fats, and other compounds, and as these groups of biochemical substances are constructed from relatively few simple organic compounds, it is not surprising that the composition of living organisms varies only a little, (see tables in Jørgensen et al., 1991, 2000). For example, if we know the uptake rate of nitrogen for phytoplankton, then we can find the approximate uptake rate of phosphorus because the uptake rates must result in a nitrogen-to-phosphorus ratio between 5:1 and 12:1, an average 1:7.

The biochemical reaction pathways are also general, which is demonstrated in all textbooks on biochemistry. The utilization of the chemical energy in the food components is basically the same for microorganisms and mammals. It is, therefore, possible to calculate approximately the energy, E1, released by digestion of food, when the composition is known:

$$E1 = 9 \text{ fat} \% 100 + 4(\text{Carbohydrates} + \text{proteins}) \% 100 \tag{2.17}$$

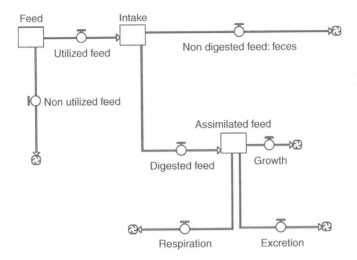

FIGURE 2.17 The principle of a fish growth model. The feed is either utilized or not utilized. The utilized food = the intake is either digested or assimilate and at steady-state intake = nondigested feed (feces) + the assimilated feed. The assimilated feed is used for either growth, excretion, or respiration and at steady state assimilated feed = growth + respiration + excretion (see Jørgensen, 2000).

The law of energy conservation is also valid for a biological system (see Figure 2.17). The chemical energy of the food components is used to cover the energy needs for respiration, assimilation, growth (increase of biomass included reproduction), and losses. As it is possible to set up relations between these needs on the one side with some fundamental properties of the species on the other, it is possible to put a number on the items in Figure 2.17 for different species. This is a general but valid approach to parameter estimation in ecological modelling.

Species surface area is a fundamental property indicating quantitatively the size of the boundary to the environment. Loss of heat to the environment must be proportional to this area and to the temperature difference, according to the law of heat transfer. The rate of digestion, the lungs, and hunting ground, are all dependent on the size of the animal and are determinant for a number of parameters.

Therefore, it is not surprising that many parameters for plants and animals are highly related to the size of the organism, which implies that it is possible to get very good first estimates for most parameters based only upon the size. Naturally, the parameters are also dependent on several characteristic features of the species, but their influence is minor compared with the organism size, and the good estimates provide at least a starting value in the calibration phase.

The conclusion of these considerations is that many parameters are related to simple properties, such as size of the organisms, and that

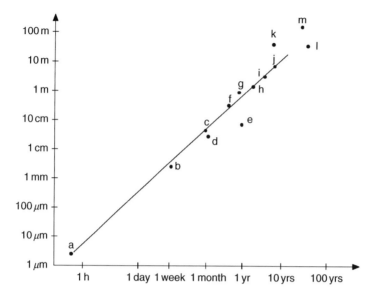

FIGURE 2.18 Length and generation time plotted on log-log scale: (a) pseudomonas, (b) daphnia, (c) bee, (d) house fly, (e) snail, (f) mouse, (g) rat, (h) fox, (i) elk, (j) rhino, (k) whale, (l) birch, (m) fir. See also Peters (1983).

such relations are based upon fundamental biochemistry and thermo-dynamics. Above all, there is a strong positive correlation between size and generation time, T_g, ranging from bacteria to the biggest mammals and trees (Bonner, 1965). This relationship is illustrated in Figure 2.18. This relationship can be explained using the relationship between size (surface) and total metabolic action per unit of body weight. It implies that the smaller the organism, the greater the specific metabolic activity (= activity/weight). The per capita rate of increase, r, defined by the exponential or logistic growth equations:

$$dN/dt = rN \tag{2.18}$$

respectively,

$$dN/dt = rN(1 - N/K) \tag{2.19}$$

is inversely proportional to the generation time.

This implies that r is related to the organism size, but, as shown by Fenchel (1974), it actually falls into three groups of organisms: unicellular, poikilotherms, and homeotherms (see Figure 2.19). Thus, the metabolic rate per unit of weight is related to the size. The same basis is expressed in the following equations, giving the respiration, feed consumption, and ammonia excretion for fish when the weight, W, is known:

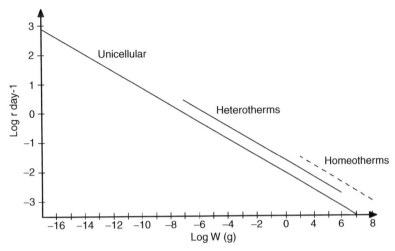

FIGURE 2.19 Intrinsic rate of natural increase against weight for various animals. See also Peters (1983).

$$\text{Respiration} = \text{constant} * W^{0.80} \tag{2.20}$$

$$\text{Feed Consumption} = \text{constant} * W^{0.65} \tag{2.21}$$

$$\text{Ammonia Excretion} = \text{constant} * W^{0.72} \tag{2.22}$$

This is also expressed in Odum's equation (E. P. Odum, 1969, 1971):

$$m = kW^{-1/3} \tag{2.23}$$

where k is roughly a constant for all species, equal to about 5.6 kJ/g$^{2/3}$ day, and m is the metabolic rate per weight unit.

Similar relationships exist for other animals. The constants in these equations might be slightly different due to differences in shape, but the equations are otherwise the same.

All of these examples illustrate the fundamental relationship in organisms between size (surface) and the biochemical activity. The surface determines the contact with the environment quantitatively along with the possibility of taking up food and excreting waste substances.

The same relationships are shown in Figures 2.20–2.22, where rates of biochemical processes involving toxic substances are plotted versus size. They are reproduced from Jørgensen (1997, 2002). In these figures, the excretion rate, uptake rate, and concentration factor (for aquatic organisms) follow the same trends as the growth rate. This is not surprising, as excretion is strongly dependent on metabolism and the direct uptake dependent on the surface.

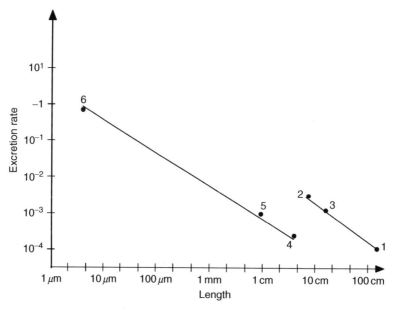

FIGURE 2.20 Excretion of Cd $(24h)^{-1}$ plotted versus the length of various animals: (1) Homosapiens, (2) mice, (3) dogs, (4) oysters, (5) clams, and (6) phytoplankton.

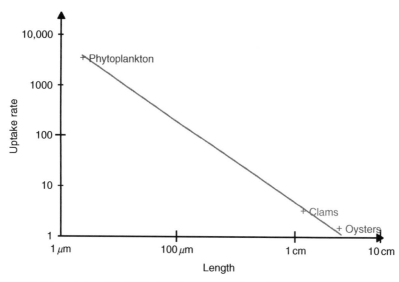

FIGURE 2.21 Uptake rate (μg Cd/g 24 h) plotted against the length of various animals' phytoplankton, clams, and oysters.

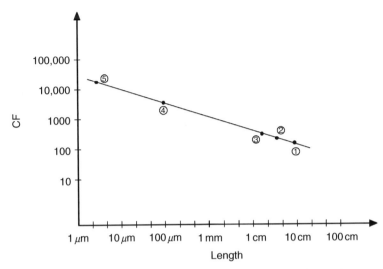

FIGURE 2.22 CF (concentration factor >= to the ratio concentration in organism to the concentration in water>) for Cd versus size: (1) goldfish, (2) mussels, (3) shrimps, (4) zooplankton, (5) algae (brown-green).

In spite of all these methods to estimate parameters, it may still be necessary to accept that a parameter is only known within some unacceptable large range. In such cases, applying a Monte Carlo simulation of the parameter within the known range should be considered. The concentration factor indicating concentration in the organism vis-á-vis concentration in the medium also follows the same lines (see Figure 2.20). By equilibrium, the concentration factor can be expressed as the ratio between the uptake rate and the excretion rate, as shown in Jørgensen (1979). As most concentration factors are determined by the equilibrium, the relationship found in Figure. 2.20 seems reasonable. Intervals for concentration factors are indicated here for some species according to the literature (Jørgensen et al., 1991, 2000).

The *allometric principles,* illustrated in Figures2.18–2.22, can be generally applied to find process rates, provided these parameters are available for the element or compound under consideration (because the slope is known). However, it is preferable to know several species to control the validity of the graph. When plots similar to Figures 2.18–2.22 are constructed, it is possible to read unknown parameters when the size of the organism is known.

It was mentioned earlier that model constraints could be used to estimate unknown parameters. The chemical composition of an organism was applied to illustrate this principal method. The topic model constraints are covered further in Section 2.12. The Darwinian survival of the fittest is used in thermodynamic translation as a goal function to find the change in properties resulting from adaptation and a shift in species composition presented. This constraint has also been applied to estimate unknown parameters, as shown in Chapter 10, after the more basic theory has been presented.

This presentation of parameter estimation methods can be summarized in the following overview and recommendations:

A. Examine the literature to find the range of as many model parameters as possible. It is recommended to use Jørgensen et al. (2000), which contains about 120,000 parameters.
B. Examine processes *in situ* or in the laboratory to assess unknown parameters.
C. Apply an intensive observation period to reveal the dynamic processes in the model. Use the method described in Figures 2.15 and 2.16 to find unknown parameters. This method often makes it possible to indicate parameters within relatively narrow ranges.
D. Apply allometric principles to find unknown parameters for the organisms included in the model as well as for other organisms. The allometric principles may also be used as a control of a parameter that is found by estimations or calibration.
E. Ecotoxicological parameters can be estimated by a network of methods based on a translation of the chemical structure to the properties of the compound. This method will be presented in detail in Chapter 10, Section 10.6.
F. Use the model constraints to estimate an unknown parameter or to control an uncertain parameter (e.g., how exergy can be used to determine parameters in Chapter 10, Section 10.3).
G. Apply calibration of submodels and/or the entire model. The better the data, the more certain and reliable results the calibration will offer.

2.11. Ecological Modelling and Quantum Theory

How can we describe such complex systems as ecosystems in detail? The answer is that it is impossible if the description must include all details, including all interactions between all the components in the entire hierarchy, as well all feedbacks, adaptations, regulations, and the entire evolutionary process.

Jørgensen (1997, 2002) introduced the application of the uncertainty principle of quantum physics in ecology. In nuclear physics, uncertainty is caused by the observer of the incredibly small nuclear particles, while uncertainty in ecology is caused by the enormous complexity of ecosystems.

For instance, if we take two components and want to know the relationship between them, we would need at least three observations to show whether the relation is linear or nonlinear. Correspondingly, the relations among three components will require 3*3 observations for the shape of the plane. If we have 18 components we would correspondingly need 3^{17} or approximately 10^8 observations. At present, this is probably an approximate practical upper limit to the number of observations that can be invested in one project aimed at one ecosystem. This could be used to formulate a practical uncertainty relation in ecology (Jørgensen, 1990):

$$10^5 * \Delta x / \sqrt{3^{n-1}} \leq 1 \qquad (2.24)$$

where Δx is the relative accuracy of one relation, and n is the number of components examined or included in the model.

The 100 million observations could also be used to give a very exact picture of one relation. Costanza and Sklar (1985) talked about the choice between the two extremes: knowing "everything" about "nothing" or "nothing" about "everything" (see Section 2.9). The former refers to the use of all the observations on one relation to obtain a high accuracy and certainty, while the latter refers to the use of all observations on as many relations as possible in an ecosystem. How we can obtain a balanced complexity in the description will be further discussed in the next section.

Equation (2.18) formulates a practical uncertainty relation, but, the possibility that the practical number of observations may be increased in the future cannot be excluded. More and more automatic analytical equipment is emerging on the market. This means that the number of observations invested in one project may be one, two, three, or even several magnitudes larger in the future. Yet, a theoretical uncertainty relation can be developed. If we go to the limits given by quantum mechanics, then the number of variables will still be low compared to the number of components in an ecosystem.

One of Heisenberg's uncertainty relations is formulated as follows:

$$\Delta s * \Delta p \geq h/2\pi \qquad (2.25)$$

where Δs is the uncertainty in determining the position, and Δp is the uncertainty of momentum. According to this relation, Δx of Eq. (2.24) should be in the order of 10^{-17} if Δs and Δp are about the same. Another of Heisenberg's uncertainty relations may now be used to give the upper limit of the number of observations:

$$\Delta t * \Delta E \geq h/2\pi \qquad (2.26)$$

where Δt is the uncertainty in time and ΔE in energy.

If we use all the energy that Earth has received during its existence of 4.5 billion years, then we get:

$$173 * 10^{15} * 4.5 * 10^9 * 365.3 * 24 * 3600 = 2.5 * 10^{34} J, \qquad (2.27)$$

where $173 * 10^{15}$ W is the energy flow of solar radiation. Δt would, therefore, be in the order of 10^{-69} seconds. Thus, an observation will take 10^{-69} seconds, even if we use all the energy that has been available on Earth as ΔE, which must be considered the most extreme case. The hypothetical number of observations possible during the lifetime of the Earth would therefore be:

$$4.5 * 10^9 * 365.3 * 3600 / 10^{-69} \approx of 10^{85}. \qquad (2.28)$$

This implies that we can replace 10^5 in Eq. (2.24) with 10^{60} since

$$10^{-17}/\sqrt{10^{85}} \approx 10^{-60}$$

If we use $\Delta x = 1$ in Eq. (2.28) we get:

$$\sqrt{3^{n-1}} \leq 10^{60} \qquad (2.29)$$

or $n \leq 253$.

From these very theoretical considerations, we can clearly conclude that we will never have enough observations to describe even one ecosystem in complete detail. An ecosystem is a middle number system, which means that the number of components are not as high as the number of gas molecules in a room, but that it may be as high as 10^{15}–10^{20}. Unlike the gas molecules in a room, all of these components are different, while there may be only 10 to 20 different types of gas molecules in a room.

These results agree with Niels Bohr's complementarity theory, which he expressed as follows: "It is not possible to make one unambiguous picture (model) of reality, as uncertainty limits our knowledge." The uncertainty in nuclear physics is caused by the inevitable influence of the observer on the nuclear particles; in ecology it is caused by the enormous complexity and variability.

No map of reality is completely correct. There are many maps (models) of the same area of nature, and the various maps or models reflect different viewpoints. Accordingly, one model (map) does not give all the information and far from all the details of an ecosystem. Applying the theory of complementarity in ecology, we see that it is important to view the ecosystem from different, complementary angles.

As stated previously, the use of maps in geography is a good parallel to the use of models in ecology. As we have road maps, airplane maps, geological maps, maps in different scales for different purposes, we have many models in ecology of the same ecosystems. We need them all if we want to get a comprehensive view of ecosystems (see Sections 1.1 and 2.9). Furthermore, a map can give an incomplete picture. We can always make the scale larger and larger and include more details, but we cannot get all the details. An ecosystem also has too many dynamic components to enable us to model all the components simultaneously, and even if we could, the model would be invalid a few seconds later after the dynamics of the system have changed the "picture."

Another good example comes from physics, in which we need a pluralistic view to consider light as waves as well as particles. The situation in ecology is similar. Because of the immense complexity, we need a pluralistic view to describe an ecosystem. We need many models covering different viewpoints. This is consistent with Gödel's Theorem from 1931 (Gödel, 1986) that the infinite truth can never be condensed in a finite theory. There are limits to our insight; we cannot produce a map of the world with every possible detail because that would be the world itself.

Ecosystems must be considered irreducible systems, because it is not possible to make observations and then reduce the observations to more or less complex laws of nature; for instance mechanics. Too many interacting components force us to consider ecosystems as irreducible systems. It is necessary to use what is called experimental mathematics or modelling to cope with irreducible systems.

Quantum theory may have an even wider application in ecology. Schrödinger (1944) suggested, that the "jump like changes" you observe in the properties of species are comparable to the jump-like changes in energy by nuclear particles. Schrödinger was inclined to call De Vries' mutation theory (published in 1902) the quantum theory of biology because the mutations are due to quantum jumps in the gene molecule.

Patten (1982a, 1985) defined an elementary "particle" of the environment, called an environ — previously Koestler (1967) used the word holon — as a unit that can transfer an input to an output. Patten suggested that a characteristic feature of ecosystems is the network of connections. Input signals go into the ecosystem components and they are translated into output signals. Such a "translator unit" is an environmental quantum according to Patten. The term comes from the Greek "holos" = whole, with the suffix "on" as in proton, electron, and neutron to suggest a particle or part.

Stonier (1990) introduced the term infon for the elementary particle of information. He envisaged an infon as a photon whose wavelength has been stretched to infinity. At velocities other than c, its wavelength appears infinite, its frequency zero. Once an infon is accelerated to the speed of light, it crosses a threshold, which allows it to be perceived as having energy. When that happens, the energy becomes a function of its frequency. Conversely at velocities other than c, the particle exhibits neither energy nor momentum, yet it could retain at least two information properties: its speed and its direction. In other words, at velocities other than c, a quantum of energy becomes converted into a quantum of information. This concept has still not found any application in ecological modelling.

2.12. Modelling Constraints

A modeller is very concerned about the application of the right description of the components and processes in his models. The model equations and their parameters should reflect the properties of the model components and processes as correctly as possible. The modeller must, however, also be concerned with the right description of the system properties, and too little research has been done in this direction. A continuous development of models as scientific tools will need to consider how to apply constraints on models according to the system properties. Several possible modelling constraints are mentioned next. The sequence reflects decreasing relations to physical properties and increasing relations to biological properties of the ecosystems. The ecological modelling constraints will only be mentioned briefly in this context. A further discussion will take place in Chapter 10 where the application of these constraints is the basis for development of what may be called next generation models.

The *conservation principles* are often used as modelling constraints. Biogeochemical models must follow the conservation of mass, and bioenergic models must equally obey the laws of energy and momentum conservation.

Energy and matter are conserved according to basic physical concepts that are also valid for ecosystems. This requires that energy and matter are neither created nor destroyed.

The expression "energy *and* matter" is used, as energy can be transformed into matter and matter into energy. The unification of the two concepts is possible by Einstein's law:

$$E = mc^2 (ML^2T^{-2}), \tag{2.30}$$

where E is energy, m is mass, and c is the velocity of electromagnetic radiation in vacuum ($= 3*10^8$ m sec^{-1}). The transformation from matter into energy and vice versa is only of interest for nuclear processes and does not need to be applied to ecosystems; therefore, we might break the proposition down to two more useful propositions, when applied in ecology:

1. Ecosystems conserve matter.
2. Ecosystems conserve energy.

The conservation of matter may mathematically be expressed as follows:

$$dm/dt = input - output (MT^{-1}) \tag{2.31}$$

where m is the total mass of a given system. The increase in mass is equal to the input minus the output. The practical application of the statement requires that a system is defined, which implies that the boundaries of the system must be indicated.

Concentration, c, is used instead of mass in most models of ecosystems:

$$Vdc/dt = \text{input} - \text{output} \ (MT^{-1}) \tag{2.32}$$

where V is the volume of the system under consideration and assumed constant.

If the law of mass conservation is used for chemical compounds that can be transformed to other chemical compounds, then Eq. (2.32) must be changed to:

$$V * dc/dt = \text{input} - \text{output} + \text{formation} - \text{transformation} \ (MT^{-1}) \tag{2.33}$$

The principle of mass conservation is widely used in the class of ecological models called biogeochemical models. Equation (2.26) is set up for the relevant elements, for example, for eutrophication models for C, P, N, and perhaps Si (see Jørgensen, 1976a,b, 1982; Jørgensen et al., 1978).

For terrestrial ecosystems, mass per unit of area is often applied in the mass conservation equation:

$$A * dma/dt = \text{input} - \text{output} + \text{formation} - \text{transformation} \ (MT^{-1}) \tag{2.34}$$

here A = area and ma = mass per unit of area.

The Streeter-Phelps model (see Chapter 7) is a classical model of an aquatic ecosystem that is based upon conservation of matter and first-order kinetics. The model uses the following central equation:

$$dD/dt + K_a * D = L_o * K_1 * K_T(T - 20) * e^{-K1*t}(ML^{-3}T^{-1}) \tag{2.35}$$

where $D = C_s - C(t)$ C_s = concentration of oxygen at saturation; C(t)= actual concentration of oxygen; t = time; K_a = reaeration coefficient (dependent on the temperature); L_o = BOD_5 at time = 0; K_1 = rate constant for decomposition of biodegradable matter; and K_T = constant of temperature dependence.

Equation (2.29) states that change (decrease) in oxygen concentration + input from reaeration is equal to the oxygen consumed by decomposition of biodegradable organic matter according to a first-order reaction scheme.

Equations according to (2.27) are also used in models describing the fate of toxic substances in the ecosystem. Examples can be found in Thomann (1984) and Jørgensen (1991, 2000).

The mass flow through a food chain is mapped using the mass conservation principle. The food taken in by one level in the food chain is used in respiration, waste food, undigested food, excretion, and growth, including reproduction (see Figure 2.17). If the growth and reproduction are considered as the net production, then it can be stated that:

$$\text{net production} = \text{intake of food} - \text{respiration} - \text{excretion} - \text{waste food} \qquad (2.36)$$

The ratio of the net production to the intake of food is called the net efficiency; it is dependent on several factors, but is often as low as 10–20%. Any toxic matter in the food is unlikely to be lost through respiration and excretions because it is much less biodegradable than the normal components in the food. Because of this, the net efficiency of toxic matter is often higher than for normal food components, and as a result some chemicals, such as chlorinated hydrocarbons including DDT and PCB, will be magnified in the food chain.

This phenomenon is called biological magnification and is illustrated for DDT in Table 2.7. DDT and other chlorinated hydrocarbons have an especially high biological magnification because they have a very low biodegradability and are excreted from the body very slowly, due to dissolution in fatty tissue. These considerations also explain why pesticide residues observed in fish increase with the increasing weight of the fish (see Figure 2.23). As humans are the last link of the food chain, relatively high DDT concentrations have been observed in the human body fat (see Table 2.8).

Table 2.7 Biological Magnification

Trophic Level	Concentration of DDT (mg/kg dry matter)	Magnification
Water	0.000003	1
Phytoplankton	0.0005	160
Zooplankton	0.04	~13,000
Small fish	0.5	~167,000
Large fish	2	~667,000
Fish-eating birds2	5	~8,500,000

Source: Data after Woodwell et al., 1967.

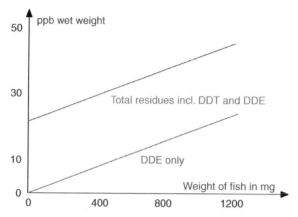

FIGURE 2.23 Increase in pesticide residues in fish as weight of the fish increases. Top line = total residues; bottom line = DDE only. *(After Cox, 1970).*

Understanding the principle of conservation of energy, called the first law of thermodynamics, was initiated in 1778 by Rumford. He observed a large quantity of heat appeared when a hole is bored in metal. Rumford assumed that the mechanical work was converted to heat by friction. He proposed that heat was a type of energy transformed at the expense of another form of energy; in his case mechanical energy. It was left to J.P. Joule in 1843 to develop a mathematical relationship between the quantity of heat developed and the mechanical energy dissipated.

Two German physicists, Mayer and Helmholtz, working separately, showed that when a gas expands the internal energy of the gas decreases in proportion to the amount of work performed. These observations led to the first law of thermodynamics: energy can neither be created nor destroyed.

If the concept internal energy, then dU, is introduced:

$$dQ = dU + dW (ML^2T^{-2}) \tag{2.37}$$

where dQ = thermal energy added to the system, dU = increase in internal energy of the system, and dW= mechanical work done by the system on its environment.

Then the principle of energy conservation can be expressed in mathematical terms as follows: U is a state variable which means that $\int dU$ is independent on the pathway 1 to 2. The internal energy, U, includes several forms of energy: mechanical, electrical, chemical, magnetic

Table 2.8   Concentration of DDT (mg per kg dry matter)

Atmosphere	0.000004
Rain water	0.0002
Atmospheric dust	0.04
Cultivated soil	2.0
Fresh water	0.00001
Sea water	0.000001
Grass	0.05
Aquatic macrophytes	0.01
Phytoplankton	0.0003
Invertebrates on land	4.1
Invertebrates in sea	0.001
Fresh-water fish	2.0
Sea fish	0.5
Eagles, falcons	10.0
Swallows	2.0
Herbivorous mammals	0.5
Carnivorous mammals	1.0
Human food, plants	0.02
Human food, meat	0.2
Man	6.0

energy, and so forth. The transformation of solar energy to chemical energy by plants conforms to the first law of thermodynamics (see Figure 2.24):

$$\text{Solar energy assimilated by plants} = \text{chemical energy of plant tissue growth} + \text{heat energy of respiration} \tag{2.38}$$

For the next level in the food chains, the herbivorous animals, the energy balance also can be set up as:

$$F = A + UD = G + H + UD, (ML^2T^{-2}) \tag{2.39}$$

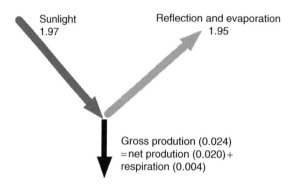

FIGURE 2.24 Fate of solar energy incident upon the perennial grass-herb vegetation of an old field community in Michigan. All values in GJ m-2 y-1.

where F = the food intake converted to energy (Joule), A = the energy assimilated by the animals, UD = undigested food or the chemical energy of feces, G = chemical energy of animal growth, and H = the heat energy of respiration.

These considerations pursue the same lines as those mentioned in context with Eq. (2.36) and Figure 2.17, where the mass conservation principle is applied. The conversion of biomass to chemical energy is illustrated in Table 2.9. The energy content per g ash-free organic material is surprisingly uniform, as is illustrated in Table 2.9. Table 2.9, part D shows ΔH, which symbolizes the increase in enthalpy, defined as H = U + p*V. Biomass can be translated into energy, and this is also true of transformations through food chains. Ecological energy flows are of considerable environmental interest as calculations of biological magnifications are based on energy flows.

Many biogeochemical models are within *narrow bands of the chemical composition of the biomass*. Eutrophication models are either based on a constant stoichiometric ratio of elements in phytoplankton or on an independent cycling of the nutrients, where the phosphorus content may vary from 0.4 to 2.5%, the nitrogen content from 4 to 12%, and the carbon content from 35 to 55%.

Some modellers have used *the second law of thermodynamics and the concept of entropy* to impose thermodynamic constraints on models; see Mauersberger (1985), who has used this constraint to assess process equations, too. Since the second law of thermodynamics is also valid for ecosystems, it raises the question: How does it apply to ecological processes?

Table 2.9*

A. Combustion Heat of Animal Material

Organism	Species	Heat of Combustion (kcal/ash-free gm)
Ciliate	*Tetrahymena pyriformis*	-5.938
Hydra	*Hydra littoralis*	-6.034
Green hydra	*Chlorohydra viridissima*	-5.729
Flatworm	*Dugesia tigrina*	-6.286
Terrestrial flatworm	*Bipalium kewense*	-5.684
Aquatic snail	*Succinea ovalis*	-5.415
Brachiipode	*Gottidia pyramidata*	-4.397
Brine shrimp	*Artemia sp.(nauplii)*	-6.737
Cladocera	*Leptodora kindtii*	-5.605
Copepode	*Calanus helgolandicus*	-5.400
Copepode	*Trigriopus californicus*	-5.515
Caddis fly	*Pycnopsyche lepido*	-5.687
	P. guttifer	-5.706
Spit bug	*Philenus leucopthalmus*	-6.962
Mite	*Tyroglyphus lintneri*	-5.808
Beetle	*Tenebrio molitor*	-6.314
Guppie	*Lebistes reticulates*	-5.823

B. Energy Values in an *Andropogus virginicus*, Old-Field Community in Georgia

Component	Energy Value (kcal/ash-free gm)
Green grass	-4.373
Standing dead vegetation	-4.290
Litter	-4.139
Roots	-4.167
Green herbs	-4.288
Average	-4.251

C. Combustion Heat of Migratory and Non-migratory Birds

Sample	Ash-Free Material (kcal/gm)	Fat Ratio (% dry weight as fat)
Fall birds	-8.08	71.7
Spring birds	-7.04	44.1
Non-migrants	-6.26	21.2
Extracted bird fat	-9.03	100.0
Fat extracted: fall birds	-5.47	0.0
Fat extracted: spring birds	-5.41	0.0
Fat extracted: non-migrants	-5.44	0.0.

D. Combustion Heat of Components of Biomass

Material	ΔH Protein (kcal/gm)	ΔH Fat (kcal/gm)	ΔH Carbohydrate (kcal/gm)
Eggs	-5.75	-9.50	-3.75
Gelatin	-5.27	-9.50	

Continued

Table 2.9*—cont'd

Material	ΔH Protein (kcal/gm)	ΔH Fat (kcal/gm)	ΔH Carbohydrate (kcal/gm)
Glycogen			-4.19
Meat, fish	-5.65	-9.50	
Milk	-5.65	-9.25	-3.95
Fruits	-5.20	-9.30	-4.00
Grain	-5.80	-9.30	-4.20
Sucrose			-3.95
Glucose			-375
Mushroom	-5.00	-9.30	-4.10
Yeast	-5.00	-9.30	-4.20

*Source: Morowitz, 1968.

Ecological models contain many parameters and process descriptions and at least some interacting components, but the parameters and processes can hardly be given unambiguous values and equations, even by using the previously mentioned model constraints. It means that an ecological model in the initial phase of development has many degrees of freedom. It is necessary to limit the degrees of freedom to develop a workable model.

Many modellers use a comprehensive data set and calibration to limit the number of possible models. Nonetheless, this is a cumbersome method if it is not accompanied by some realistic constraints on the model. Calibration is therefore often limited to give the parameters realistic and literature-based intervals, within which the calibration is carried out, as mentioned in Section 2.10.

But far more would be gained if it were possible to give the models more ecological properties and/or test the model from an ecological point of view to exclude those versions of the model that are not ecologically possible. For example: How could the hierarchy of regulation mechanisms be accounted for in the models? Straskraba (1979, 1980) classified models according to the number of levels that the model includes from this hierarchy. He concluded that we need experience with the models of the higher levels to develop structural dynamic models. This is the topic for Chapter 10.

We know that evolution has created very complex ecosystems with many feedback mechanisms, regulations, and interactions. The coordinated co-evolution means that rules and principles for the cooperation

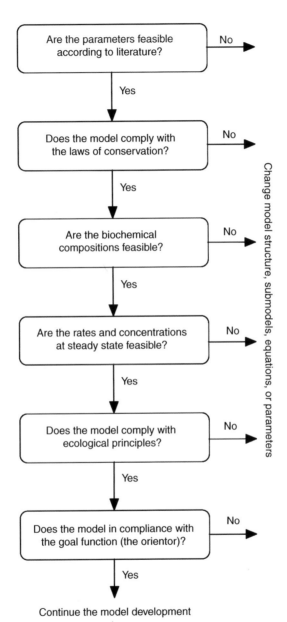

FIGURE 2.25 Considerations on using various constraints by development of models. The range of parameter values is particularly limited by the procedure shown.

among the biological components have been imposed. These rules and principles are the governing laws of ecosystems, and our models should follow these principles and laws.

It also seems possible to limit the number of parameter combinations by using what could be named "ecological" tests. The maximum growth rates of phytoplankton and zooplankton may have realistic values in a eutrophication model, but when the two parameters do not fit to each other because they will create chaos in the ecosystem, it is inconsistent with the actual or general observations. Such combinations should be excluded at an early stage of the model development.

Figure 2.25 summarizes the considerations of using various constraints to limit the number of possible values for parameters, possible descriptions of processes, and possible submodels to facilitate the development of a feasible and workable model. The two last steps of the procedure will be presented in Chapter 10, where the next generation models are developed.

It requires the introduction of variable parameters, governed by a goal function (an orientor). Several possible goal functions have to be introduced before a presentation of structural dynamic models can take place.

Problems

1. Which class of models would you select for the following problems:
   a. Protection of a lion population in a national park?
   b. Optimization of fishery in marine environment?
   c. Construction of a wetland for denitrification of nitrate input from agriculture?
2. Explain the importance of verification, calibration, and validation. Can models without these three steps be developed at all?
3. Find the concentration factor of cadmium for a whale, estimated to have a length of 20 m.
4. The ammonia excretion for a fish of 500 g is 200 mg/24h. Estimate the ammonia excretion for a fish of 4 kg. What is the excretion rate of a shark of 2000 kg?
5. Set up an adjacency matrix for the models shown in Figure 2.10 and 2.11.

6. Improve the model in Figure 2.5 by adding two more state variables. Which two state variables would probably be most important to add to the present model focused on eutrophication?

7. How often would you determine the phytoplankton concentration, if a model for the diurnal variations of primary production during a month was supposed to be modelled? Would the number of observations be dependent on the season? If yes, why?

8. Set up the equations for a model explaining the accumulation of DDT in fish according to Figure 2.23.

9. How many state variables could a model have, if all the relationships are based entirely on 10,000,000 observations?

10. Develop a model for the biomagnification of a toxic substance through a food chain with primary producers, primary consumers, and secondary consumers.

3

An Overview of Different Model Types

3.1. Introduction

In Chapter 2, a modelling procedure was presented and how to select close to optimum complexity was discussed. In the 1970s, when ecological models started to be applied in environmental management and as a scientific tool in systems ecology, most of the applied models were of three types: population dynamic, bioenergetic, and biogeochemical. For the first type, conservation of the number of individuals in a population was applied to set up the equations. For the second and third types, the conservation of energy and/or mass was the key principle applied for the development of the equations (see Table 2.3). For all three types, there can be both a dynamic version using the differential equations and a steady-state version using algebraic equations. The steady-state versions are used when the problem can be solved by presentation of an average situation or a worst-case situation, presuming both situations are steady state. In data-poor situations, it is often beneficial to apply the steady-state versions, because the quality and the quantity of the data are not sufficient to develop a dynamic model. For all three types, the development of a conceptual model is the first step to visualizing how the state variables are connected by processes. Sometimes the conceptual model is considered an independent model type when it is developed only to get

Fundamentals of Ecological Modelling. DOI: 10.1016/B978-0-444-53567-2.00003-X an overview of the model components and how they are connected through processes such as transfer of mass, energy, and/or information. This does not prevent the conceptual model from being used to further develop the model. Section 3.3 presents conceptual models that can be visualized by several different methods.

In the last 20 to 30 years, several new types of models have emerged to solve a wide spectrum of problems that cannot be solved by the application of bioenergetic models, biogeochemical models, population dynamic models, or even conceptual models. This makes the selection of model type even more complicated. It is therefore crucial to have a good overview of all the available model types and their characteristics to be able to choose the model type that best meets the model objectives. In this chapter, the information needed to be able to make the best selection of model type will be presented.

3.2. Model Types — An Overview

The new model types, developed during the last couple of decades, have been created to answer a number of relevant modelling problems or questions that arose as a result of the increasing use of ecological models in the 1970s. Seven relevant modelling questions formulated around 1980 as a result of this model experience are listed below:

1. How can we describe the spatial distribution that is often crucial to understand ecosystem reactions and to select the best environmental strategy?
2. Ecosystems are middle number systems (Jørgensen, 2002). Since all of the components are different, what is the proper description of the ecosystem reactions when considering the differences in properties among individuals?
3. The species are adaptable and may change their properties to meet the changes in the prevailing conditions, which means forcing functions. Furthermore, the initial species may be replaced by other species better fitted to the combinations of forcing functions. How should we account for these changes? Even the networks may change if more biological components with very different properties are replaced by other species. How should we account for these structural changes?

4. Can we model a system that has a poor database — a few data of only low quality?
5. The forcing functions and several ecological processes are in reality stochastic. How do we account for the stochasticity?
6. Can we develop a model when our knowledge is mainly based on a number of rules, properties, and propositions?
7. Can we develop a model based on data from a wide spectrum of different ecosystems, which means that we have only a very heterogeneous database?

These problems could not be solved by the three "old" model types mentioned in Section 3.1, but they have all found a solution with the new model types.

Spatial models often based on the use of Geographical Information System (GIS) have been developed to answer question 1. Individual-based models (IBMs) are able to answer question 2. Software that can be used to develop IBMs is even available to facilitate IBM development. This software can also be utilized to cover spatial distribution (see question 1). Structurally dynamic models (SDM) have been developed to solve the problem expressed in question 3. Fuzzy models can be used to make models based on a poor or semiquantitative database. Stochastic models were not often applied in the 1970s, but they are still used today. The application remains infrequent, probably because an urgent need to include stochastic processes in ecological models does not happen often. IBMs can often meet the demands expressed in question 6. Artificial neural networks (ANN) are a good solution to the problem formulated in question 7.

Ecotoxicological models, discussed in Chapter 8, are sometimes considered a special model type. They are developed similar to other biogeochemical models, and have been widely used, particularly the last 10–15 years, because they are needed for environmental risk assessment of chemicals. It is therefore relevant to devote a special chapter to ecotoxicological models.

This book presents the development of biogeochemical models (both dynamic and steady-state types are discussed), population dynamic models, spatial models, ecotoxicological models, structurally dynamic models, IBMs, fuzzy models, and application of ANN. These types are the most common types (see Jørgensen, 2008b; Jørgensen, Chon, & Recknagel, 2009). Different methods used to present conceptual models considered as the first model step for any of the nine types of models will be presented in the next section.

Table 3.1 gives a summary of model statistics based on the number of publications in the journal *Ecological Modelling*. The percentage application of most general model types from the 1975 to 1980 are compared to the period from 2001 to 2006. Ecotoxicological models are included as a model type, although they are constructed similar to biogeochemical models. In Table 3.1, we have distinguished between nine types of models:

1. Dynamic biogeochemical
2. Steady-state biogeochemical
3. Population dynamics
4. Spatial
5. Structurally dynamic
6. Individual-based
7. Ecotoxicological
8. Fuzzy
9. Artificial Neural Networks

Table 3.1 Application of the Most General Model Types From 1975 to 1980 with the Model Types from 2001 to 2006

	% Application 1975–1980	% Application 2001–2006
Dynamic biogeochemical models	62.5	32.0
Steady state biogeochemical models	0	1.8
Population dynamic models	31.0	24.9
Spatial models	0	19.1
Structurally dynamic models	1.5	8.0
Individual-based and cellular automata	0	5.2
Artificial Neural Networks and use of artificial intelligence	0	4.9
Fuzzy models	0.5	1.8
Ecotoxicological models	0	2.2

The data in Table 3.1 are shown graphically in Figures 3.1 and 3.2 and are reproduced from Jørgensen (2008b).

The number of papers published from 2001 to 2006 is about nine times the number of papers published from 1975 to 1980. This means that the number of dynamic biogeochemical model papers published recently is more than 4.5 times the number published during the late 1970s, and that the number of papers on structurally dynamic modelling has increased by a factor of almost 50 during the last 35 years. A comparison of Figures 3.1 and 3.2 also shows that the spectrum of model types applied today is much wider than applied about 30 years ago. This is not surprising as the new types of models were developed because there was an urgent need to answer the seven modelling problems previously listed.

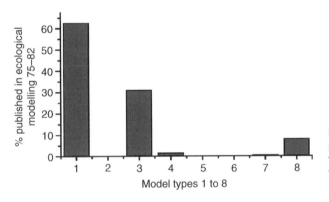

FIGURE 3.1 Percentage of papers published about the eight model types in *Ecological Modelling* from 1975 to 1982 previously listed and in Table 3.1.

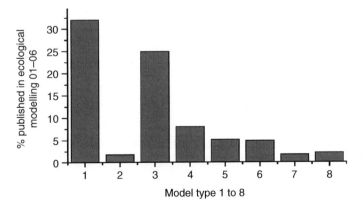

FIGURE 3.2 Percentage of papers published about the eight model types in *Ecological Modelling* from 2001 to 2006 previously listed and in Table 3.1.

With the present spectrum of model types, it is possible to address the major modelling problems from the 1970s. This development has increased the application of ecological models in general, particularly the use of the new model types. However, it is also clear that all the problems cannot be solved completely. We still have a number of problems that may not be possible to solve by use of a single model type. The very complicated problems require the use of hybrid models — a combination of the model types presented in this chapter.

New model types may be developed in the future to solve the complicated problems that today require the use of hybrid models. It is, however, agreed among ecological modellers that we currently have a sufficient toolbox of model types to address many ecological modelling problems we now face, although some modelling case studies are still need to guarantee their feasibility in real situations.

3.3. Conceptual Models

A conceptual model has a function of its own. If flows and storage are given by numbers, then the diagram gives an excellent survey of a steady-state situation. It can give a picture of the changes in flows and storages if one or more forcing functions are changed and another steady-state situation emerges. If first-order reactions are assumed, then it is even easy to compute other steady-state situations which might prevail under other combinations of forcing functions (see Chapter 6). Conceptualization is one of the early steps in the modelling procedure (see Chapter 2), but it can also have a function of its own, as will be illustrated in this section.

A *conceptual model* can be considered as a list of state variables and forcing functions of importance to the ecosystem and the problem in focus, but it also shows how these components are connected by processes. It is employed as a tool to create abstractions of reality in ecosystems and to delineate the level of organization that best meets the objectives of the model. A *wide spectrum* of conceptualization approaches is available and will be presented in this chapter. Some conceptual models give only the components and the connections; others imply the first steps toward a mathematical description.

It is almost impossible to model without a conceptual diagram to visualize the modeller's concepts and the system. The modeller usually plays with the idea of constructing various models of different complexity at this stage in the modelling procedure, making the first assumptions and selecting the complexity of the initial model or alternative models. It requires intuition to extract the applicable parts of the knowledge about the ecosystem and the problem involved. Models attempt to make a synthesis of what we know, and the conceptual diagram is the first step of this synthesis.

Construction of a conceptual diagram is system and developer dependent, but it is often better at this stage to use a slightly too complex model rather than an approach that is too simple. In the later stage of modelling, it is easy to exclude redundant components and processes. On the other hand, it makes the modelling too cumbersome if an overly complex model is used even at this initial stage. Generally, good knowledge about the system and the problem facilitates the conceptualization step and increases the chance to get closer to the right complexity for the initial model. The questions to be answered include: What components and processes of the real system are essential to the model and the problem? Why? How? In this process a suitable balance is sought between elegant simplicity and realistic detail.

Identification of the level of organization and selection of the needed complexity of the model are not trivial problems. Miller (1978) indicated 19 hierarchical levels in living systems. To include all of them in an ecological model is impossible, mainly due to lack of data and a general understanding of nature. Usually, it is not difficult to select the focal level — where the problem is or where the components of interest operate. The step below the focal level is often relevant for a good description of the processes; for instance, photosynthesis is determined by the processes occurring in the individual plants. One step higher than the focal level determines many of the constraints. These considerations are visualized in Figure 3.3.

In most cases it is not necessary to include more than a few or even only one hierarchical level to understand a particular behavior of an ecosystem at a particular level (see Patten, 1971, 1976; Wilson, 2000; Miller, 1978; Allen, 1976; Allen & Starr, 1982). Figure 3.4 illustrates a model with three hierarchical levels, which might be needed if a multi-goals model is constructed. The first level could be a hydrological model, the next level a eutrophication model, and the third level a model of phytoplankton growth considering the intracellular nutrients concentrations.

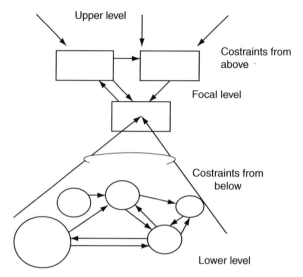

FIGURE 3.3 The focal level has constraints from both lower and upper levels. The lower level determines to a large extent the processes, and the upper level determines many of the constraints on the ecosystem.

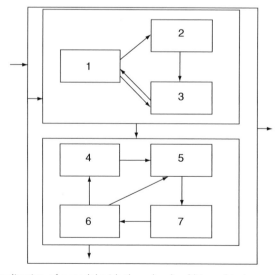

FIGURE 3.4 Conceptualization of a model with three levels of hierarchical organization.

Each submodel has its own conceptual diagram; for example, the conceptual diagram of the phosphorus flows in a eutrophication model (see Chapter 7). In the latter submodel there may be a sub-submodel considering the growth of phytoplankton by use of intracellular nutrients concentrations which is shown as a conceptual diagram in Figure 3.5. The nutrients are taken up by phytoplankton at a rate determined by the temperature and nutrient concentration in the cells and in the water. The closer the nutrient concentrations in the cells are to the minimum, the faster the uptake. The growth, on the other hand, is determined by solar radiation, temperature, and the concentration of nutrients in the cell. The closer the nutrient concentration is to the maximum concentration, the faster the growth. This description is according to phytoplankton physiology and will be presented in Chapter 7.

The modeller can choose among several conceptualization methods for the development of the conceptual diagram. Six of the most applied methods are presented next. Which one to choose depends on how much information the modeller wants to include in the conceptual diagram. The more information the modeller includes, the more informative the diagram is, but it becomes more difficult to interpret and manage.

Picture conceptual models use components seen in nature and place them within a framework of spatial relationships. Figure 3.6 is a simple example.

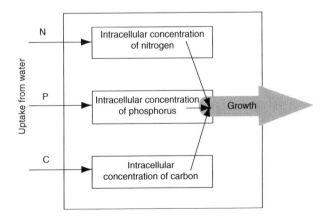

FIGURE 3.5 A phytoplankton growth model with two hierarchical levels: the cells that determine the uptake of nutrients and the phytoplankton population production (growth) determined by the intracellular nutrient concentrations. This model is applied in Chapter 7.

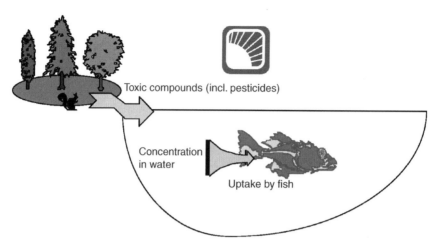

FIGURE 3.6 Example of a picture model: pesticides are coming from the littoral zone, resulting in a certain concentration in the water. Fish take up the toxic compounds directly from the water. The model attempts to answer the crucial question: What is the concentration of the toxic substance in the fish?

Box conceptual models are simple and commonly used conceptual designs for ecosystem models. Each box represents a component in the model and arrows between boxes indicate processes. Figures 2.1, 2.10, and 2.11 show examples of this model type. The conceptual diagrams show the nutrient flows (nitrogen and phosphorus) in a lake. The arrows indicate mass flows caused by processes. Some modellers prefer other geometric shapes, for example, Wheeler et al. (1978) preferred circles to boxes in their conceptualization of a lead model. This results in no principal difference in the construction and use of the diagram. A box model for predicting the carbon dioxide concentration in the atmosphere and the consequences for the climatic changes will be presented in Chapter 7, Section 7.7.

The term *black-box model* is used when the equations are set up based on an analysis of input and output relations, for example, by statistical methods. The modeller is not concerned with the causality of these relations, and such a model might be very useful provided the input and output data are of sufficient quality. Yet, the model can only be applied to the case study for which it has been developed. New case studies will require new data, a new analysis of the data, and, consequently, new relations. *White-box* models are constructed based on causality for all processes. This does not imply that these models can be applied to all similar case studies, because, as previously discussed, a model inevitably reflects ecosystem characteristics. In general, a white-box model will be applicable to other case studies with some minor or major modifications. In practice, *most models are gray,* as they contain some causalities but also often apply empirical expressions to account for some of the processes.

Input/output models differ only slightly from box models; they can be considered as box models with numerical indications of inputs and outputs. An example of this type of model is shown in Figure 3.7, which is an oyster community model developed by Dame and Patten (1981).

The *feedback dynamics diagrams* use a symbolic language introduced by Forrester (1961) (Figure 3.8). Rectangles represent state variables, parameters or constants are small circles, sinks and sources are cloud-like symbols, flows are arrows, and rate equations are the pyramids that connect state variables to the flows. Several modifications have been developed and they differ from the Forrester diagrams by giving more

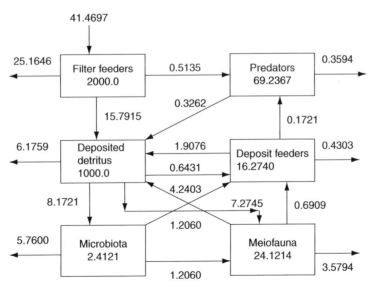

FIGURE 3.7 Input/output model for energy flow (cal m^{-2} d^{-1}) and storage (kcal m^{-2}) in an oyster reef community. Matrix representation: 1. filter feeders, 2. deposited detritus, 3. microbiota, 4. meiofauna, 5. deposit feeders, and 6. predators

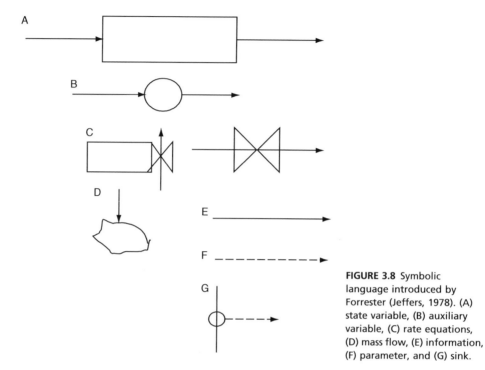

FIGURE 3.8 Symbolic language introduced by Forrester (Jeffers, 1978). (A) state variable, (B) auxiliary variable, (C) rate equations, (D) mass flow, (E) information, (F) parameter, and (G) sink.

information about the processes. The conceptualization used in the model development software STELLA (see Figure 2.3) uses symbols similar to the Forrester diagram (compare Figures 2.3 and 3.8).

Energy circuit diagrams, developed by H. T. Odum (1983), are designed to give information on thermodynamic constraints, feedback mechanisms, and energy flows. The most commonly used symbols in this language are shown Figure 3.9. As the symbols have an implicit mathematical meaning, it gives an abundance of information about the mathematics of the model. Furthermore, it is rich in conceptual information and hierarchical levels can easily be displayed. Numerous other examples can be found in the literature (Odum, 1983; Odum & Odum, 2000). A review of these examples reveals that energy circuit diagrams are very informative, but they are difficult to read and survey when the models become a little more complicated. On the other hand, it is easy to set up energy models from energy circuit diagrams. Sometimes it is even sufficient to use the energy circuit diagrams

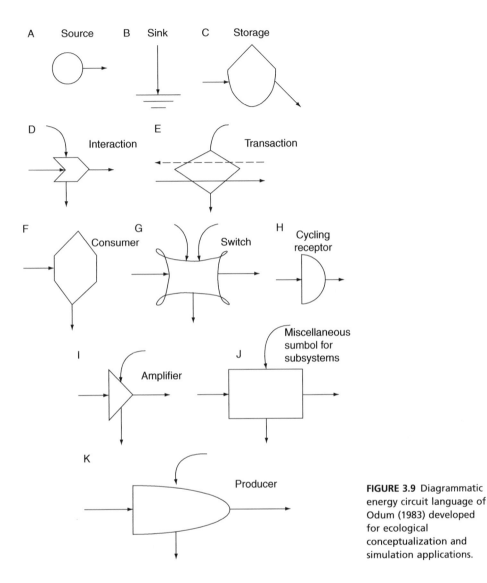

FIGURE 3.9 Diagrammatic energy circuit language of Odum (1983) developed for ecological conceptualization and simulation applications.

directly as energy models. These diagrams have found a wide application for development of ecological/economic models, where the energy is used as the translation from economy to ecology and vice versa. H.T. Odum has used the approach for developing models for entire countries.

3.4. Advantages and Disadvantages of the Most Applied Model Types

The characteristics, advantages, and disadvantages (mostly expressed as a limitation of the application) for all of the available main model types are given in this section. The applicability of the various model types is discussed in the next section. The application of catastrophe theory and chaos theory are not included in the overview, because they can be considered mathematical tools that, in principle, can be applied as mathematical tools in the development of several different model types. Furthermore, statistical models in this textbook are not considered as a particular model type, but as a tool that can be applied in ecological modelling to give a better process description. If a model is based entirely on application of statistics, then it is denoted as a black-box model, because it has no causality. Black-box models are not used to uncover new ecological knowledge where the focus is on causality. A short review of the most applied model types based on Jørgensen (2008b) is given in the following list.

1. **Biogeochemical and bioenergetic dynamic models.** This model type is widely used, as can be seen in Table 1.1. It applies differential equations to express the dynamics. Change in state variables are expressed as the results of the ingoing minus the outgoing processes and the model is therefore based on conservation principles. The process equations are usually based on causality. The model type has some clear advantages that make it attractive to use for the development of many models.
   Advantages:
   * Most often based on causality
   * Based on mass or energy conservation principles
   * Easy to understand, interpret, and develop
   * Software is available (e.g., STELLA)
   * Easy to use for predictions
   Disadvantages:
   * Not used for heterogeneous data
   * A relatively good database is required
   * Difficult to calibrate when they are complex and contain many parameters

- Do not account for adaptation and changes in species composition

The advantages and disadvantages define the area of application such as the description of the state of an ecosystem when a good data set is available. This model type has been extensively applied in environmental management as a powerful tool to understand the reactions of ecosystems to pollutants and to set up prognoses.

2. **Steady-state biogeochemical models.** Due to the limitations of this model type, it has not been used in more than 1.8% of the publications in *Ecological Modelling* from 2001 to 2006. This model type is a biogeochemical or bioenergetic dynamic model where the differential equations all are set to zero to obtain the values of the state variables corresponding to the static situation.
   Advantages:
   - Require generally smaller databases than most other types
   - Excellent for worst-case or average situations
   - Results are easily validated (and verified)
   Disadvantages:
   - Do not give any information about dynamics and changes over time
   - Prediction with time as independent variable is not possible
   - Only give average or worst-case situations
   This model type is often used when a static situation is sufficient to give a proper description of an ecological system or to make environmental management decisions.

3. **Population dynamic models**. This model type is rooted in the Lotka-Volterra model developed in the 1920s. Numerous papers have been published about the mathematics behind this model and a number of deviated models developed. The mathematics of these equation systems are not very interesting from an ecological modelling point of view, where the focus is a realistic description of ecological populations. Population dynamic models may include age structure, which in most cases is based on matrix calculations. The number of population dynamic papers is 5 times as much today as in the late 1970s, which illustrates that ecological modelling has developed significantly over the past 30 years. The

minor reduction in percentage is due to the application of a much wider spectrum of different model types.

Advantages:

- Able to follow the development of a population
- Age structure and impact factors can easily be considered
- Easy to understand, interpret, and develop
- Most often based on causality

Disadvantages:

- Conservation principles are sometimes not applied, although it is easy in most cases
- Application is limited to population dynamics
- Require a relatively good and homogenous database
- Difficult to calibrate in some situations

This model type is typically applied to keep track of the development of a population. The number of individuals is the most applied unit, but it can easily be translated into biomass or bioenergy. Effects of toxic substances on the development of populations can easily be covered by increasing the mortality and decreasing the growth corresponding to the effect of the toxic substance. This model type is extensively used in the management of fisheries and other natural resources and national parks.

4. **Structurally Dynamic Models**. This model type can change the parameters, corresponding to the properties of the biological modelling components to account for adaptation and changes in species composition. It is possible either to use knowledge or artificial intelligence to describe the changes in the parameters. Most often a goal function is used to find the parameter changes. The thermodynamic variable, eco-exergy, is a commonly used goal functions in structurally dynamic models. Minor changes of the parameters may be due to adaptation to the changed conditions, but for major changes, it is most probably due to a change in the state variables — that is, a shift in the species composition — that causes the changed parameters. This approach can be used for a major change in the ecological network, although no reference to this application of the structurally dynamic modelling approach is yet available. SDMs are applied much more today than 25 to 30 years ago.

Advantages:
- Able to account for adaptation
- Able to account for shift in species composition
- Can be used to model biodiversity and ecological niches
- Parameters determined by the goal functions do not need to be calibrated
- Relatively easy to develop and interpret

Disadvantages:
- Selection of a goal function or use of artificial intelligence is needed
- Computer use is time-consuming
- Information about structural changes is needed for a proper calibration and validation
- No available software; programming needed (in most cases C++ has been used)

This model type should be applied whenever it is known that structural changes take place. It is also recommended for models that are used in environmental management to make prognoses resulting from major changes in the forcing functions (impacts).

5. **Fuzzy models.** This type of model may either be knowledge-based (the Mamdani type) or data-based (the Sugeno type). Mamdani-type models are based on a set of linguistic expert formulations, and they are applied when no data are available. The Sugeno-type model applies an optimization procedure and is applied when only uncertain data are available.

Advantages:
- Can be applied on a fuzzy data set
- Can be applied on semiquantitative (linguistic formulations) information
- Can be applied for development of models where a semiquantitative assessment is sufficient

Disadvantages:
- Not usable for more complex model formulations
- Cannot be used where numeric indications are needed
- No software available to run this type of model, although there are facilities in Matlab to run fuzzy models

This model type is applied when the data set is fuzzy or only semiquantitative expert knowledge is available, provided that the semiquantitative results are sufficient for the ecological description or the environmental management.

6. **Artificial Neural Networks.** This model type is able to show relationships between state variables and forcing functions based on a heterogeneous database. In principle, it is a black-box model and therefore not based on causality. It is very useful when applied for prognoses, provided that the model has been based on a sufficiently large database that allows the discovery of relationships and to test these discoveries on an independent data set. This model type was not applied in ecological modelling before 1982. It can be developed by using available software or Matlab.
Advantages:
- May be used where other methods fail
- Easy to apply
- Give a good indication of the certainty due to the application of a test set
- Can be used on a heterogeneous data set
- Give a close to optimum use of the data set

Disadvantages:
- No causality unless algorithms are introduced or a hybrid between ANN and another model type is applied
- Cannot replace biogeochemical models based on the conservation principles
- Accuracy of predictions is sometimes limited, although validation is almost always used.

The advantages and disadvantages of this model type indicate where it would be advantageous to apply ANN; namely where ecological descriptions and understandings are required on the basis of a heterogeneous database, such as data from several different ecosystems of the same type. It is also often applied beneficially when the database is more homogeneous; for instance, when the focus is on a specific ecosystem. The modeller should seriously consider using biogeochemical dynamic models due to their causality. ANN is, however, faster to use and the time-consuming calibration that is part of the biogeochemical models is not necessary.

7. **Spatial models.** Spatial differences of the forcing functions and the nonbiological and biological state variables may be decisive for model results, and are often required to obtain model results that reveal spatial differences. They are often urgently needed to understand the ecological reactions or to make a proper environmental management strategy. Models that produce spatial differences must also consider the spatial differences in the processes, forcing functions, and state variables. Due to the urgent need for a proper description of the spatial differences, it is not surprising that the journal *Ecological Modelling* has published almost 250 papers about spatial modelling from 2001 to 2006 and that the number of models that focus on spatial distribution is increasing rapidly. There are a number of ways to cover the spatial differences in the development of an ecological model. It is not possible to review them all here, but this important model type is presented in more detail in Chapter 11. For aquatic ecosystems, the ultimate spatial model is a 3D description of the processes, forcing functions, and state variables. When studying this ecosystem, there are often questions regarding a good description of hydrodynamics. There has been an increasing use of models that couple 3D hydrodynamic models and ecological models.

   Advantages:
   • Cover spatial distribution, which is often important in ecology
   • Results can be presented in many informative ways, for instance, GIS

   Disadvantages:
   • Require a huge database
   • Calibration and validation are difficult and time-consuming
   • A very complex model is usually needed to properly describe the spatial patterns

   Spatial models are applied whenever it is required that the results include the spatial distribution, because it is decisive or the spatial distribution is crucial to the model results. Landscape models covering the exchange of matter among several different ecosystems in a landscape have been developed.

8. **Individual Based Models**. This model type was developed because all the biological components in ecosystems have different

properties, which is not considered in biogeochemical or population dynamic models. Within the same species the differences are minor and are therefore often neglected in biogeochemical models, but the differences among individuals of the same species may sometimes be important for ecological behavior. For instance, individuals may have different sizes, which gives a different combinations of properties from the allometric principles (see Chapter 2, Section 2.9). The right property may be decisive for growth and/or survival in certain situations. Consequently, a model that ignores the differences among individual species could produce a completely wrong result. Advantages:

- Able to account for individuality
- Able to account for adaptation within the spectrum of properties
- Software is available; although the choice is more limited than software used by biogeochemical dynamic models
- Spatial distribution can be covered

Disadvantages:

- When a number of properties are considered, the models get very complex
- Cannot always cover mass and energy transfer based on the conservation principle
- Require a large amount of data to calibrate and validate the models

As mentioned earlier, we know that the individuals have different properties that may sometimes be crucial for model results. In such cases, IBMs are absolutely needed.

9. **Ecotoxicological models**. Ecotoxicological models, in principle, do not represent a separate model type. Biogeochemical models or population dynamic models are applied widely in ecotoxicology. It is, however, preferable to treat ecotoxicological models as a separate model type, because they are characterized by the following:

- Our knowledge of the parameters is limited so estimation methods are needed to a much larger extent than for other model types. Fortunately, many estimation methods are available in ecotoxicology to estimate process rates.

- Due to the use of safety factors and the limited knowledge of the parameters, ecotoxicological models are often quite simple; particularly, the so-called fugacity models.
- They should often include an effect component.

Advantages:
- Tailored to ecotoxicological problems
- Usually simple to use
- Often includes an effect component or can easily be interpreted to quantify the effect

Disadvantages:
- The number of parameters needed to develop models for all toxic substances is very high and we know approximately 1% of these parameters
- It implies that we need estimation methods that inevitably have a high uncertainty; model results therefore have a high uncertainty
- Inclusion of an effect component requires knowledge of the effect, which is also limited

The area of application for this model is to solve ecotoxicological research and management problems and perform environmental risk assessment for the application of chemicals.

10. **Stochastic models**. This model type is characterized by an element of randomness. The randomness could be in the forcing functions, particularly the climatic forcing functions, or it could be in the model parameters. In both cases, it is caused by a limitation in our knowledge. For instance, we may not know the temperature on May 15 next year at a given location, but we know the normal distribution of the temperature over the last hundred years and can use it to represent the temperature on this date. Similarly, many of the parameters in our models are dependent on random forcing functions or on factors that we cannot include in our model without making it too complex. Using Monte Carlo simulations based on this knowledge, it is possible to consider the randomness. By running the model many times, it becomes possible to obtain the uncertainty of the model results. A stochastic model may be a biogeochemical/ bioenergetic model, a spatial model, a structural dynamic model,

an IBM, or a population dynamic model. In principle, a model can become a stochastic model regardless of its type.

Advantages:

* Able to consider the randomness of forcing functions or processes
* Uncertainty of the model results are easily obtained by running the model many times

Disadvantages:

* Must know the distribution of the random model elements
* High complexity and requires many hours of computer time

It is recommended to apply stochastic models whenever the randomness of forcing functions or processes are significant.

11. **Hybrid models**. In principle, hybrid models are any combination of two of the previously listed ten model types; but only few hybrid models have been developed. It is expected that many more will be developed in the future to combine some of the advantages and eliminate some of the disadvantages of the existing models. *Ecological Modelling* has published several hybrid models that combine a biogeochemical/bioenergetic dynamic model with other model types. The result of combining a biogeochemical dynamic model and an ANN is a hybrid model that may have causality and is able to squeeze as much information out of the database as an ANN.

3.5. Applicability of the Different Model Types

Which model types are recommended to solve which problems? What are the data requirements of the different model types? These are questions answered in the first step of modelling development (see Figure 2.2). As mentioned in the introduction to this chapter, new model types were needed to solve specific problems that emerged during the late 1970s, when ecological modelling started to be applied more extensively as a tool in ecological research and environmental management. Biogeochemical/bioenergetic dynamic models and population dynamic models have shortcomings that ecological modellers have tried to solve for the last 30 years by developing new model types. Today, the shortcomings have at least been partially eliminated by the development of new model types, particularly spatial models, IBMs, and SDMs. It is possible with the available model types to make the best choice in a given model situation, which is defined by the available data and the combination of problem and system. It is possible to recommend a particular model type from the eleven model types presented in Section 3.4 based on (1) available data sets and the (2) combination of problem and system.

The core question, "Which model type should be applied in which context?", is answered in Tables 3.2 and 3.3, which cover, respectively, different data sets and different problem/system combinations.

Now we have a wide spectrum of model types available to solve a wide spectrum of relevant ecological problems, which include a description of shifts in species compositions, ecotoxicological effects and spatial distributions, and the use of heterogeneous data sets and uncertain data sets. This wide spectrum of models is richly represented in the ecological modelling literature (Table 3.4).

Table 3.2 Selection of Model Type Based on the Available Data Set

Data Set	Recommended Model Type
High quality, homogeneous	Biogeochemical dynamic models and population dynamic models
Medium-high quality, heterogeneous	ANN
Low quality, homogeneous	Steady-state model
Uncertain data	Fuzzy models
No data only rules	Fuzzy models

Table 3.3 Selection of Model Type Based on Problem/System

Problem/System	Recommended Model Type
Exchange of matter and/or energy	Biogeochemical dynamic model
Population dynamics	Population dynamic model
Toxic substances, distribution and effect	Ecotoxicological model
Individuality important for the results	IBM
Structural changes occur	SDM
Adaptation significant	SDM
Spatial differences	Spatial model
Stochasticity important for the results	Stochastic model

Table 3.4 Case Studies Illustrating the Application of the Various Model Types

Model Type	Description of Case Study
Dynamic biogeochemical models	Eutrophication of a lake; a relatively good database available
Steady-state biogeochemical models	Eutrophication of a lake; only three annual average values available
Population dynamic models	Management of deer in a national park; a relatively good database available
Spatial models	Distribution of nutrients in a landscape; a relatively good database available
Structurally dynamic models	Oxygen deficiency in a stream; significant changes of control functions are expected, and knowledge about shifts in species composition available
IBMs	Growth of trees in a forest; the trees have very different conditions (sun, exposure to wind, soil, etc.) and a good database with different growth pattern under different conditions available
ANN and use of artificial intelligence	The presence of different fish species in a wide spectrum of different streams; a huge but heterogeneous database available
Fuzzy models	Presence or absence of 5 species of songbirds in 20 different wetlands
Ecotoxicological models	The fate of an insecticide used in agriculture; an agricultural area, a wetland, and a stream are considered

Problems

1. Which type of model would you select for the following problems?
   a. Protection of a lion population in a national park
   b. Optimization of a fishery in a marine ecosystem like the North Sea
   c. Construction of a wetland for removal of nitrate mainly by denitrification
   d. Adaptation (change in the rate of evapotranspiration) of plants to a dry climate

    e. Interpretation of a database with 12 stations in drainage areas, rivers, and many observations as function of time for the stations of (i) water quality, (ii) fish diversity, (iii) dominant fish species, and (iv) the use and composition of the land adjacent to the rivers

2. Consider a shallow lake with a surface of 100 ha and a depth of 2 m has an initial phosphorus concentration of 0.1 mg/L. The loading is 100 kg/year. The water retention time is 4 months. No input of phosphorus from the sediment is considered.

    a. What would be the concentration of phosphorus in the lake by steady state if no settling of phytoplankton or suspended matter takes place?

    b. What would be the concentration of phosphorus in the lake at steady state if it is considered that phosphorus is settled by a rate of 10 m/24h?

    c. What would be the phytoplankton concentration in the two cases if phytoplankton contains 1% phosphorus?

    The differential equation needed to answer question (a) and (b) should be indicated and the steady-state solution should be found.

3. Draw a Forrester and energy circuit diagram for Figure 2.9.

4. Develop a STELLA diagram of the picture model in Figure 3.6. Set up an adjacency matrix for the model.

5. Set up an adjacency matrix for the model in Figure 3.7.

6. Give an example of a case study that is best solved by use of the nine model types listed in Table 3.1. Describe the case study by the problem, the ecosystem, and the data needed. Present the answer by use of a table.

4

Mediated or Institutionalized Modelling

CHAPTER OUTLINE

4.1. Introduction: Why Do We Need Mediated Modelling?

The following questions arise as result of the proposed modelling procedure (Figure 2.2):

- How is it possible to consider the many different aspects of an environmental problem including natural science aspects such as geology, zoology, botany, and chemistry as well as the economic and social aspects?
- The answer to this question is to implement a very wide spectrum of expertise in the modelling team, but it gives rise to the next question: How do you ensure good cooperation from team members when they represent many different disciplines and have many different opinions and "languages"?
- How is it possible to consider *all* relevant ecosystem properties at the same time?
- How is it possible to integrate these insights?
- How can we ensure that all important stakeholders are included in the modelling process?
- How is it possible to integrate impacts and knowledge at different scales?

Fundamentals of Ecological Modelling. DOI: 10.1016/B978-0-444-53567-2.00004-1

121

- How is it possible to understand the very root of the problems and their sources and have this understanding reflected in the modelling and the final model result?
- How is it possible to build the best consensus among the different opinions and disciplines?

Institutionalized or mediated modelling (IMM) can address these questions. The main idea of IMM is to represent without exception *all* stakeholders, policymakers, managers, and scientists with knowledge and ideas about the problem, the system, and possible solutions for the modelling procedure. The model is developed as a result of integrated brainstorming where all ideas, opinions, disciplines, and knowledge are represented. For the development of most mediated models, depending on the complexity of the problem and the system, several days of intense interaction among participants are required to reach a satisfactory basis for model development. The advantages of IMM (partly taken from van den Belt, 2004) are that the:

1. Level of shared understanding increases;
2. Consensus is built about the structure of a complex problem for a complex system, because all interests are represented in the stepwise model development;
3. Result of the modelling process, the model, serves as a tool to disseminate the insights gained by the modelling procedure;
4. Effectiveness of the decision making is increased, because the mediated model makes it possible for policymakers and the stakeholders to see the consequences of the action plans over longer time scales;
5. Team building is developed parallel to the model development;
6. Process is emphasized over the product;
7. State-of-the-art knowledge is captured, organized, and synthesized.

When a team develops a mediated model "groupiness" is increased because:

1. Individual members perceive clearly that they are a part of the group.
2. Members become oriented toward a common goal.
3. Interaction between group members takes place.
4. Interdependence is realized and acknowledged.
5. A structure of roles/status and norms is built.

4.2. The Institutionalized Modelling Process

The first step in the development of an institutionalized model is to invite representatives for *all* possible groups and stakeholders that have an interest in the focal problem to a brainstorming workshop focused on model development. This could include green organizations, social organizations, policymakers, managers, ecologists, engineers, economists, sociologists, and so on. It is crucial that *all* groups with a well-founded interest in the problem or in the (eco)system are represented. All scientific disciplines associated with the problem and all knowledge bases must also have a voice.

The *first* stage of the workshop is to introduce the objectives; namely to develop a model that can be used as a common reference for all the participating groups, and to understand and hopefully, on a long-term basis, to solve a well-defined problem of common interest. The advantages and disadvantages of modelling, particularly mediated modelling, are presented at this stage, together with the basic ideas behind the system. The various teams participating in the brainstorming must introduce themselves and clearly present their interest in solving the problem, as well as give an overview of their knowledge about the problem.

An IMM is coordinated by the following individuals:

- Facilitator that prepares the meeting and guides the discussion;
- Mediator that plays the role of the facilitator during group meetings;
- Modeller that tries stepwise to conclude the discussion in the form of a model; the model is changed currently to follow stepwise the conclusions made as a result of the discussion and group meetings.

The *second* stage of the workshop focuses on a clear definition of the problem and the scale in terms of spatial system boundaries, time horizon, and time step. The problem can eventually be defined by an ecological risk assessment (ERA), but under all circumstances it will include these crucial questions:

- What has caused the problem or problem complex?
- What are the impacts of the problem?
- If the problem complex consists of several problems, then how are these problems interrelated?

The geographical boundaries are usually already determined by the stakeholders. The time horizon and time step will inevitably lead the participants to focus on some questions, while other aspects are ignored.

The focus can be designed to be narrow or wider in its inclusiveness of economic and social problems, often determined by the roots of the problems. The focus will be very clear as a result of the three crucial questions listed above.

An envisioning exercise attempts to describe the future the participants want and the future that they would settle for. This vision should not be considered a static picture; it has to be redefined over time. Finally, a survey of what we know about the system and the problem is presented, including a list of data and observations.

During the *third* stage of the workshop, a qualitative model is built. The modeller is translating the discussion into state variables, processes, and forcing functions. Simultaneously, he is explaining the meaning of these modelling components, and what it means when the model presents a relationship between forcing functions and state variables. The possibilities for changing the forcing functions and making simulations accordingly will inevitably become a part of the debate in this phase of the model development. It may be beneficial to break up into smaller groups to discuss submodels. Causalities, interacting processes, or possible change of forcing functions should be discussed as well.

The *fourth* stage of the workshop focuses on the quantitative model. The quantitative process description requires an extensive discussion among the participants. It is crucial that the quantitative description of processes adheres to the known ecosystem dynamics. Another topic, open for the discussion at this stage, is the use of indicators. Which indicators best express the system quality and can be used in the follow-up phase when the model results are implemented to pursue the best possible environmental strategy? Jørgensen et al. (2005, 2010) provided a good overview of possible indicators.

When the quantitative model is prepared, the observations are compared to the model simulations and the possibilities for calibrations are discussed. In some cases, it may be beneficial to close the workshop and leave the calibration and validation to a modelling team and re-open the workshop when the calibration and validation are ready. This is recommended when the calibration and validation are very time-consuming because the model is very complex or because the number of observations is high.

The *fifth* stage of the workshop encompasses the testing of various selected scenarios and their conclusions. The simulated scenarios are made after the calibration and validation and may be carried out after the workshop has been reconvened. The model is foreseen to be adaptive, because if the basic conditions for the model have been changed (we are living in a dynamic world), the model should be changed correspondingly. A follow-up workshop should be agreed upon during this stage.

The follow-up workshop, perhaps one to three years after the first workshop, should adjust the model according to the observed "mistakes" by the model and the changing basic conditions. To what extent the previous conclusions should be changed also needs to be discussed.

During the follow-up workshop it is recommended to examine whether the IMM has been a success or failure. This can be determined by answering the following questions (see van den Belt, 2004):

1. Did the participants establish or reach common goals?
2. Did the participants contribute their knowledge and creative thinking toward innovative solutions?
3. Is the model considered a common reference for the participants?
4. Does the model use a common language when the different aspects are discussed?
5. Is the model expressing all the different opinions and knowledge of the stakeholders?
6. Has a cooperative climate emerged?
7. Have all participants accepted the model as an acceptable learning tool?
8. Is there an increased sense of interdependence among the participants?

4.3. When Do You Apply Institutionalized or Mediated Modelling (IMM)?

All of the models presented in this book could, in principle, be developed as non-institutionalized models, but they still require a workshop as a part of the modelling procedure. Not all models need to be developed as IMMs; for example, if they focus on a less complex problem that only touches on a few aspects. When the problem is complex and many different interests, interactions, and aspects are integrated within the problem, it is strongly recommended to use an IMM. Examples where institutionalized modelling is almost mandatory include: water quality or ecosystem models of important lakes, rivers, coastal areas, lagoons, bays, landscapes, wetlands, recreational areas (national parks, sanctuaries) and so on, where many problems have many sources and there are many conflicting interests. The previously listed ecosystems all have different applications that may be in conflict; for instance, a lake, which is often used simultaneously for recreation, production of drinking water, fisheries, and to ensure recycling of important elements. The cost of wastewater treatment is increasing with increasing water quality of the treated water, but the required water quality is not necessarily the same for each application. The willingness to pay for a better water quality is therefore dependent on the use of the lake, which could lead to conflicts.

Institutionalized models applied to ecosystem management can conclude in an environmental management policy and cost strategies. The model conclusions should be accepted by the population and all interest groups, because they have participated in the process, the simulations, and the conclusions and understand the details and basis for the conclusions.

There are many examples of noninstitutionalized models that have failed without a workshop. It is difficult to collect all the knowledge about the problem and the system, the roots of the problems, and all of the different interests in solving the problems without representation from the different groups. It can also be difficult to understand all of the different aspects of the core problem without a brainstorming session. Complex problems are like icebergs, only 10% is visible.

Most of the crucial problems humans face are very complex. Consider the difference between the problem of climate changes due to global warming and the problem of putting the first human on the moon. The global warming problem interferes with an enormously wide spectrum of other problems involving agriculture, industries, developing versus developed countries, sufficient drinking water of an acceptable quality to all citizens, poverty, and so on. The realization of human lunar exploration was entirely a question about very advanced technology with a much narrower spectrum of interdisciplinary issues.

Using IMM is recommended for all complex problems and modellers are encouraged to use this method for their environmental planning problems, which may often be the most complex problem they have. van den Belt (2004) gave several examples where IMM actually resulted in a good planning strategy for a complex problem. The most illustrative case studies are typical environmental management problems such as:

- Watershed management in Wisconsin
- Planning of Banff National Park
- Coastal zone management

Problems

1. Give examples of problems where the development of IMM would be a good solution.
   List the stakeholders interested in the problem. Which group would you invite to a brainstorming meeting? Which type of model do you expect will be developed?
2. Who would be interested in an IMM focusing on the wildlife in a national park, which has enormous income value for a district due to tourism? The wildlife is damaging the surrounding agriculture and negatively impacting the quality of the drinking water in a lake close to the national park.

5

Modelling Population Dynamics

CHAPTER OUTLINE

5.1. Introduction

This chapter covers population dynamic models where state variables are the number or biomass of individuals or species. The growth of one population is used — see Sections 5.2 and 5.3 — to present the basic concepts. Afterward, the interactions between two or more populations are presented. The famous Lotka-Volterra model and several more realistic predator-prey and parasitism models, are shown. Age distribution is introduced and computations with matrix models are illustrated, including the relations to biological growth. Finally, the last three sections illustrate the use of fishery/harvest models, metapopulation dynamics, and infection models.

5.2. Basic Concepts

This chapter deals with biodemographic models, which are population models characterized by numbers of individuals or kilograms of biomass of individuals or species as typical units for state variables. As early as the

1920s, Lotka and Volterra developed the first population model, which is still widely used (Lotka, 1956; Volterra, 1926). So many population models have been developed, tested, and analyzed since that it would not be possible to give a comprehensive review of these models here. This chapter mainly focuses on models of age distribution, growth, and species interactions. Only deterministic models will be mentioned. Those interested in stochastic models can refer to Pielou (1966, 1977), which gives a very comprehensive treatment of this type of population dynamic model.

A *population* is defined as a collective group of organisms of the same species. Each population has several characteristic properties, such as population density (population size relative to available space), natality (birth rate), mortality (death rate), age distribution, dispersion, growth forms, and so forth.

A population changes over time, and we are interested in its size and dynamics as it grows or shrinks. If N represents the number of organisms and t the time, then dN/dt = the rate of change in the number of organisms per unit time at a particular instant (t), and $dN/(Ndt)$ = the rate of change in the number of organisms per unit time per individual at a particular instant (t). If the population is plotted against time, then a straight line tangential to the curve at any point represents the growth rate.

Natality is the number of new individuals appearing per unit of time and per unit of population.

We have to distinguish between absolute natality and relative natality, denoted B and B_r, respectively:

$$B = \Delta N / \Delta t \tag{5.1}$$

$$B_r = B/N \tag{5.2}$$

where ΔN = production of new individuals in the population.

Mortality refers to the death of individuals in the population. The absolute mortality rate, M, is defined as:

$$M = \Delta M / \Delta t \tag{5.3}$$

where ΔM = number of organisms in the population that died during the time interval, Δt, and the relative mortality rate, M_s, is defined as:

$$M_s = M/N \tag{5.4}$$

5.3. Growth Models in Population Dynamics

The simplest growth models consider only one population. Its interactions with other populations are taken into consideration by the specific growth rate and the mortality, which might be dependent on the magnitude of the considered population but independent of other populations. In other words, we consider only one population as a state variable. The simplest growth model assumes unlimited resources and exponential population growth. A simple differential equation can be applied:

$$dN/dt = B_s N - M_s N = rN \qquad (5.5)$$

where B_s is the instantaneous birth rate per individual, M_s is the instantaneous death rate, $r = B_s - M_s$, N is population density, and t is time. Equation (5.5) represents first-order kinetics (see exponential growth in Chapter 2, Section 2.3, equation 2.2a). If r is constant, then we get after integration:

$$N_t = N_0 e^{rt}, \qquad (5.6)$$

where N_t is the population density at time t and N_0 is the population density at time 0. A logarithmic presentation of Eq. (5.6) is given in Figure 5.1.

The net reproductive rate, R_0, is defined as the average number of age class zero offspring produced by an average newborn organism during its entire lifetime. Survivorship, l_x, is the fraction surviving at age x. It is the probability that an average newborn will survive to that age, designated x. The number of offspring produced by an average

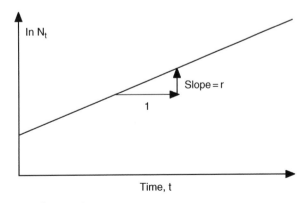

FIGURE 5.1 In N_t is plotted versus time, t.

organism of age x during the age period is designated m_x. This is termed *fecundity,* while the product of l_x and m_x is called the *realized fecundity.* According to its definition, R_0 can be found as:

$$R_0 = \int_0^\infty l_x m_x dx \tag{5.7}$$

A curve that shows l_x as function of age is called a survivorship curve. Such curves differ significantly for various species, as illustrated in Figure 5.2.

The so-called *intrinsic rate of natural increase,* **r,** is like l_x and m_x, dependent on the age distribution, and it is only constant when the age distribution is stable. When R_o is as high as possible, that is, under optimal conditions and with a stable age distribution, the maximal rate of natural increase is realized and designated r_{max}. Among various animals it ranges over several orders of magnitude (Table 5.1).

Exponential growth is a simplification, which is only valid over a certain time interval. Sooner or later every population must encounter the limitation of food, water, air, or space, as the world is finite. To account for this we introduce the concept of *density dependence;* that is, vital rates, like r, depend on population size, N (while we now ignore

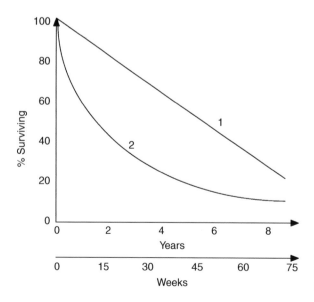

FIGURE 5.2 Survivorships of (1) the lizard Uta (the lower x axis) and (2) the lizard Xantusia (the upper x axis). *(After Tinkle, 1967).*

Table 5.1 Estimated Maximal Instantaneous Rate of Increase (r_{max}, per Capita per Day) and Mean Generation Times (in Days) for a Variety of Organisms

Taxon	Species	r_{max}	Generation Time
Bacterium	*Escherichia coli*	ca. 60.0	0.014
Algae	*Scenedesmus*	1.5	0.3
Protozoa	*Paramecium aurelia*	1.24	0.33–0.50
Protozoa	*Paramecium caudatum*	0.94	0.10–0.50
Zooplankton	*Daphnia pulex*	0.25	0.8– 2.5
Insect	*Tribolium confusum*	0.120	ca. 80
Insect	*Calandra oryzae*	0.110(0.09–.011)	58
Insect	*Rhizopertha Dominica*	0.085(0.07–0.10)	ca. 100
Insect	*Ptinus tectus*	0.057	102
Insect	*Gibbium psylloides*	0.034	129
Insect	*Trigonogenius globules*	0.032	119
Insect	*Stethomezium squamosum*	0.025	147
Insect	*Mezium affine*	0.022	183
Insect	*Ptinus fur*	0.014	179
Insect	*Eurostus hilleri*	0.010	110
Insect	*Ptinus sexpunctatus*	0.006	215
Insect	*Niptus hololeucus*	0.006	154
Octopus	—	0.01	150
Mammal	*Rattus norwegicus*	0.015	150
Mammal	*Microtus aggrestis*	0.013	171
Mammal	*Canis domesticus*	0.009	ca. 1000
Insect	*Magicicada septendecim*	0.001	6050
Mammal	*Homosapiens*	0.0003	ca. 7000

differences caused by age). Let the *carrying capacity,* K, be defined as the density of organisms at which r is zero. At zero density, R_o is maximal and r becomes r_{max}. The logistic growth equation has already been mentioned in Section 2.3, equation 2.4. The application of the logistic growth equation requires three assumptions:

1. All individuals are equivalent.
2. K and r are immutable constants independent of time, age distribution, and so forth.
3. There is no time lag in the response of the actual rate of increase per individual to changes in N.

All three assumptions are unrealistic and can be strongly criticized. Nevertheless, several population phenomena can be nicely illustrated by using the logistic growth equation.

Illustration 5.1

An algal culture shows a carrying capacity due to a self-shading effect. In spite of "unlimited" nutrients, the maximum concentration of algae in a chemostat experiment was measured to be 120 g/m^3. At time 0, 0.1 g/m^3 of algae was introduced and 2 days after a concentration of 1 g/m^3 was observed. Set up a logistic growth equation for these observations.

Solution

During the first 5 days, we are far from the carrying capacity and we have with good approximations:

$$\ln 10 = r_{max} 2$$

$$r_{max} = 1.2 \, day^{-1}$$

and since the carrying capacity is 120 g/m^3 (C = algae concentration), we have:

$$dC/dt = 1.2C(120 - C/120)$$

Integration and use of the initial condition C(0)= 0.1 yield

$$C = 120/(1 + e^{(a-1.2t)})$$

where

$$a = \ln((120 - 0.1)/0.1) = 7.09.$$

This simple situation in which there is a linear increase in the environmental resistance with density, that is, logistic growth is valid, seems to hold well only for organisms that have a very simple life history.

In populations of higher plants and animals that have more complicated life histories, there is likely to be a delayed response. Wangersky and Cunningham (1956, 1957) have suggested a modification of the logistic equation to include two kinds of time lag: (1) the time needed for an organism to start increasing, when conditions are favorable, and (2) the time required for organisms to react to unfavorable crowding by altering birth and death rates. If these time lags are $t - t_1$ and $t - t_2$, respectively, then we get:

$$dN/dt = rN_{t-t_1}(K - N_{t-t_2})/K \tag{5.8}$$

Population density tends to fluctuate as a result of seasonal changes in environmental factors or due to factors within the populations themselves (so-called intrinsic factors). We will not go into details here, but will just mention that the growth coefficient is often temperature dependent and since temperature shows seasonal fluctuations, it is possible to explain some seasonal population fluctuations in density in that way.

5.4. Interaction Between Populations

The growth models presented in Section 5.3 might have a constant influence from other populations reflected in the selection of parameters. It is, however, unrealistic to assume that interactions between populations are constant. A more realistic model must therefore contain the interacting populations (species) as state variables. For example, in the case of two competing populations, we can modify the logistic model and use the following equations, often termed the Lotka-Volterra equation:

$$dN_1/dt = r_1N_1(K_1 - N_1 - \alpha_{12}N_2)/K_1 \tag{5.9}$$

$$dN_2/dt = r_2N_2(K_2 - N_2 - \alpha_{21}N_1)/K_2 \tag{5.10}$$

where α_{12} and α_{21} are competition coefficients. K_1 and K_2 are carrying capacities for species 1 and 2. N_1 and N_2 are numbers of species 1 and 2, while r_1 and r_2 are the corresponding maximum intrinsic rate of natural increase.

The steady-state situation is found by setting Eqs. (5.9) and (5.10) equal to zero. We get:

$$N_1 = K_1 - \alpha_{12}.N_2$$
$$N_2 = K_2 - \alpha_{21}.N_1, \tag{5.11}$$

These two linear equations are plotted in Figure 5.3 giving dN/dt isoclines for each species. Below the isoclines, populations will increase, above them, they decrease. So, four cases result, as illustrated in Figure 5.3 and summarized in Table 5.2.

Lotka-Volterra also wrote a simple pair of *predation equations*:

$$\frac{dN_1}{dt} = r_1.N_1 - p_1N_1.N_2 \tag{5.12}$$

$$\frac{dN_2}{dt} = p_2.N_1.N_2 - d_2.N_2 \tag{5.13}$$

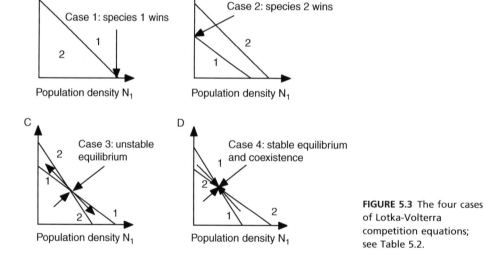

FIGURE 5.3 The four cases of Lotka-Volterra competition equations; see Table 5.2.

Table 5.2 Summary of the Four Possible Cases of Lotka-Volterra Competition Equations

	Species 1 Can Contain Species 2 ($K_2/\alpha_{21} < K_1$)	Species 2 Cannot Contain Species 2 ($K_2/\alpha_{21} < K_1$)
($K_1/\alpha_{12} < K_2$)	Either species may win (Case 3)	Species 2 always wins (Case 2)
($K_1/\alpha_{12} > K_2$)	Species 1 always wins (Case 1)	Stable coexistence (Case 4)

where N_1 is prey population density, N_2 is predator population density, r_1 is the intrinsic (maximal) rate of increase of the prey population (per head), d_2 is the mortality of the predator (per head), and p_1 and p_2 are predation coefficients. Each population is limited by the other and in absence of the predator, the prey population increases exponentially. By setting the two right-hand sides equal to zero, we find, respectively,

$$N_2 = \frac{r_1}{p_1} \tag{5.14}$$

$$N_1 = \frac{d_2}{p_2} \tag{5.15}$$

Thus each isocline of the two species corresponds to a particular density of the other species. Below the threshold prey density, the predator population will always decrease, whereas above that threshold, it will increase. Similarly, the prey population will increase below a particular predator density but decrease above it (Figure 5.4). A joint equilibrium exists where the two isoclines cross, but prey and predator densities do not generally converge to this point; instead any given pair of initial densities results in oscillations of a certain magnitude. The amplitude of fluctuations depends on the initial conditions. These equations are unrealistic since most populations encounter either self-regulations, density-dependent feedbacks, or both. The addition of a simple self-damping term to the prey equation results either in a rapid approach to equilibrium or in damped oscillations. Perhaps a more realistic pair of simple equations for modelling the *prey-predator relationship* is

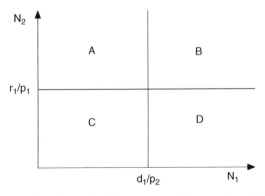

FIGURE 5.4 Prey-predator isoclines for Lotka-Volterra prey-predator equation. (A) both species decrease; (B) predators increase, prey decrease; (C) prey increase, predators decrease; (D) both species increase.

$$\frac{dN_1}{dt} = r_1.N_1 - z_1.N_1^2 - \beta_{12}.N_1.N_2 \qquad (5.16)$$

$$\frac{dN_2}{dt} = \gamma_{21}.N_1.N_2 - \beta_2.\frac{N_2^2}{N_1} \qquad (5.17)$$

where r_1, z_1 and so on are coefficients.

The prey equation is a logistic expression combined with the effect of the predator, while the predator expression considers a carrying capacity dependent on the prey concentration.

The literature of ecological modelling contains many papers focusing on modified Lotka-Volterra equations, but the equations can also be criticized for not following the conservation principle. The increase in the biomass of the predator is less than the decrease in the biomass of the prey. Kooijman (2000) developed many population dynamic models based on the energy conservation principles; they give new and emerging properties of the energy flow in ecosystems. His approach is recommended when energy is in focus or if a more complex food web is considered.

However, Eqs. (5.16) and (5.17) can also easily be criticized. The growth term for the predator is a linear function of the prey concentration of density. Other possible relations are shown in Figure 5.5. The first relation (a) corresponds to a Michaelis-Menten expression (see Section 2.3, equation 2.5), while the second relation (b) only approximates a Michaelis-Menten expression by using a first-order expression in one interval and a zero order expression in another. The third relation (c) shown in Figure 5.5 corresponds to a logistic expression: With increasing prey density the predator density first grows exponentially and afterward a damping takes place. This relation is observed in nature and might be explained as follows: The energy and time used by the predator to capture a prey is decreasing with increasing density of the prey. This implies that the predator can capture more prey due to increasing density, and less of the energy consumed is used to capture the next prey.

Thus, the density of the predator increases more than proportionally to the prey density in this phase. Yet, there is a limit to the food (energy) that the predator can consume and at a certain density of the prey, a further decrease in the energy used to capture the prey cannot be obtained. So the increase in predator density slows down as it reaches a saturation point at a certain prey density.

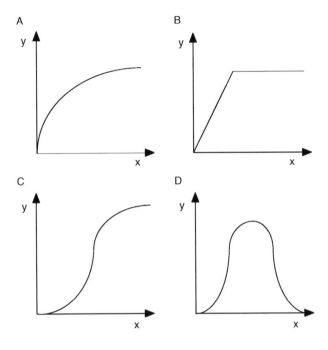

FIGURE 5.5 Four functional responses (Holling, 1959) where y is number of prey taken per predator per day and x is the prey density.

The fourth relationship (d) is similar to the relation between growth and pH or temperature. It is characteristic here that the predator density decreases above a certain prey density. This response might be explained by the effect of the waste produced by the prey on the predator. At a certain prey density the concentration of waste is sufficiently high to have a pronounced negative effect on predator growth.

Holling (1959, 1966) developed more elaborate models of prey-predator relationships. He incorporated time lags and hunger levels to attempt to describe the situation in nature. These models are more realistic, but they are also more complex and require knowledge of more parameters. Besides these complications, we have coevolution of predators and prey. The prey will develop better and better techniques to escape the predator and the predator will develop better and better techniques to capture the prey. To account for the convolution, it is necessary to have a current change of the parameters according to the selection taking place. The effect of *parasitism* is similar to that of predation, but is different because members of the host species affected are seldom killed, but may live for some time after becoming parasitized.

This is accounted for by relating the growth and the mortality of the prey, N_1, to the density of the parasites, N_2. Furthermore, the carrying capacity for the parasites is dependent on the prey density.

The following equations account for these relations and include a carrying capacity of the prey:

$$\frac{dN_1}{dt} = \frac{r_1}{N_2} N_1 \left(\frac{K_1 - N_1}{K_1} \right) \tag{5.18}$$

$$\frac{dN_2}{dt} = r_2.N_2 \left(\frac{K_2.N_1 - N_2}{K_2.N_1} \right) \tag{5.19}$$

Symbiotic relationships are modelled with expressions similar to the Lotka-Volterra competition equations simply by changing the signs for the interaction terms:

$$\frac{dN_1}{dt} = r_1.N_1 \left(\frac{K_1 - N_1 + \alpha_{12}N_2}{K_1} \right) \tag{5.20}$$

$$\frac{dN_2}{dt} = r_2.N_2 \left(\frac{K_2 - N_2 + \alpha_{21}N_1}{K_2} \right) \tag{5.21}$$

Another criticism of the Lotka-Volterra prey-predator model is that it isolates two entities out of their larger contextual web of interactions. In reality, a complex food web both provides and constrains the behavior of species comprising it. The control is much more distributed and decentralized than is evident from the Lotka-Volterra model, which packs all causation into lumped parameters of natality, mortality, and interference.

In nature, interactions among populations often become intricate. The expressions (5.20) and (5.21) might be of great help in understanding population reactions in nature, but when it comes to the problem of modelling entire ecosystems, they are in most cases insufficient. Investigations of stability criteria for Lotka-Volterra equations are an interesting mathematical exercise, but can hardly be used to understand the stability properties of real ecosystems or even of populations in nature.

The experience from investigations of population stability in nature shows that it is necessary to account for many interactions with the environment to explain observations in real systems (e.g., Jørgensen & Fath, 2007).

The stability concept was widely discussed during the 1970s, but today almost all ecologists agree that the stability of an ecosystem is a very complex problem that cannot be solved by simple methods and at least not by examinations of the stability of two coupled differential equations. It is also acknowledged that there is no simple relationship between stability and diversity (May, 1974, 1975, 1977). Stability must be considered a multidimensional concept because the stability is dependent on the particular changes we are concerned with. Some changes the ecosystem might easily adsorb, some other changes can cause drastic reorganization in the ecosystem by minor changes in the forcing function. The buffer capacity introduced in Section 2.6 (see Figure 2.12) may be a relevant concept to use, because it is multidimensional. There is a buffer capacity for each combination of state variable and forcing function.

Illustration 5.2

This illustration concerns an anaerobic cultivation of two species of yeast first described by Gause (1934). The two species are *Saccharomyces cerevisiae* (Sc) and *Schizosaccharomyces pombe* (Kephir; K). Gause cultivated both species in mono-cultures and in mixture, and the results suggest that the two species have a mutual effect upon each other. His hypothesis was that a production of harmful waste products (alcohols) was the only cause of interactions.

A conceptual diagram for this model is shown Figure 5.6. The model has three state variables: the two yeast species and the waste products. The amount of waste products depends on the growth of yeast. The growth of the yeast species depends on the amount of yeast and the growth rate of the yeast, which is again dependent on the species and a reduction factor. This accounts for the influence of the waste products on the growth. The observed and computed values for growth of the two yeast species are shown in Table 5.3. The fit between observed and

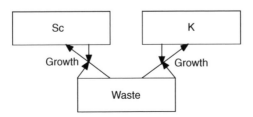

FIGURE 5.6 Conceptual diagram of the model presented in Illustration 5.2. Waste is alcohol that affects the growth of two yeast species Sc and K.

Table 5.3 Observed and Calculated Values for the Growth of Two Species of Yeasts in Mono-Cultures and Mixtures

Schizosaccharomyces "Kephir"				
Volume of Yeast (arbitrary units)				
	Mono-culture		*Mixed*	
Hours	**Observed**	**Calculated**	**Observed**	**Calculated**
0	0.45	0.45	0.45	0.45
6	—	0.60	0.291	0.59
16	1.00	0.95	0.98	0.81
24	—	1.34	1.47	0.88
29	170	1.64	1.46	0.89
48	2.73	3.04	1.71	0.89
53	—	3.44	1.84	0.89
72	4.87	4.72	—	—
93	5.67	5.51	—	—
117	5.80	5.86	—	—
141	5.83	5.96	—	—
Saccharomyces cerevisiae				
Hours	**Observed**	**Calculated**	**Observed**	**Calculated**
0	0.45	0.45	0.45	0.45
6	0.37	1.72	0.375	1.70
16	8.87	8.18	3.99	7.56
24	10.66	11.83	4.69	10.86
29	12.50	12.46	6.15	11.47
40	13.27	12.73	—	11.75
48	12.87	12.74	7.27	11.77
53	12.70	12.74	8.30	11.77

calculated values is acceptable for the mono-culture experiments, but it is completely unacceptable for the mixed culture experiments. It can be concluded that the two species do not interfere solely through the production of alcohol. Additional biological knowledge about the interference between the two species must be introduced to the model to explain the observations.

Illustration 5.3

This illustration is a summary of an example presented by Starfield and Bleloch (1986) in their book on population dynamics titled *Building*

Models for Conservation and Wildlife Management. The example illustrates a very common and generally applicable approach to use population dynamic models in wildlife management. This illustration also demonstrates how an analysis of the focal problem can be used to construct a model. The equations are all based on semiquantitative to quantitative known relationships between determining factors on the one side and the influence on the state variables on the other. It is a clear illustration of how "down to earth" considerations might be used to construct models. As many interacting species are involved, the model is rather complex by including many different relationships between the different state variables of the model. The illustration is concerned with a spectrum of herbivores, while no significant predators are present. The principal grazers are warthog, wildebeest, zebra, and the white rhinoceros. The principal browsers are giraffe, kudu, and the black rhinoceros. Impala and nyala are the two most important mixed feeders.

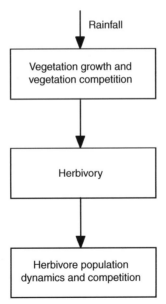

FIGURE 5.7 Conceptualization of the problem in Illustration 5.3. The influence of rainfall on the vegetation, the competition among the different forms of vegetation, the food availability for the herbivorous state variables, and the competition among the herbivores should all be considered in the model.

The problem is illustrated in Figure 5.7. It implies that the model should consider the interactions between rainfall and vegetation, between vegetation and herbivores, and the competition among the herbivores for food.

The first question to consider is How many classes of species do we need? Clearly the giraffe should be a class of its own, as only this animal can browse on tall trees. The black rhinoceros and the kudu browse on shrubs and short trees. Both the white rhinoceros and zebra are grazers that can use relatively tall, coarse grass, while wildebeest and warthog are grazers that require short grass. Finally, impala and nyala are mixed feeders, utilizing short grass, shrubs, and short trees. By this short analysis we have suggested how to reduce the number of state variables of herbivores from nine to five. Converting one variable to another is made by using the concept of equivalent animal units (EAU), defined as the daily food intake of a domestic cow. The black rhinoceros is about 2 EAU, while a kudu is only about 0.4. When we lump the two animals in one group, each black rhinoceros is equivalent to 5 kudu. The same considerations are made for the other species.

The next problem concerns the food preferences. Here Starfield and Bleloch (1986) have suggested setting up the preferences in table form (see Table 5.4). This implies that we have to increase the number of herbivore types from five to six, as shown in the table. For example, Impala will first choose palatable grass, then palatable shrubs, and as last resort, less palatable grass. Kudu, on the other hand, have only two preferences: fist palatable shrubs, then unpalatable shrubs. The effect of switching to a

Table 5.4 Food Preferences of the Herbivores

Species	Preference 1	Preference 2	Preference 3
Giraffe	Palatable tall trees	Palatable shrubs	Unpalatable trees
Impala	Grass, palatability > 0.8	Palatable shrubs	Less palatable grass
Kudu	Palatable shrubs	Unpalatable shrubs	
Warthog	Grass, palatability > 0.8	Less palatable grass	
Wildebeest	Grass, palatability > 0.8	Less palatable grass	
Zebra	Grass, palatability > 0.6	Less palatable grass	

second or third preference is accounted for by a condition index with an arbitrarily chosen scale from 1 to 6. A value of 1 corresponds to the peak condition, while a 6 means extremely poor condition. It is important whether an animal class has an inadequate diet for just one month or for a number of consecutive months. The scale is therefore used to consider the cumulative effect and it is used stepwise. The condition index influences the mortality, particularly the juvenile mortality, which will increase sharply as the condition index approaches 6.

For each of the five classes, we consider two subclasses: adults and juveniles. We estimate, for example, that an adult kudu requires B kg and a juvenile b kg of food per month, which is selected as the time step of the model. If there are K adult kudu and k juveniles, then the kudu population in that park will potentially eat KB + kb kg of leaves in the next month. The model calculates a demand for food, first assuming that every species eats only its first preference. If there is sufficient for all, then the food is shared accordingly, but if there is a shortage, the model allocates a share of each animal's second preference, which determines a possible change of the condition index.

Except for zebra, all births take place during the first months of the summer. It is assumed that zebra produce their young throughout the year. The annual birthrate varies from 0.2 for giraffe to 0.95 for warthog.

Six types of vegetation are considered in the model: A grass, B shrubs + small trees, and C tall trees; each with a palatable and unpalatable subclass. The growth in leaf biomass for the two subclasses of B and C are modelled by using the following equation:

$$dl/dt = r * f * S * [1 - L/(q * S)] - b \qquad (5.22)$$

where L denotes the leaf biomass, r a growth parameter, f is a rainfall correction factor, S the woody component, q the maximum leaf mass that one unit of wood mass normally can support, and b is calculated from the herbivore module as the food requirement. Equation (5.22) is based on the following assumptions:

1. New leaf growth depends on how many bushes/trees, S, there are.
2. Rainfall will influence production.
3. Herbivores will consume some biomass each month.
4. There is an inhibitory effect of existing leaf biomass, which is considered in the expression [1 − L/(q*S)].

The application of Eq. (5.22) implies that we have to model the wood mass, S. This is made by using:

$$dS/dt = r_s * f_s * S * [1 - (\Sigma S)/T_{max} * C] \qquad (5.23)$$

where r_s is the growth parameter for woody biomass, f_s is the rainfall correction factor for the woody biomass of shrubs and trees, S is the present total wood mass, T_{max} is the saturation level for woody biomass, and C is the competition from grass. C is found from:

$$C = \exp(-[p * c * A * h + \Sigma I])/U \qquad (5.24)$$

where p is a competition factor (must be calibrated), c is converting grass volume to biomass, A is the grass area, h the height of the grass, ΣI is the total leaf biomass, and U is the saturation level for green production.

A and h are state variables, too. Equations for the grass area (m^2), A, and for the grass height (m), h, are included in the model:

$$dA/dt = r_a * f_g * A * C \qquad (5.25)$$

$$dh/dt = r_h * f_g * h[1 - h/h_{max}] - G/(c * A) \qquad (5.26)$$

where r_a and r_h are the growth parameters for A and h, f_g is the rainfall correction factor for grass area and grass height, h_{max} is the saturation height for grass, and G is the grass biomass consumed by herbivores (kg/month). All of these variables are obtained from the herbivore module. Empirical tables are available for f. For instance, f_g is dependent on the rainfall, whether it is low, medium, and high, and it is dependent on the season.

Figures 5.8 and 5.9 show some of the simulations carried out by the model. The number of kudu versus the number of years is plotted in Figure 5.8, while Figure 5.9 gives the palatable browse on shrubs in the same period. The condition index will roughly be opposite this curve. When the palatable browse is high, the condition index is low and vice versa.

Rain is — not surprisingly — of very great importance for the herbivorous populations, as is seen in Figure 5.7, where the indirect effect from rain on herbivores is obvious. This effect is seen by the violent fluctuations in palatable browse on shrubs, which can be explained by fluctuations in rainfall.

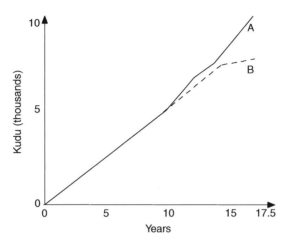

FIGURE 5.8 The kudu population is plotted versus the number of years. A corresponds to cropping of the impala, whenever their population exceeds 6000. B corresponds to no cropping of impala under otherwise similar conditions.

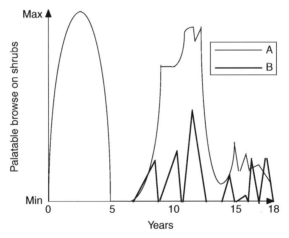

FIGURE 5.9 The amount of palatable browse on shrubs and short trees is plotted versus the time. A corresponds to cropping of the impala, whenever their population exceeds 6000. B corresponds to no cropping of impala under otherwise similar conditions.

5.5. Matrix Models

Another important aspect of modelling population dynamics is the influence of the age distribution, which shows the proportion of the population belonging to each age class. If a population has unchanged

l_x and m_x schedules, then it will eventually reach a stable age distribution. This means that the percentage of organisms in each age class remains the same. Recruitment into every age class is exactly balanced by its loss due to mortality and aging.

The growth equations (5.6) and (5.8) assume that the population has a stable age distribution. The intrinsic rate of increase, r, the generation time, T, and the reproductive value, vx, is conceptually independent of the age distribution, but might be different for populations of the same species with different age distributions. Therefore, the models presented in Sections 5.2 and 5.3 did not need to consider age distribution, although the parameters in actual cases reflect the actual age distribution.

A model predicting the future age distribution was developed by Lewis (1942), Leslie (1945), and Levine (1980). The population is divided into n + 1 equal age groups — group 0, 1, 2, 3, ..., n. The model is then presented by the following matrix equation:

$$\begin{matrix} f_0 & f_1 & f_2 & \ldots & f_{n-1} & f_n & n_{t,0} & n_{t+1,0} \\ p_0 & 0 & 0 & \ldots & 0 & 0 & n_{t,1} & n_{t+1,1} \\ 0 & p_1 & \ldots & \ldots & 0 & 0 & n_{t,2} & n_{t+1,2} \\ \ldots & \ldots & \ldots & \ldots & \ldots & \ldots & . & . \\ \ldots & \ldots & \ldots & \ldots & \ldots & \ldots & . & . \\ 0 & 0 & 0 & \ldots & p_{n-1} & 0 & n_{t,n} & n_{t+1,n} \end{matrix} \qquad (5.27)$$

The number of organisms in the various age classes at time t + 1 is obtained by multiplying the numbers of animals in these age classes at time t by a matrix, which expresses the fecundity and survival rates for each age class. $t_{,o}$, f_1, f_2 ... f_n give the reproduction in the ith age group and P_0, P_1, P_2, P_3, P_4 ... P_n represent the probability that an organism in the ith age group will still be alive after promotion to the $(i + 1)^{th}$ group.

The model can be written in the following form:

$$A * a_t = a_{t+1} \qquad (5.28)$$

where A is the matrix, a_t is the column vector representing the population age structure at time t, and a_{t+1} is a column vector representing the age structure at time t + 1. This equation can be extended to predict the age distribution after k periods of time:

$$a_{t+k} = A^k * a_t \qquad (5.29)$$

Matrix A has n possible eigenvalues and eigenvectors. Both the largest eigenvalues, λ, and the corresponding eigenvectors are ecologically meaningful. λ gives the rate at which the population size is increased:

$$A * v = \lambda * v \tag{5.30}$$

where v is the stable age structure. In λ is the intrinsic rate of natural increase. The corresponding eigenvector indicates the stable structure of the population.

Illustration 5.4

Usher (1972) gave a very illustrative example on the use of matrix models. This model is based upon data for the blue whale before its sharp changes in survival rates.

The eigenvalue can be used to find the number of individuals that can be removed from a population to maintain the same number in each age class. It can be shown that the following equation is valid:

$$H = 100(\lambda - 1)/\lambda$$

where H is the percentage of the population that can be removed.

The blue whales reach maturity at between four and seven years of age. They have a gestation period of about one year. A single calf is born and is nursed for about seven months. On average, not more than one calf is born to a female every two years. The male-to-female sex ratio is approximately equal. Survival rates are about 0.7 each 2 years for the first 10 years and 0.78 for whales above 12 years. We divide the population into 7 groups with a 2-year period for the first 6 groups and the age of 12 years and above as the seventh group. The fecundity for the first two groups is according to the information about zero. The third group has a fecundity of 0.19, and the fourth group, 0.44. The maximum fecundity of 0.50 is reached between ages of 8 and 11 years. The fecundity of the last group is 0.45.

Find the intrinsic rate of natural increase, the stable structure of the whale population, and the harvest, which can be taken to maintain a stable population size.

Solution

The eigenvalue can be found either by an iterative method or by plotting the number of whales (totally or for each age class separately) versus the period of time. The slope of this plot will, after a stabilization period, correspond to r, the intrinsic rate of increase, or ins. We find that r = 0.0036 1/year or λ or l = antilog 0.0036 = 1.0036 (for one year) or

1.0036^2 = 1.0072 for two years. Using Eq. (5.30), the corresponding eigenvector is found to be a = [1000, 764, 584, 447, 341, 261, 885] as the Leslie matrix is

$$
\begin{array}{ccccccc}
0 & 0.19 & 0.44 & 0.50 & 0.50 & 0.45 \\
0.77 & 0 & 0 & 0 & 0 & 0 & 0 \\
0 & 0.77 & 0 & 0 & 0 & 0 & 0 \\
0 & 0 & 0.77 & 0 & 0 & 0 & 0 \\
0 & 0 & 0 & 0 & 0.77 & 0 & 0 \\
0 & 0 & 0 & 0 & 0 & 0.77 & 0 \\
0 & 0 & 0 & 0 & 0 & 0.77 & 0.78 \\
\end{array}
$$

The harvest that can be taken from the population is estimated to be

$$H = 100(\lambda - 1)/\lambda = 0.71\%$$

every two years or about 0.355% every year.

If the harvest exceeds this value, then the population will decline. Population models of r-strategies generally cause some difficulties when developing models of K-strategies due to the high sensitivity of the fecundity. The number of offspring might be well known, but the number of survivors to be included in the first age class and the number of recruits is difficult to predict. This is the central problem of fish population dynamics, since it represents nature's regulation of population size (Beyer, 1981).

5.6. Fishery Models

Figure 5.10 shows the growth rate dN/dt versus the biomass or the number for the logistic growth equation. It is a parabolic shape in accordance with the s-shape of the logistic growth equation. The slope has maximum at an intermediate value of N, but is zero for N = 0 and for N = K.

It is also possible to include harvest, H, which is of interest in fishery and forest models. The following expression is used:

$$dN/dt = rN(1 - N/K) - H$$

The harvest H is proportional to N and to the fish effort E:

$$dN/dt = rN(1 - N/K) - fEN \qquad (5.31)$$

where f is a proportional constant.

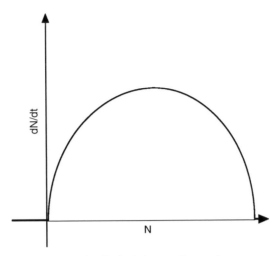

FIGURE 5.10 dN/dt is plotted versus N for the logistic growth equation.

This expression has two equilibriums corresponding to $dN/dt = 0$ $N_1 = 0$ and $N_2 = K (1 - fE/r)$.

N_2 can be found graphically as shown in Figure 5.11.

If the specific fishing mortality fE is $> r$, then there is no equilibrium value N_2, only the equilibrium value $N_1 = 0$. For a sustainable harvest

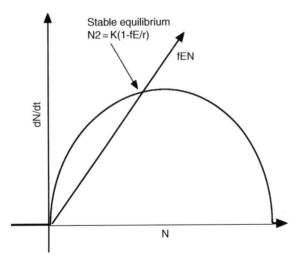

FIGURE 5.11 The growth rate of the logistic growth as function of N and the fishing mortality as function of N are both plotted. A stable equilibrium is obtained where the two functions are equal. A sustainable fishery will therefore require that the fishing mortality is equal or less than the increase of N due to the logistic growth.

fE/r < 1.0. The sustainable yield = fEN_2 can be found as function of the fishing effort by using the previously shown expression for N_2:

$$\text{Yield} = fEN_2 = fEK(1 - fE/r) \qquad (5.32)$$

This graph yield 0 f(E) is shown in Figure 5.12. The optimal effort is found by:

$$d\text{Yield}/dE = fK - 2f^2EK/r = 0, \qquad (5.33)$$

which leads to:

$$E = r/(2f) \text{ and the maximum yield} = rK/4. \qquad (5.34)$$

In populations of higher plants and animals with more complicated life histories, there is likely to be a delayed response. Wangersky and Cunningham (1957) suggested a modification of the logistic equation to include two kinds of time lag: (1) the time needed for an organism to start increasing under favorable conditions, and (2) the time required for organisms to react to unfavorable crowding by altering birth and death rates. If these time lags are $t - t_1$ and $t - t_2$ respectively, then we get:

$$\frac{dN}{dt} = rN_{t-t_1}\left(1 - \frac{N_{t-t_2}}{K}\right) \qquad (5.35)$$

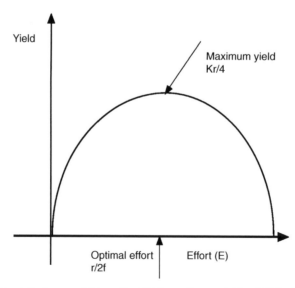

FIGURE 5.12 Yield is plotted versus fishing effort E. The optimum yield = Kr/4 is obtained by E = r/2f. A sustainable fishery will therefore require that E < r/2f.

Population density tends to fluctuate as a result of seasonal changes in environmental factors or due to factors within the actual populations (so-called intrinsic factors). We will not go into details here, but will just mention that the growth coefficient is often temperature dependent and since temperature shows seasonal fluctuations, it is possible to explain at least some seasonal population fluctuations in density as temperature changes.

The simple fishery model presented earlier focuses on one species only, and it is insufficient for setting up an optimal fishery strategy. It is necessary to include several species, because all species interact and influence each other. A fishery policy based on one species will inevitably fail. Consequently, the European fishery policy for the North Sea is based on a multi-species fishery model used to assess the optimal fishery strategy. The fishery is, however, not optimal because the politicians are not following the recommendations given by the model.

5.7. Metapopulation Models

A regional set of local populations that occupy isolated habitat patches but are interconnected by dispersal movements are denoted metapopulations (see an example in Figure 5.13). All of the local populations have a finite possibility of becoming extinct. Even if the local population is fairly large, extinction may still occur through catastrophic events.

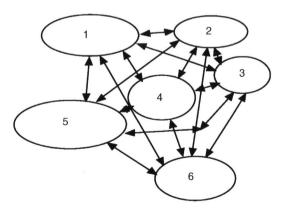

FIGURE 5.13 Conceptual model of metapopulation interactions. The populations occupy isolated patch habitats (1-6) that are connected by dispersal corridors.

The dispersal movements are essential to reestablish populations that have faded or crashed. Species that are widely distributed in many local populations have reduced likelihood to be extinct regionally.

Landscape fragmentation has increased due to human expansion. Populations that were formerly continuously distributed have become broken into separate localized groupings. Dispersal may even be inhibited by hazards in traversing the human-transformed areas separating suitable habitats. Metapopulation models assess the risks of species extinctions as a consequence of such fragmentations and identify how actions such as providing dispersal corridors can reduce the risks.

The metapopulation concept was formulated by Levins (1969) and further modified by Hanski (1994, 1999). P is the proportion of sites occupied by populations, E is the extinction rate of these populations, and C is the colonization rate of vacant sites by migrants from occupied patches. The change over time in the proportion of patches, dP/dt, occupied is a matter of balance between colonization and extinction:

$$dN/dt = CP(1 - P) - EP \qquad (5.36)$$

The equilibrium proportion P_{eq} is given by

$$P_{eq} = 1 - E/C \qquad (5.37)$$

As seen from Eq. (5.36) the patch occupancy will become zero if the extinction rate exceeds the colonization rate.

Figure 5.14 shows a STELLA diagram for a metapopulation model based on Eq. (5.36) plus the introduction of temporal disturbances.

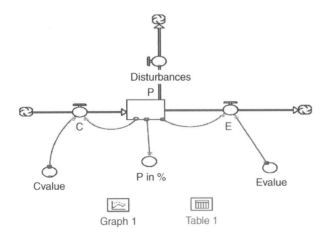

FIGURE 5.14 A STELLA diagram of a metapopulation model with disturbances (removal of patches).

The model can be applied to give the consequences of disturbances such as road construction that interrupts connections among population sites. The populations in the threatened population sites may become extinct due to unbalanced natality and mortality that require immigration to be in balance. A simple simulation of the threatened populations will be able to assess the E-value as a consequence of the disturbances.

5.8. Infection Models

Population models of disease dynamics have the proportion of the host population that is infected as the focal state variable. A simple infection model is shown in Figure 5.15. Susceptible hosts, S, become infected at rate b. After a period of time, the infected hosts either recover, maybe with long-lasting immunity, or die. The number of infected hosts is reduced at a rate corresponding to m + r, where m is the mortality rate and r is the recovery rate. The disease spreads as a result of contact between infected and susceptible hosts. The following equation can be used to express the number of infected hosts, I:

$$dI/dt = bSI - vI \qquad (5.38)$$

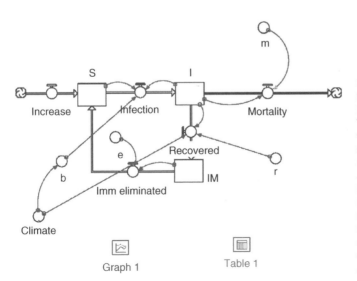

FIGURE 5.15 A conceptual STELLA diagram for an infection model is shown. The model has three state variables: the number of susceptible hosts, the number of infected hosts, and the number of immune hosts. The infection rate is b, the mortality rate is m, the recovery rate is r, and the rate of immunity elimination is e. IM is ...., and Imm is ....

The duration of recovery, D, is the inverse of the rate of recovery from the infection:

$$D = bN/v \tag{5.39}$$

where N is the susceptible hosts, which initially may be equal to the total population, denoted N.

Notice that Eq. (5.38) implies that the infection will spread very fast because it is at a rate that is S × I. Figure 5.15 shows a conceptual diagram of an infection model with three state variables: susceptible hosts, infected hosts, and immune hosts. The spreading of the infection follows Eq. (5.38). The infected hosts either die at a rate m or recover at a rate r. The immunity is eliminated at a rate e, which implies that immune hosts are transferred to the susceptible hosts at a rate e. The state variables are S, I, and IM. It is sometimes necessary to distinguish between recovered hosts that still can transmit the infection and completely immune hosts. In this case, a fourth state variable is introduced to represent the recovered hosts that can still transmit the infection.

The influence of the climate is included in the model. The auxiliary variable "climate" (see Figure 5.15) follows a sine-like curve with higher values during the winter time (maximum in February with an approximate value of 2.0) and lower during the summer time, with an approximate value of 0.3. Both b and the recovery rate are influenced by the climate. B is dependent on a number of factors. Vaccination or isolation of infected individuals will decrease b. For some diseases, the transmission is dependent on the proportion of susceptible individuals within the population rather than on their absolute number. This is the case for sexually transmitted diseases, where spread is frequency dependent.

The equations of the model are shown in Table 5.5. The result of a simulation with the duration of 1000 days is shown in Figure 5.16. As expected, the number of infected hosts increased very rapidly, although b is only 0.000001. The number of susceptible hosts is 1,000,000 at time = 0, and the number of infected hosts has as an initial value 1.0. These numbers could be realistic for an influenza epidemic. Notice that the number of infected hosts after the peak has been reached is decreasing, but with fluctuations according to the auxiliary variable climate. The fluctuations of the number of immune hosts and infected hosts are opposite with maximum for infected hosts when the number of immune hosts are in minimum.

Table 5.5   Equations Using STELLA for the Model Shown in Figure 5.15

```
I(t) = I(t - dt) + (infection - mortality - recovered) * dt
INIT I = 1
INFLOWS:
infection = b*S*I
OUTFLOWS:
mortality = m*I
recovered = I*r*climate
IM(t) = IM(t - dt) + (recovered - imm_eliminated) * dt
INIT IM = 0
INFLOWS:
recovered = I*r*climate
OUTFLOWS:
imm_eliminated = IM*e
S(t) = S(t - dt) + (imm_eliminated + increase - infection) * dt
INIT S = 1000000
INFLOWS:
imm_eliminated = IM*e
increase = 200
OUTFLOWS:
infection = b*S*I
b = 0.000001*climate
e = 0.025
m = 0.002
r = 0.05
climate = GRAPH(TIME)
(0.00, 1.40), (20.4, 1.91), (40.8, 2.00), (61.2, 1.80),
(81.6, 1.35), (102, 0.85), (122, 0.61), (143, 0.4), (163,
0.32), (184, 0.29), (204, 0.36), (224, 0.43), (245, 0.81),
(265, 1.08), (286, 1.26), (306, 1.46), (327, 1.60), (347,
1.75), (367, 1.86), (388, 1.96), (408, 2.00), (429, 1.87),
(449, 1.53), (469, 0.86), (490, 0.62), (510, 0.44), (531,
0.35), (551, 0.3), (571, 0.34), (592, 0.5), (612, 0.73),
(633, 0.97), (653, 1.39), (673, 1.75), (694, 1.94), (714,
2.00), (735, 1.97), (755, 1.86), (776, 1.68), (796, 1.43),
(816, 1.18), (837, 0.9), (857, 0.6), (878, 0.42), (898,
0.3), (918, 0.35), (939, 0.55), (959, 1.01), (980, 1.28),
(1000, 1.45)
```

Problems

1. Set up a STELLA model representing Lotka-Volterra equations. How is it possible to consider the conservation principles, which is a prerequisite for the application of STELLA?
2. Express the model in Illustration 5.1 by STELLA.
3. Make a conceptual diagram of a four species model based on Eq. (5.10).

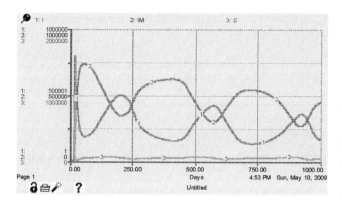

FIGURE 5.16 The simulation results of the model shown in Figure 5.15. The equation is applied in Table 5.5. Notice the rapid increase due to the equation $dI/dt = bIS$. The peak of infection is after ten days. The fluctuations of all three state variables, particularly for I and IM, are due to the auxiliary variable "climate."

4. Mention at least 3 reasons for the unrealistic nature of the Lotka-Volterra model.

5. A fish culture has a carrying capacity of 50 g/L. Set up a logistic growth equation for the fish culture when the initial concentration at day 0 is 1 g/L and after 10 days the concentration 2 g/L is obtained. How long does it take to increase the concentration from 24 g/L to 48 g/L? Find an equation that expresses the doubling time as a function of the time.

6. Explain under which conditions the four functional responses may occur.

7. Set up a matrix model for a bird population that has the following characteristics:
   a. Life span 7 years
   b. 4 eggs from the second year per pair, increasing to 5 eggs the third year, and 6 eggs the following years
   c. The mortality is 30% the first year, 20% the following years, except the last year where it is 100%. What is the steady-state age distribution?

8. Give an overview of factors that may be able to limit the carrying capacity of a population.

9. Make a conceptual diagram of a four-species model based on population interactions representing prey-predator-top predators and two competing top predators.

10. The following equation is valid for a fish population: $dN/dt = 0.025*$ number of fish* $(1 - \text{fish}/1.5*10^7)$. The fishing effort, E, is 0.22 and $f = 0.66$. By using dN/dt find the stable equilibrium. What is the maximum yield? What is the optimal effort?

6

Steady-State Models

6.1. Introduction

Steady-state models presume a condition in which the values of the model state variables do not change over time. A steady-state model corresponds to a dynamic model where all the derivates dx/dt, dy/dt, and so forth are equal to zero. Observations, giving the time variations of the ecological components, are therefore not needed. Instead, average values are sufficient. Average values (e.g., annual average values) often correspond to steady-state values because they are only changed on a longer time basis. The model is only able to give information about average values of the modelled components or about specific steady-state conditions.

This chapter presents three different approaches to develop steady-state models:

1. Chemostat models, which are often used for aquatic ecosystems with a moderate to long retention time such as ponds, lakes, estuaries, lagoons, and the open sea
2. Using downloadable software Ecopath, which has been widely used to develop steady-state models of aquatic ecosystems
3. Using steady-state models to analyze ecological networks

These three approaches illustrate the advantages and disadvantages of steady-state models and how they can be used to obtain a good picture of the interactions among ecological components in an ecosystem.

Fundamentals of Ecological Modelling. DOI: 10.1016/B978-0-444-53567-2.00006-5

6.2. A Chemostat Model to Illustrate a Steady-State Biogeochemical Model

A chemostat model is a mixed flow reactor, which implies that the concentrations of all the components are the same in the entire reactor. Physical, chemical, and biological processes take place in the reactor, but the reactor is sufficiently well mixed to maintain the same concentration throughout the entire reactor so there are no gradients. A well-mixed aquatic ecosystem can be considered a mixed flow reactor or a chemostat (Figure 6.1). The concentrations of the various components can be found as a function of time using a developed system of equations. It is easy to find the steady-state concentrations that adequately describe the conditions in the aquatic ecosystem.

Let us consider one component in a chemostat with the concentration, C. The following differential equation is valid:

$$\bullet dC/dt = (\text{input} - \text{output} - \text{decomposition} - \text{settling} - \text{evaporation})/V \qquad (6.1)$$

where V is the volume (m^3). The processes can be expressed by algebraic equations, which change the differential equation to:

$$\bullet dC/dt = (I - Q*C)/V - k*C - sr*C - A*v*C/V) \ \text{mg}/(\text{m}^3/24\text{h}) \qquad (6.2)$$

where I is the input in mg/24h, Q is the flow rate (m^3/24h), k is the first-order decomposition rate coefficient for the component, sr is the settling rate (1/24h), A is the surface area (m^2), and v is the evaporation

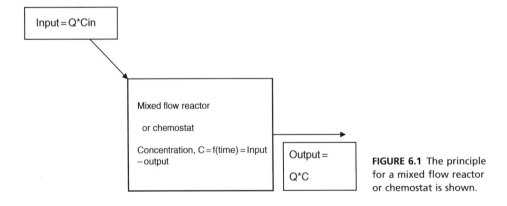

FIGURE 6.1 The principle for a mixed flow reactor or chemostat is shown.

rate (m/24h) of the component. At steady state, the following equation is valid:

• Steady state : $dC/dt = 0$, which implies that

$$\bullet I = (Q * C + V * k * C + V * sr * C + A * v * C) \tag{6.3}$$

If I, Q, V, A, k, sr, and v are known, C can be expressed as function of time and the steady-state value can be found:

$$C = (I - Q * C)/V - k * C - sr * C + A * v * C/V) \tag{6.4}$$

The solution of Eq. (6.2) or (6.3) will give C as a function of time. The steady state is, in principle, never reached, but C approaches asymptotically the steady-state value. A typical and possible plot of C= f(time) is shown in Figure 6.2.

It is possible to follow the concentration of a toxic substance in an organism as a function of time by a similar simple model, which will yield a similar result: The toxic substance concentration as f(time) is approaching the steady state asymtotically as shown in Figure 6.2. The applied equations are:

$$dtx/dt = (daily)intake - kTx \tag{6.5}$$

At steady state:

$$dTx/dt = 0$$

$$kTx = input \text{ or } Tx = input/k \tag{6.6}$$

$$Concentration = Tx/biomass$$

where Tx is the total amount of toxic substance in the organism, k is the excretion coefficient (units 1/24h), which is very dependent on the toxic

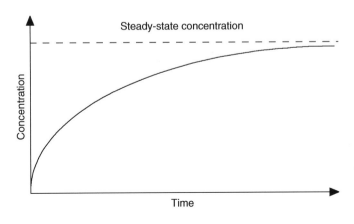

FIGURE 6.2 C as function of time. C is approaching asymptotically the steady-state concentration.

compounds and the organism. The daily intake may consist of uptake from the medium (air or water) and uptake from the food.

Illustration 6.1

A pesticide, Pz, is discharged to a lake that can be considered a mixed flow reactor. 400 m^3 of wastewater with a Pz concentration of 4 mg/L is discharged to the lake per 24 hours. The natural inflowing stream has a rate of 1000 m^3/24h. The pesticide has a biological half-life of 40 days. The settling and evaporation are negligible. The volume of the lake is 200,000 m^3. What is the steady state concentration of Pz?

Solution

The specific decomposition rate k can be found from the first-order differential equation for the decomposition:

dPz/dt = k* Pz, which means since the half-life is 40 days:

ln 2 = 0.693 = k*40 or k = 0.0173 (1/24h)

The differential equation: dPz/dt = (input − output − decomposition)/V, which implies that:

dPz/dt = (400*4 − Pz*1400)/200,000 − 0.0173*Pz = (at steady state)

or

0.008 − 0.007*Pz − 0.0173Pz = 0; or Pz = 0.329 mg/m^3

6.3. Ecopath Models

Ecopath was designed to help the user construct trophic network models of aquatic ecosystems. Ecopath is downloadable, public domain software (see www.ecopath.org). Several hundred Ecopath models have been developed, and it has been extensively applied for marine ecosystems and fishery models; see and Christensen and Pauly (1992, 1993). Many Ecopath models have been published in the journal *Ecological Modelling*. This software also provides useful procedures for parameter estimation and for balancing the system of equations for mass and energy conservation. The latest versions, which have been accessible for more than a decade, have introduced accumulation and depletion of biomass by any organisms during the considered time period. This addition allows us to refrain from the restrictive steady-state conditions.

The required input data can be different types depending on the available data set. The software accepts as input biomass values (standing stock at presence or means of the period considered) and flow values. The latter should be given together with metabolic parameters (food uptake, respiration, excretion rate, etc.). The eventually unknown parameters are automatically determined by means of energy balance equations. An estimation of the diet composition of the various organisms is always asked for as input. The necessary input ratios of fundamental metabolic parameters are:

- Production/biomass ratio (P/B)
- Consumption/biomass ratio (Q/B)
- Gross efficiency, GE = production/consumption = (P/B)*(Q/B)
- Unassimilated part of the food

It is sufficient to know two out of the three ratios (P/B), (Q/B), and GE.

Figure 6.3 shows an example of an Ecopath model. As seen from the figure, the model provides the quantitative information about the trophic network, including flows between compartments and storages within the compartments. Note that all inflows are balanced by outflows so that the model is at steady state.

6.4. Ecological Network Analysis

Ecological network analysis (ENA) is a methodology used to study objects as part of a larger system. It starts with the assumption that a system can be represented as a network of nodes (compartments, components, etc.) and the connections between them (flows of energy or matter). Network analysis, by design, provides a systems-oriented perspective because it is based on uncovering patterns and influence among all the objects in a system. Therefore, it illustrates how system components are tied to a larger web of interactions. Ecological network analysis is included in this chapter because the current methodologies are developed for models in which input and output are balanced for each compartment, but the approach is not conceptually limited to steady-state models as time-varying methodologies are being considered and developed (Hippe, 1983; Shevtsov et al., 2009).

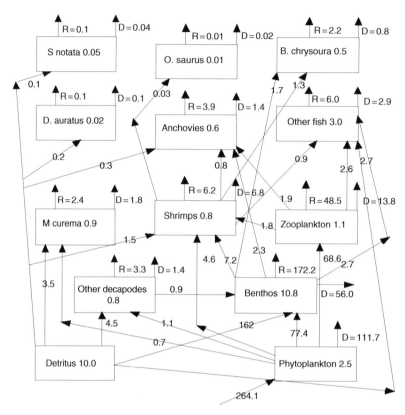

FIGURE 6.3 Example of a marine Ecopath model, taken from Christensen and Pauly (1993). R means respiration and D the transfer to detritus. Notice that all the components are in steady state: input = output.

Ecological network analysis is based on economic input-output analysis (Leontief, 1951) and was first introduced into ecology by Hannon (1973). His models were linked by the energy flow through the food web to determine interdependence of organisms in an ecosystem based on their direct and indirect energy flows. In ecological systems, the connections are often based on the flow of conservative units such as energy, matter, or nutrients between the system compartments. If such a flow exists, then there is a direct transaction between the two connected compartments. These direct transactions give rise to both direct and indirect relations between all of the objects in the system. Several formulizations of ENA have arisen including embodied energy analysis (Herendeen, 1981), ascendency analysis (Ulanowicz, 1980, 1986, 1997, 2009), and network environ analysis (Fath & Patten, 1999; Patten, 1978a, 1981, 1982a). There are several available software packages that carry out the network analysis, such as NETWRK (Ulanowicz, 1982), WAND (Allesina & Bondavalli, 2004), NEA (Fath & Borrett, 2006), EcoNet (Kazanci, 2007), and R.ENA (Scotti & Bondavalli, 2010).

An ecological network flow model is essentially an ecological food web (energy–matter flow of who eats whom), which also includes nonfeeding pathways such as dissipative export out of the system and pathways to detritus. The first step is to identify the system of interest and place a boundary (real or conceptual) around it. Energy–matter transfers within the system boundary comprise the network; transfers crossing the boundary are either input or output to the network; and all transactions starting and ending outside the boundary without crossing it are external to the system and are not considered. Once the system boundary has been established, it is necessary to compartmentalize the system into the major groupings. The most aggregated model could have only two compartments: producers and consumers (where decomposers are included in consumers). A slightly more disaggregated model could have the following functional groups: producers, herbivores, carnivores, omnivores, decomposers, and detritivores (Fath, 2004). Most ecological network models will have many more compartments.

Mathematically, network analysis is built on the formalisms of graph theory and matrix algebra. The most basic realization of network analysis consists of a graph in which an edge or arc links two or more nodes together. An isomorphic mapping allows for representation of the graph as a matrix, which is called an adjacency matrix (Figure 6.4). When the edges are directed, that is, flow in a particular direction, the model can

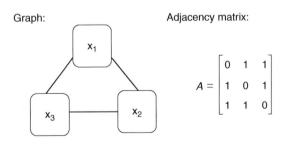

FIGURE 6.4 Graph and adjacency matrix for simple network.

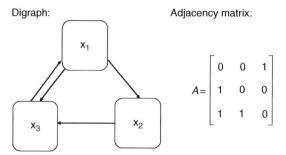

FIGURE 6.5 Digraph and adjacency matrix for simple network.

be considered as a directed graph or digraph (Figure 6.5). A digraph-based adjacency matrix can be used for investigating the structural properties of the network such as number of pathways, connectivity, and rate of path proliferation (Borrett et al., 2007). The next level of analysis occurs when weighted flow values can be determined for each arc including (since an ecosystem is an open environmental system) the input and output boundary exchanges (Figure 6.6). This should also include information regarding the storage values, in which case the full suite of network analysis properties can be investigated and calculated.

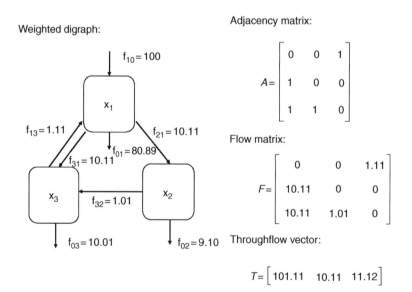

FIGURE 6.6 Weighted digraph, adjacency, and flow matrix.

Lastly, it is useful to normalize and nondimensionalize the flows between compartments relative to a reference condition, typically, using the component's throughflow or storage values (Figure 6.7). These values are used in the network environ analysis. This involves the power series analysis and subsequent transitive closure matrix, which yield the combined direct and indirect flows in the network.

An adjacency matrix element, $a_{ij} = 1$ if a direct arc from j to i exists and zero if no arc exists (note that some authors orient the adjacency matrix from rows to columns $a_{ij} = 1$ if there is a direct arc from i to j). In other words, the A matrix gives the direct connectivity of a model. A row sum would give the total number of arrows emanating out of the compartment (called out degree), and a column sum gives the total number of arrows entering the compartment (called in degree). In ecological models, the interpretation is clear and relevant because the in degree is the number of diet sources (prey items) and the out degree is the number of predators or sinks for the compartment. Whereas the adjacency matrix indicates the presence or absence of a direct connection, matrix A^2 gives the number of pathways that take exactly two steps between two compartments, and matrix A^3 gives the number of pathways that take exactly two steps, and so forth. Therefore, A^m gives the

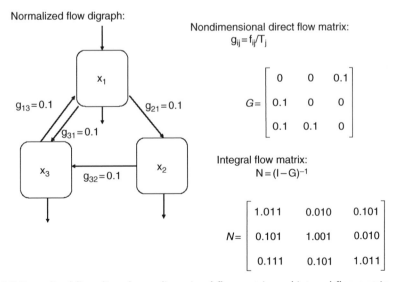

FIGURE 6.7 Normalized flow digraph, nondimensional flow matrix, and integral flow matrix.

number of pathways of length m. For example, A^2, A^3, and A^4 for Figure 6.7 are given as:

$$A^2 = \begin{bmatrix} 1 & 1 & 0 \\ 0 & 0 & 1 \\ 1 & 0 & 1 \end{bmatrix} A^3 = \begin{bmatrix} 1 & 0 & 1 \\ 1 & 1 & 0 \\ 1 & 1 & 1 \end{bmatrix} A^4 = \begin{bmatrix} 1 & 1 & 1 \\ 1 & 0 & 1 \\ 2 & 1 & 1 \end{bmatrix}$$

The reachability matrix $A^{[R]}$ is given by:

$$A^{[R]} = A + A^2 + \ldots + A^{n-1}$$

Two nodes are reachable if and only if all elements of $A^{[R]} \geq 1$, and not reachable if it is 0. In this example, all compartments are reachable to each other. Furthermore, by A^4 we see that there are multiple pathways from compartment 1 to compartment 3 ($a_{31} = 2$). These are: $1\rightarrow2\rightarrow3\rightarrow1\rightarrow3$ and $1\rightarrow3\rightarrow1\rightarrow2\rightarrow3$, which are distinct paths carrying energy through the network. Note that in this case the elements will continue to increase as the path length increases. In fact, the rate of increase in the limit for each pairwise combination is equal to the magnitude of the maximum eigenvalue:

$$\frac{a_{ij}^{(m+1)}}{a_{ij}^m} = \lambda_{max}$$

as $m \rightarrow \infty$

Not all networks have this property of increasing pathways with increasing path length, referred to as path proliferation, in which the pathways increase without bound. It depends on the connectivity and is an important feature for ecological systems that cycle material and energy. Three classes of connectivity and therefore cycling, can be defined using the eigenvalues as follows (Fath and Halnes, 2007):

$$\lambda_{max} = 0; \text{no cycling}$$

$$\lambda_{max} = 1; \text{weak cycling}$$

$$\lambda_{max} > 1; \text{strong cycling}$$

As previously stated there are several ENA approaches applied. Here, for illustrative purposes, we present an example using EcoNet software for a common network model of an oyster reef community (Dame & Patten, 1981, see Fig. 3.7). EcoNet 2.1 is an online, user-friendly, interactive domain that allows the easy calculation of many of the network properties (http://eco.engr.uga.edu/). The Web site provides sufficient background information for new users, including modelling information, theoretical background, and preloaded examples. The user is prompted to enter the model structure in the provided window, which will include flows, flow types, coefficients, initial conditions, and comments. The model is currently set up to accept three types of flow: donor controlled, donor-recipient controlled, and Michaelis-Menten taking the following forms, respectively, for a flow from compartment A to compartment B:

$$\text{Donor controlled}: \text{ Flow } A \rightarrow B = c * A$$

$$\text{Donor} - \text{recipient controlled}: \text{ Flow } A \rightarrow B = r * A * B$$

$$\text{Michaelis} - \text{Menten flow}: \text{ Flow } A \rightarrow B = A * B/(v + A)$$

The symbols c, r, and v for the coefficient values indicate the type of flow used in the model. Multiple flow types are permissible in this model.

The oyster reef is one of the examples preloaded in the EcoNet software, and is formulated as follows:

```
# Intertidal Oyster Reef Ecosystem Model, by Dame and
Patten.
# Model flows are in kcal mo-2 dayo-1; storage data is
# kcal mo-2.
# This model is based on the Matlab model written by Fath
# and Borrett (2004).
# Dame, R. F., and B. C. Patten. 1981. Analysis of energy
# flows in an intertidal oyster reef. Marine Ecology
Progress
# Series 5:115-124.
# Patten, B. C. 1985. Energy cycling, length of food
chains,
# and direct versus indirect effects in ecosystems. Can.
# Bull. Fish. Aqu. Sci. 213:119-138.
* -> Filter_Feeders c=41.4697
Filter_Feeders -> Dep_Detritus c=0.0079
Filter_Feeders -> Predators c=0.0003
Dep_Detritus -> Microbiota c=0.0082
Dep_Detritus -> Meiofauna c=0.0073
Dep_Detritus -> Dep_Feeders c=0.0006
Microbiota -> Meiofauna c=0.5
Microbiota -> Dep_Feeders c=0.5
Meiofauna -> Dep_Detritus c=0.1758
Meiofauna -> Dep_Feeders c=0.0274
Dep_Feeders -> Dep_Detritus c=0.1172
Dep_Feeders -> Predators c=0.0106
Predators -> Dep_Detritus c=0.0047
Filter_Feeders -> * c=0.0126
```

```
Dep_Detritus -> * c=0.0062
Microbiota -> * c=2.3880
Meiofauna -> * c=0.1484
Dep_Feeders -> * c=0.0264
Predators -> * c=0.0052
Filter_Feeders = 2000; Dep_Detritus = 1000;
Microbiota = 2.4121; Meiofauna = 24.121;
Dep_Feeders = 16.274; Predators = 69.237
# Model flows are in kcal mo-2 dayo-1; storage data is
# kcal mo-2.
```

The model diagram is given in Figure 6.8. This model can be run using four numerical methods: (1) adaptive time-step, Runge-Kutta-Fehlberg is the default, and also allows for (2) fixed time step, 4th order Runge-Kutta;

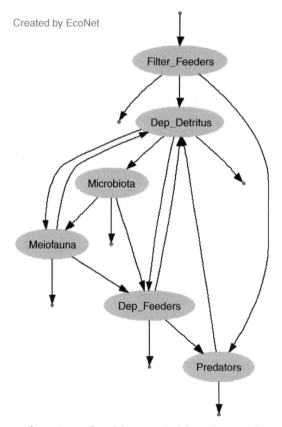

FIGURE 6.8 Flow diagram for oyster reef model constructed from EcoNet software.

(3) stochastic, fixed time step, Langevin equation which generates different solutions at each run, and (4) stochastic, adaptive time step, Gillespie's method, which is a variety of a dynamic Monte Carlo method. The default approach has maximum time and sensitivity as parameters. The sensitivity is related to the amount of error between the actual and numerical solution, which provides a trade-off between accuracy and computational time.

After the model is run, the results are presented, starting with the diagram, followed by a plot of the state variables during the simulation (called the time course figure). For the oyster reef example, which is balanced, there is no change in any of the state variables during the duration of the simulation. Next, the compartmental properties are provided including the initial and final storage values (essentially the same in this case) as well as the boundary input and output from each compartment. Also included in the compartmental properties are the input throughflow (and equivalently the output throughflow) and the residence time for energy in each compartment. Input throughflow is the amount of flow through each compartment that arrives from boundary flow without cycling. This is also referred to as first passage flow in the literature (Higashi et al., 1993; Fath et al., 2001). Values for input throughflow and residence time are given in Table 6.1.

Next the analysis gives results for a suite of system-wide properties that are described in Box 6.1 and the values given in Table 6.2.

Lastly, in the "short" version of the output, the software completes the throughflow, storage, and utility analyses. As these are the key environ analysis results, they are presented here in full.

Table 6.1 Values for Input Throughflow and Residence Time

	Input Throughflow	Residence Time
X1- Filter_feeders	41.4697	48.0769
X2-Dep-Detritus	22.2257	44.843
X3-Microbiota	8.1779	0.2952
X4-Meiofauna	8.4872	2.8441
X5-Dep_Feeders	2.4667	6.4851
X6-Predators	0.7679	101.01

Notes: This is also referred to as first passage flow in the literature
Source: Fath et al., 2001; Higashi et al., 1993.

BOX 6.1   DESCRIPTION OF THE SYSTEM-WIDE PROPERTIES CALCULATED IN ECONET SOFTWARE

1. **Link Density:** Number of intercompartmental links (d) per compartment: d/(number of compartments).
2. **Connectance:** Ratio of the number of actual intercompartmental links (d) to the number of possible intercompartmental links: d/(Number of compartments)^2.
3. **Total System Throughflow (TST):** The sum of throughflows of all compartments: TST $= T_1 + T_1 + \ldots + T_n$.
4. **Finn's Cycling Index:** Measures the amount of cycling in the system by computing the fraction of total system throughflow that is recycled.
5. **Indirect Effects Index:** Measures the amount of flow that occurs over indirect connections versus direct connections. When the ratio is greater than one, indirect flows are greater than direct flows.
6. **Ascendency:** It quantifies both the level of system activity and the degree of organization (constraint) with which the material is being processed in ecosystems.
7. **Aggradation Index:** Measures the average path length. In other words, it is the average number of compartments a unit flow quantity (e.g., an N atom, unit biomass, energy quanta, etc.) passes through before exiting the system: TST/$(z_1 + z_2 + \ldots + z_n)$.
8. **Synergism Index:** Based on utility analysis, it provides a system-wide index for pairwise compartment relations. Values larger than 1 indicate a shift toward quantitative positive interactions (synergism). It is computed as the ratio of the sum of positive entries over the sum of negative entries in the utility analysis matrix U.
9. **Mutualism Index:** Similar to synergism index, mutualism index provides a system-wide index for pairwise compartment relations. Values larger than 1 indicate a shift toward qualitative positive interactions (mutualism). It is computed as the ratio of number of positive entries over the number of negative entries in the mutual relations matrix.
10. **Homogenization Index:** Quantifies the action of the network making the flow distribution more uniform. Higher values indicate that resources become well mixed by cycling in the network, giving rise to a more homogeneous distribution of flow.

Source: EcoNet Web site: http://eco.engr.uga.edu/DOC/econet4.html#scalar)

Nondimensional, direct flow matrix:

$$G = \begin{bmatrix} 0 & 0 & 0 & 0 & 0 & 0 \\ 0.38 & 0 & 0 & 0.5 & 0.76 & 0.475 \\ 0 & 0.367 & 0 & 0 & 0 & 0 \\ 0 & 0.327 & 0.148 & 0 & 0 & 0 \\ 0 & 0.027 & 0.148 & 0.0779 & 0 & 0 \\ 0.014 & 0 & 0 & 0 & 0.0687 & 0 \end{bmatrix}$$

Table 6.2 Analysis Gives Results for a Suite of System-wide Properties Described in Box 6.1 and the Values Given in Table 6.2.

Link density	2
Connectance	0.3333
Total system throughflow	83.5959
Finn's cycling index	0.1097
Indirect effects index	1.5297
Ascendency	115.329
Development capacity	188.573
Aggradation index	2.0158
Synergism index	6.5379
Mutualism index	2
Homogenization index	1.8905

Integral flow matrix:

$$N = \begin{bmatrix} 1.000 & 0 & 0 & 0 & 0 & 0 \\ 0.536 & 1.386 & 0.277 & 0.779 & 1.01 & 0.658 \\ 0.197 & 0.510 & 1.102 & 0.286 & 0.404 & 0.242 \\ 0.205 & 0.529 & 0.253 & 1.297 & 0.419 & 0.251 \\ 0.059 & 0.154 & 0.190 & 0.164 & 1.122 & 0.073 \\ 0.0185 & 0.0106 & 0.013 & 0.0113 & 0.077 & 1.005 \end{bmatrix}$$

Partial turnover rate matrix:

$$C = \begin{bmatrix} -0.0208 & 0 & 0 & 0 & 0 & 0 \\ 0.0079 & -0.0223 & 0 & 0.1758 & 0.1172 & 0.0047 \\ 0 & 0.0082 & -3.388 & 0 & 0 & 0 \\ 0 & 0.0073 & 0.5 & -0.3516 & 0 & 0 \\ 0 & 0.0006 & 0.5 & 0.0274 & -0.1542 & 0 \\ 0.0003 & 0 & 0 & 0 & 0.0106 & -0.0099 \end{bmatrix}$$

Integral storage matrix:

$$S = \begin{bmatrix} 48.0769 & 0 & 0 & 0 & 0 & 0 \\ 24.0392 & 62.172 & 12.423 & 34.929 & 49.283 & 29.516 \\ 0.0582 & 0.150 & 0.325 & 0.0845 & 0.119 & 0.071 \\ 0.582 & 1.505 & 0.721 & 3.690 & 1.193 & 0.714 \\ 0.386 & 0.997 & 1.231 & 1.066 & 7.276 & 0.473 \\ 1.870 & 1.068 & 1.318 & 1.141 & 7.790 & 101.517 \end{bmatrix}$$

Mutualism relations:

$$
\text{sgn}(U) = \begin{bmatrix}
+ & - & + & + & - & - \\
+ & + & - & - & + & - \\
+ & + & + & - & - & + \\
+ & + & + & + & - & + \\
- & + & + & + & + & - \\
+ & - & + & + & + & +
\end{bmatrix}
$$

Utility analysis:

$$
U = \begin{bmatrix}
0.832 & -0.221 & 0.0699 & 0.0126 & -0.027 & -0.014 \\
0.424 & 0.599 & -0.193 & -0.0357 & 0.065 & -0.0012 \\
0.394 & 0.547 & 0.741 & -0.200 & -0.061 & 0.0071 \\
0.208 & 0.287 & 0.0014 & 0.946 & -0.056 & 0.0054 \\
-0.0035 & 0.065 & 0.446 & 0.169 & 0.911 & -0.0615 \\
0.473 & -0.449 & 0.251 & 0.066 & 0.156 & 0.975
\end{bmatrix}
$$

The software does have extended results that include control analysis and an eco-exergy calculator. The analysis data may be downloaded in Matlab format (.m file) or in spreadsheet format (.csv file). Overall, the EcoNet software provides a very easy to use application for researchers or students interested in conducting network analysis.

Problems

1. Construct the adjacency matrix of the oyster reef model in Figure 6.8.
2. How can you determine if the compartments are at steady state in a network model?
3. What are typical units of flow in a network model? Why is it necessary to normalize the flow values by the throughflows before taking the powers of the matrix?
4. How could one conduct a network analysis for time-varying data?
5. What is network mutualism? What role do indirect influences play in determining it?

7

⠿

Dynamic Biogeochemical Models

7.1. Introduction

This chapter gives detailed examples of typical dynamic biogeochemical models and discusses the considerations that have to be made when selecting the model complexity and equations. The past 30 years have witnessed a pronounced development and application of biogeochemical models. The models are often formulated as a set of differential equations combined with some algebraic equations and a parameter list. The differential equations require the definition of an initial state.

The following biogeochemical models are included in this chapter to illustrate and demonstrate their wide applicability in ecological and environmental modelling:

Fundamentals of Ecological Modelling. DOI: 10.1016/B978-0-444-53567-2.00007-7

1. Classical Streeter-Phelps river BOD/DO model
2. Simple eutrophication models based on up to only 2–4 state variables
3. Complex eutrophication model that has been applied to 25 case studies with modifications from case-to-case
4. Wetland model used for design and construction of wetlands for the treatment of drainage water or wastewater
5. Model for the prediction of global warming

The two eutrophication models (2 and 3) are used to show the complexity spectrum of available models. The selection of model complexity will be discussed with reference to Chapter 2, Section 2.6. Furthermore, the generality of models and their ability to develop predictions will be discussed using the eutrophication models as examples. All five of these models are discussed in detail; the reader will get a good impression of how to develop and use biogeochemical models and how to assess the advantages and disadvantages of each model. Hopefully, the reader will learn to be critical and understand the considerations involved in modelling, including the selection of balanced model complexity.

Wetland models have been very much in focus recently due to an increasing interest for these ecosystems as habitats for birds and amphibians. Wetland restoration or wetland construction is an effective method of abatement of nutrient pollution from nonpoint sources (agricultural pollution). This has increased the demand for good management models in this area. The presented wetland model has been widely applied to design and construct wetlands.

Biogeochemical models are widely used to solve a number of concrete management problems:

- Optimization of biological treatment
- Groundwater contamination
- Atmospheric acidification (see Rains model in Alcamo et al., 1990)
- Forest growth and yield (Vanclay, 1994)
- Air pollution problems (Gryning & Batchvarova, 2000)
- Agricultural production (France & Thornley, 1984)

7.2. Application of Biogeochemical Dynamic Models

Ecosystems are dynamic systems and biogeochemical models attempt to capture the dynamics and cycling of biochemical and geochemical compounds in the ecosystems. When models are used as an instrument in environmental management, they must account for the fate and distribution of both pollutants and of nature's own compounds. This requires the application of biogeochemical models, since they focus on the processes and transformation of various compounds in the ecosystem. As pointed out in Chapter 2, Section 2.3, the construction of dynamic models requires data, which can elucidate the dynamics of the processes included in the model. Generally, a more comprehensive database is required to build a dynamic model than a static model. Therefore, in a data-poor situation, it might be better to draw up an average situation under different circumstances using a static model than to construct an unreliable dynamic model, which contains uncertainty in the most crucial parameters

The first biogeochemical model constructed was the Streeter-Phelps BOD/DO model in 1925. It has been used numerous times as an illustration of biogeochemical models and of the practical use of ecological models in environmental management (Jørgensen, 2009). As a seminal example, it clearly illustrates the concepts of the biogeochemical models, and is presented in detail in the next section. The Streeter-Phelps model consists, opposite from most dynamic models, of only one differential equation that can be solved analytically. Here we use STELLA to simulate and demonstrate the applicability of the model solution.

Hydrodynamic models can be considered biogeochemical models, since they describe the fate and distribution of the important compound water in ecosystems. Output from hydrodynamic models is often used as forcing function in ecological models. If only the hydrology is modelled, then hydrodynamic models are not ecological models, as they do not account for biological processes. However, they are often used in conjunction with ecological models, as the distribution of chemical compounds and living organisms is dependent on the hydrodynamics. During the 1990s, 3-D hydrodynamic models were applied more frequently, and today, well-developed ecological models such as eutrophication models are coupled with 3-D hydrodynamic models. It is important to emphasize that coupling simple ecological models with 3-D models is not feasible because the standard deviations of a validation and the reliability of the predications are determined by the weakest link in the chain of calculations. Hydrodynamic models are, however, beyond the scope of this book and will not be discussed further.

The experience gained by developing many biogeochemical models over time has shown that:

1. A good knowledge of the ecosystem is required to capture the essential features, which should be reflected in the model.
2. The scope of the model determines the complexity, which again determines the quality and quantity of the data needed for calibration and validation.
3. If good data are not available, then it is better to go for a somewhat oversimplified model instead of one that is too complex.
4. Simple models are more general than complex models. If the database allows development of a more complex model, then it will contain some processes and components specific for the considered ecosystem.

During the 1970s and the early 1980s, a great deal of experience was gained in modelling many different types of ecosystems and many different aspects, including a number of problems relating to environmental degradation. The modellers also learned which modifications were necessary when a model was applied for the same situation but on a different ecosystem from which it was originally developed. It was seen that the same model could not be applied to another ecosystem without some changes, unless the model was very simple. More and more models became well calibrated and validated. The models could often be used as a practical management tool, but in most cases it was necessary to combine the use of the model with a good knowledge of general environmental issues. Also, in cases when the model could not be applied to set up accurate predictions, it was useful for the manager to qualitatively understand the ecosystem for various management strategies. Scientists who applied models found that they were very useful to indicate research priorities and also to capture the system features of ecosystems.

7.3. The Streeter-Phelps River BOD/DO Model, Using STELLA

For rivers and streams, the main environmental problem is low oxygen concentration that occurs in response to the discharge of organic matter. The questions posed for the model are:

1. What is the concentration of organic matter, expressed frequently as BOD_5 mg/L, as a function of time? BOD_5 mg/L is the amount of oxygen that the decomposition of the organic matter will consume during a period of 5 days.
2. What is the oxygen concentration as a function of the distance from the discharge point of organic matter?
3. What is the minimum oxygen concentration?

   A river model is presented in that next section that is able to answer these questions. It is developed using the STELLA software, which was introduced in Chapter 2, Section 2.3 and 5. After presentation of the classical Streeter-Phelps model, a discussion about which processes would probably be beneficial to include if the model is to be expanded to include more components and interactions will be presented.

   Organic matter decomposition can be approximated by a first-order reaction. If L is the concentration of organic matter (mg/L) and k_1 is the rate coefficient for the decomposition, then the following differential equation is valid:

$$dL/dt = -k_1 L \qquad (7.1)$$

Equation (7.1) has the following analytical solution:

$$L_t = L_0 e^{-k_1 t} \qquad (7.2)$$

where L_t is the concentration at time t and L_0 is the initial concentration.

   L is most often expressed as BOD_5 mg/L oxygen consumption during a period of 5 days. If it is expressed as mg/L (average) organic matter or detritus, then the concentration, according to the processes, has to be multiplied by 1.39. In other words, 1 g of detritus or organic matter requires an average of 1.39 g oxygen to be decomposed as much as it is possible during a period of 5 days (which is nearly 100%)

   Nitrification of ammonium also causes oxygen depletion and should be included in this process. If the ammonium concentration is denoted

NC (mg N/L in the form of ammonium) and it is presumed that nitrification follows a first-order reaction, then the following differential equation is valid:

$$dNC/dt = -k_N NC \qquad (7.3)$$

where k_N is the rate coefficient for the nitrification. Equation (7.3) has the following solution:

$$NC_t = NC_0 e^{-k_N t} \qquad (7.4)$$

where NC_t is the concentration at time t and NC_0 the initial concentration.

Notice that NC is the concentration in mg ammonium-N/L and the corresponding oxygen consumption is found from the chemical equation for the nitrification:

$$NH_4^+ + 2O_2 \rightarrow NO_3^- + H_2O + H^+ \qquad (7.5)$$

It means that 1 g of ammonium-N requires $2*32/14 = 4.6$ g of oxygen, which will be included in the model when the nitrification is "translated" to oxygen depletion. The factor is 4.3 not 4.6, due to the bacterial assimilation of ammonium by the nitrifying microorganisms.

Instead of a first-order expression for Eq. (2.5), one could apply a Michaelis-Menten equation (Jørgensen & Bendoricchio, 2001). According to the Michaelis-Menten expression used in eutrophication models (see Chapter 2), Eq. (7.1) could be multiplied by:

$$[Ox]/(k_{mo} + [Ox]) \qquad (7.6)$$

to account for the influence of oxygen as a limiting factor of the decomposition rate. Similarly, Eq. (7.3) could be multiplied by:

$$\min ([NC]/(k_{ma} + [NC]), \, [Ox]/(k_{ao} + [Ox])) \qquad (7.7)$$

to account for the influence of both ammonium and oxygen as possible limiting factors of the decomposition. When the model has to be erected, the Michaelis-Menten expressions will be applied, but it is also possible to get reasonably good results using the first-order expressions, which have the advantage that they can easily be solved analytically.

The decomposition of organic matter and nitrification are temperature dependent. A simple Arrhenius expression may be applied:

The rate coefficient at temperature (Celsius) T
$$= \text{rate coefficient at 20 degree Celsius} * K^{(T-20)} \qquad (7.8)$$

K is, with good approximation, 1.05 for organic matter decomposition while nitrification is more sensitive to temperature changes; therefore, K is 1.07–1.08 for this process (Jørgensen, 2000).

Typical values for the rate coefficients and the initial concentrations for various sources of organic matter and ammonium are shown in Table 7.1.

If the oxygen concentration is below the saturation concentration, that is, the water is in equilibrium with the atmosphere, then reaeration from the atmosphere takes place. The equilibrium concentration can be found by Henry's Law. The saturation concentration is dependent on the water temperature and salinity. In Appendix 1, Table 1, the equilibrium concentration of oxygen can be found as a function of the temperature and salinity.

Aeration is proportional to the difference between the oxygen concentration at saturation, Ox_{sat}, and the actual oxygen concentration, $[Ox]$. The driving force for the aeration is this difference. It is expressed by:

$$\text{Reaeration} = d[Ox]/dt = K_a(Ox_{sat} - [Ox]) \tag{7.9}$$

where K_a is the reaeration coefficient (1/24h).

K_a is dependent on the water temperature and flow rate. The aeration is also proportional to the surface area relatively to the volume; it is inversely proportional to the water depth. There are several hundred empirical equations that can be used to estimate the reaeration or the reaeration coefficient. One equation often applied is:

$$K_a = 2.26 * v * \exp(0.024 * (T - 20)/d), \tag{7.10}$$

Table 7.1 Characteristic Values for k_1, k_N (1/24h) and Initial Concentrations (mg/L) for Various Sources to Oxygen Depletion in Streams and Rivers

Source	k_1	k_N	L_0	NC_o
Municipal waste water	0.35–0.40	0.15–0.25	180–300	20–45
Mechanically treated waste water	0.32–0.36	0.10–0.15	100–200	18–35
Biologically treated waste water	0.10–0.25	0.05–0.20	10–40	15–32
Potable water	0.05–0.10	0.03–0.06	0–2	0–1
River water (average)	0.05–0.15	0.04–0.10	1–4	0–2
Agricultural drainage water	0.08–0.20	0.04–0.12	5–25	0–10
Waste water, food industry	0.4–0.5	0.1–0.25	200–5000	20–200

where v is the water flow rate in m/s, T is the temperature in Celsius, and d is the depth of the stream or river in m.

The oxygen concentration is determined by the difference between the consumption and the reaeration. If we only consider decomposition of organic matter and nitrification, then the oxygen concentration is determined by the following differential equation:

$$d[Ox]/dt = k_a(Ox_{sat} - [Ox]) - k_1L - k_NNC \qquad (7.11)$$

The solution of the differential equations for L and NC can be used in this differential equation to yield the following expression for the solution of [Ox] as function of time:

$$d[Ox]/dt = k_a(Ox_{sat} - [Ox]) - k_1L_0 \ e^{-k_1t} - k_NNC_0 \ e^{-k_Nt} \qquad (7.12)$$

This equation can be solved analytically, but we will use STELLA to determine [Ox], L, and NC as a function of time in accordance with the previous Michaelis-Menten equation. It is possible to add many more processes to the model, such as primary production by phytoplankton and macrophytes producing oxygen, denitrification consuming organic matter as a carbon source, the presence of organic matter with different biodegradability, changed reaeration at turbulent flow, and so on. As always, when we develop models, the problem, the available data, and the system processes should determine the model complexity.

A diagram of the STELLA model (Figure 7.1) shows three state variables: (1) organic matter, L; (2) ammonium-nitrogen, NC; and (3) oxygen, Ox. These are each covered by three different connected submodels. The oxygen is consumed by the decomposition of organic matter and nitrification. The oxygen concentration influences the decomposition rate and nitrification as presented in Eqs. (7.6) and (7.7). Time is considered the independent variable, but it could also be the distance from the discharge of wastewater. For example, if the water flow rate is 1 m/s, then the time in days will correspond to 1*3600*24 m = 86400 m. The model has a constant discharge, which can be considered to be agricultural drainage water along the stream shoreline. The point discharge of wastewater takes place at time 0 corresponding to the initial value of L and NC. The dilution has to be considered when the initial values are calculated. If 1000 m^3 waste water/h with 30 mg/L of BOD$_5$ and 13 mg/L ammonium-N is discharged to a river with 5000 m^3 of water flow/h, then the dilution factor is 6. If the river water

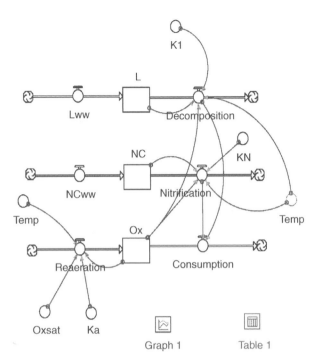

FIGURE 7.1 Conceptual diagram of the presented river model by application of the STELLA software.

has 3 mg BOD_5/L and 1 mg ammonium-N/L, then the mixture of river water and wastewater will have $(5*3 + 30*1)/6 = 7.5$ mg BOD_5/L and $(1*5 + 13*1)/6 = 3$ mg ammonium-N/L, which are applied as the initial concentrations. The model should be able to give information about how these concentrations change over time. The concentration of oxygen in treated wastewater will almost always be close to 0 mg/L. If the river water has 8.6 mg/L oxygen, then the mixture of wastewater and river water will have an oxygen concentration of about 7.2 mg/L corresponding to a dilution of the wastewater by a factor 6.

The result of running the model 90 days is shown in Figure 7.2, and the equations are presented in Table 7.3. Notice the form of the differential equation applied in STELLA. Time can be translated to distance from the discharge point by the flow rate. If the cross-sectional area is $50m^3$, then the water flow of 10,000 m^3/h corresponds to 200 m/h. Twenty-four hours therefore corresponds to 4800 m and 90 days to 432 km. The minimum oxygen concentration occurs after 8 days or 38.2 km. Table 7.2 charts the model results in table form for every 5 days.

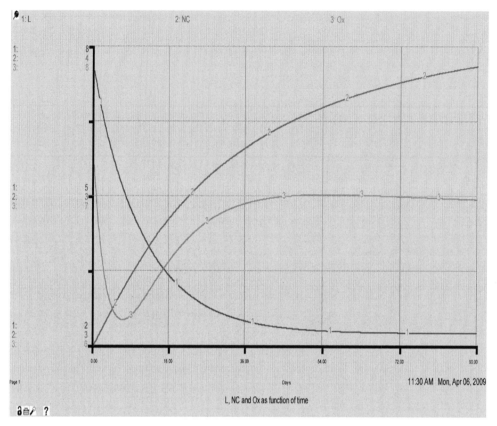

FIGURE 7.2 The result of using the following initial values: for BOD$_5$, 7.5 mg/L; for ammonium-N, 3 mg/L; and for oxygen, 7.2 mg/L.

7.4. Eutrophication Models I: Simple Eutrophication Models with 2–4 State Variables

Eutrophication is the main cause of environmental degradation in lakes and reservoirs. It results from a nutrient concentration that is too high. The core questions for this model are:

1. What is the concentration of the limiting nutrient (which is often phosphorus)?
2. What is the primary production?
3. What are the transparency and the chlorophyll concentration when the eutrophication is at maximum?

Table 7.2   Model Results

Day	BOD$_5$	NC	Oxygen	Ox consumption	Decomp.	Nitrification
.0	7.50	3.00	7.20	1.01	0.67	0.08
5.0	5.64	3.11	6.21	0.83	0.49	0.08
10.0	4.47	3.21	6.24	0.74	0.39	0.08
15.0	3.70	3.30	6.45	0.69	0.32	0.08
20.0	3.19	3.37	6.64	0.66	0.28	0.09
25.0	2.86	3.43	6.78	0.64	0.25	0.09
30.0	2.64	3.48	6.88	0.63	0.24	0.09
35.0	2.50	3.52	6.94	0.62	0.22	0.09
40.0	2.40	3.56	6.98	0.62	0.21	0.09
45.0	2.34	3.59	7.00	0.61	0.21	0.09
50.0	2.30	3.62	7.00	0.61	0.21	0.09
55.0	2.28	3.64	7.01	0.61	0.20	0.10
60.0	2.26	3.66	7.01	0.62	0.20	0.10
65.0	2.25	3.68	7.00	0.62	0.20	0.10
70.0	2.24	3.70	7.00	0.62	0.20	0.10
75.0	2.24	3.71	6.99	0.62	0.20	0.10
80.0	2.24	3.73	6.99	0.62	0.20	0.10
85.0	2.24	3.74	6.98	0.62	0.20	0.10
Final	2.24	3.75	6.98			

Notes: Concentrations (BOD$_5$ or L, NC, or ammonium-N and oxygen) are all in mg/L and process rates are in mg/L per 24h

From a thermodynamic view, a lake can be considered an open system, which exchanges material (wastewater, evaporation, precipitation) and energy (evaporation, radiation) with the environment. However, in some lakes (e.g., the Great Lakes) the material input per year does not change the concentration measurably. In such cases, the system can be considered nearly closed, which means that it exchanges energy but not material with the environment.

The flow of energy through the lake system leads to at least one cycle of material in the system (provided that the system is at a steady state,

Table 7.3 Model Equations (STELLA Format)

```
L(t) = L(t – dt) + (Lww – Decomposition) * dt
INIT L = 7.5
INFLOWS:
Lww = 0.2
OUTFLOWS:
Decomposition = (L*K1*Ox*1.05^(20-Temp)/(Ox+2.5))
NC(t) = NC(t – dt) + (NCww – Nitrification) * dt
INIT NC = 3
INFLOWS:
NCww = 0.1
OUTFLOWS:
Nitrification = NC*KN*MIN(Ox/(Ox+3), NC/(NC+1))*1.075^
(Temp-20)
Ox(t) = Ox(t – dt) + (Reaeration – Consumption) * dt
INIT Ox = 7.2
INFLOWS:
Reaeration = Ka*(Oxsat-Ox)*exp(0.024*(Temp-20))
OUTFLOWS:
Consumption = Decomposition+4.3*Nitrification
K1 = 0.1
Ka = 0.226
KN = 0.05
Oxsat = 10
Temp = 16
```

see Morowitz, 1968). As illustrated in Figures 2.1 and 7.3, important elements participate in cycles that control eutrophication.

The word eutrophy is generally known as "nutrient rich." In 1919, Nauman introduced the concepts of oligotrophy and eutrophy, distinguishing between oligotrophic lakes containing little planktonic algae and eutrophic lakes containing an abundance of phytoplankton.

The eutrophication of lakes all over the world has increased rapidly during the last decades due to increased human population growth and the consequent increase in the application of synthetic fertilizers and urbanization (Vitousek et al., 1997; Jørgensen et al., 2004). The production of fertilizers has grown exponentially in this century and the concentration of phosphorus in many lakes reflects this.

The word eutrophication is used increasingly to define an artificial addition of nutrients, mainly nitrogen and phosphorus, to waters. Eutrophication is generally considered to be undesirable, but this is not always true.

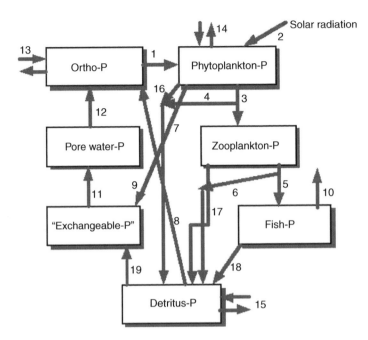

FIGURE 7.3 The phosphorus cycle. The processes are: (1) uptake of phosphorus by algae; (2) photosynthesis; (3) grazing with loss of undigested matter; (4), (5) predation with loss of undigested material; (6), (7), and (9) settling of phytoplankton; (8) mineralization; (10) fishery; (11) mineralization of phosphorous organic compounds in the sediment; (12) diffusion of pore water P; (13), (14), and (15) inputs/outputs; (16), (17), and (18) mortalities; and (19) settling of detritus.

The green color of eutrophied lakes makes swimming and boating less safe and pleasant due to the increased turbidity. From an aesthetic point of view, the chlorophyll concentration should not exceed 100 mg m^{-3}. However, the most critical effect from an ecological point of view is the reduced oxygen content of the hypolimnion caused by the decomposition of dead algae. Eutrophic lakes sometimes show a high oxygen concentration at the surface during the summer, but a low concentration of oxygen during the fall in the hypolimnion that is lethal to fish.

About 16–20 chemical elements are necessary for the growth of freshwater plants; Table 7.4 lists the relative quantities of essential elements in plant tissue. The present concern about eutrophication relates to the rapidly increasing amount of phosphorus and nitrogen, which are naturally present at relatively low concentrations. Of the two, phosphorus is considered the major cause of eutrophication in lakes, because it was formerly the growth-limiting factor for algae in the majority of lakes. But as mentioned previously, its use has increased tremendously during the last decades.

Nitrogen is a limiting factor in a number of East African lakes as a result of soil nitrogen depletion by intensive erosion. However, today nitrogen

Table 7.4 Average Freshwater Plant Elementary
Composition on a Wet Weight Basis

Element	Plant content%
Oxygen	80.5
Hydrogen	9.7
Carbon	6.5
Silicon	1.3
Nitrogen	0.7
Calcium	0.4
Potassium	0.3
Phosphorus	0.08
Magnesium	0.07
Sulfur	0.06
Chlorine	0.06
Sodium	0.04
Iron	0.02
Boron	0.001
Manganese	0.0007
Zinc	0.0003
Copper	0.0001
Molybdenum	0.00005
Cobalt	0.000002

may become limiting in lakes as a result of the tremendous increase in the phosphorus concentration caused by discharge of wastewater, which contains relatively more phosphorus than nitrogen. While algae use 4 to 10 times more nitrogen than phosphorus, wastewater generally contains only 3 times as much nitrogen as phosphorus in lakes and a considerable amount of nitrogen is lost by denitrification (nitrate→N_2).

The growth of phytoplankton is the key process of eutrophication, and it is important to understand the interacting processes that regulate growth. Primary production has been measured in great detail in a number of lakes. This process represents the synthesis of organic matter and the overall process can be summarized as follows:

$$Light + 6CO_2 + 6H_2O \rightarrow C_6H_{12}O_6 + 6O_2$$

The composition of phytoplankton is not constant (note that Table 7.4 gives only an average concentration), but reflects the concentration of the water. If the phosphorus concentration is high, then the phytoplankton will take up relatively more phosphorus — this is called luxury uptake.

As seen in Table 7.4, phytoplankton consists mainly of carbon, oxygen, hydrogen, nitrogen, and phosphorus: without these elements, no algal growth takes place. This leads to the concept of limiting the nutrients mentioned earlier in Section 2.3, which is known as the law of the minimum developed by Liebig (1840). This states that the yield of any organism is determined by the substance that in relation to the needs of the organism is least abundant in the environment (Hutchinson, 1970, 1978). However, the concept has been considerably misused due to over-simplification. First of all, growth might be limited by more than one nutrient. The composition is not constant; it varies with the composition of the environment. Furthermore, growth is not at its maximum rate until the nutrients are used, and is then stopped. But the growth rate slows down when the nutrients become depleted. Another side of the problem is the consideration of the nutrient sources. It is important to set up mass balances for the most essential nutrients.

The sequences of events leading to eutrophication have often been described as follows. Oligotrophic waters will have a ratio of N:P greater than or equal to 10, which means that phosphorus is less abundant than nitrogen for the needs of phytoplankton. If sewage is discharged into the lake, then the ratio will decrease, since the N:P ratio for municipal wastewater is 3:1; consequently, nitrogen will be less abundant than phosphorus relative to the needs of phytoplankton. In this situation, however, the best remedy for the excessive growth of algae is not the removal of nitrogen from the sewage because the mass balance might then show that nitrogen-fixing algae will release an uncontrollable input of nitrogen into the lake. It is necessary to set up mass balances for each of the nutrients as these will often reveal that the input of nitrogen from nitrogen-fixing blue-green algae, precipitation, and tributaries contribute too much to the mass balance for the removal of nitrogen from the sewage to have any effect. On the other hand, the mass balance may reveal that the phosphorus input (often more than 95%) comes mainly from sewage, which means that it is better management to remove phosphorus from the sewage than nitrogen. Thus, in environmental management it is not always important which nutrient is the most limiting, but which nutrient can most easily be made to limit algal growth.

7.4.1. Predictions of Eutrophication From Concentrations of Nutrients

Dillon and Rigler (1974) developed a relationship for estimating the average summer chlorophyll a concentration (chl.a) with the N:P ratio of the water >12:

$$\log_{10}(\text{chl.a}) = 1.45 \log_{10}[(P) \cdot 1000] - 1.14 \tag{7.13}$$

For the case where the N:P ratio is <4 the following equation, based upon eight case studies was evolved:

$$\log_{10}(\text{chl.a}) = 1.4 \log_{10}[(N) \cdot 1000] - 1.9 \tag{7.14}$$

(N) and (P) are expressed as mg/L while (chl.a) is found in mg/L^{-1}. If the N:P ratio is between 4 and 12, the smallest value of (chl.a) found on the basis of the two equations is recommended.

Many correlations between phosphorus concentrations and chlorophyll concentrations have been developed. Dillon and Kirchner (1975) set up a relationship between the Secchi disc transparency, SE, and phosphorus concentration. Kristensen et al. (1990) developed eight different equations that related the total phosphorus concentration (P_{lake}) with the average transparency depth (z_{eu}). The influence of the mean depth, z, is included in three of the equations (see Table 7.5).

The simple model presented earlier will never be as good a predictive tool as a model based on more accurate data and considering more processes. However, the semiquantitative estimations, which can be obtained by use of the simple model we have presented, are better than none at all, and in a data-poor situation it may be the only model the data can support. Furthermore, it is often an advantage to use simple models to find first estimations before a more advanced model is developed. A model with the state variables PS, NS, Psed, and Nsed, and the previously mentioned regression equations is available as a simple

Table 7.5 Relations Between Average Transparency Depth, z_{eu}, Phosphorus Concentration, P_{lake} and Mean Depth, z

Number	Equation
1	$z_{eu} = 0.44\ (+/-0.038)\ P-0.54(+/-0.031)$
2	$z_{eu} = 0.36(+/-0.029)\ P-0.29(+/-0.028)z0.51(+/-0.042)$
3	$z_{eu} = 0.39(+/-0.038)\ P-0.58(+/-0.034)$
4	$z_{eu} = 0.34(+/-0.028)\ P-0.29(+/-0.028)z0.55(+/-0.040)$
5	$z_{eu} = 0.52\ (+/-0.042)\ P-0.48(+/-0.031)$
6	$z_{eu} = 0.43\ (+/-0.026)\ P-0.20(+/-0.022)z0.55(+/-0.030)$
7	$z_{eu} = 0.40\ (+/-0.055)\ P-0.69(+/-0.064)$
8	$z_{eu} = 0.34\ (+/-0.0424\ P-0.60(+/-0.041)$

After Kristensen et al., 1990.

one-layer model in the UNEP-software Pamolare. It is relatively quick and easy to use this simple model, and it is often recommended that it be used first as a modelling approach to gain insight into the most crucial processes that determine the eutrophication, before a more complex model is developed.

The differential equations for the model are:

$$dPS/dt = Q*(Pin-PS)/V - sr*PS/D + rr*Psed*AL/D \qquad (7.15a)$$

$$dPsed/dt = k*sr*PS*D/AL - rr*Psed \qquad (7.16a)$$

$$dNS/dt = Q*(Nin-NS)/V - sr*NS/D + rr*Nsed*AL/D \qquad (7.15b)$$

$$dNS/dt = k'*sr*NS*D/AL - rr*Nsed \qquad (7.16b)$$

Q is the flow rate to and from the lake. It is presumed that precipitation and evaporation are equal and that the inflows and outflows are in balance.

PS is the total concentration of phosphorus in the water, including all forms (soluble phosphorus, detritus-phosphorus, and phytoplankton phosphorus).

NS is the total concentration of nitrogen in the water, including all forms (soluble inorganic nitrogen [ammonium, nitrate, and nitrite], soluble organic nitrogen, detritus-nitrogen and phytoplankton nitrogen).

Pin is the total phosphorus concentration in the inflowing water and Nin is the total nitrogen concentration in the inflowing water; sr is the settling rate that could be in units of m/year based on the total amount of phosphorus and nitrogen. If it is estimated that phytoplankton and detritus in average is for instance 20% per year of total phosphorus and nitrogen, then the settling rate should be calculated as settling in (m/24 h)*365/5.

AL is the active layer of sediment, D is the water depth, and rr and rr' are the release rate of phosphorus and nitrogen, respectively. They are parameters that usually are determined by model calibration.

Notice that when sediment phosphorus and sediment nitrogen are released, the sediment nutrients are diluted by the factor AL/D, and when the settled nutrients are transferred from the water column to the sediment, the concentration becomes D/AL times higher in the sediment than in the water column.

k is the fraction of exchangeable phosphorus to total phosphorus for the settled material. A part of the settled phosphorus is bound in the sediment and cannot be released again. k accounts for exchangeable phosphorus only. If, for instance, 25% of the phosphorus is bound in the sediment, then k is 0.75.

k' is the exchangeable nitrogen to total nitrogen. Usually k' is higher than k because phosphorus compounds can be bound to a higher extent than nitrogen in the sediment by formation of calcium-hydroxo-phosphate or iron (III) phosphate.

This model has been successfully applied to Lake Washington, which is close to Seattle. The model was able to approximately predict the development of the observed phosphorus concentration (for further details see the Pamolare Software, developed by United Nations Environmental Program—International Environmental Technology Center (UNEP-IETC).

7.5. Eutrophication Models II: A Complex Eutrophication Model

7.5.1. Eutrophication Models: An Overview

As expected, due to the importance of eutrophication in environmental management, numerous eutrophication models covering a wide spectrum of complexity have been developed. As for other ecological models, the right complexity of the model is dependent on the available data and the ecosystem. Table 7.6 reviews various eutrophication models.

Table 7.6 Various Eutrophication Models

Model Name	Number of St. Var. per Layer or Segment	Nutrients	Segments	Dimension 2L, 1D	CS or NC*	C and/or V**	Number of Case Studies
Vollenweider	1	P(N)	1	1L	CS	C+V	many
Imboden	2	P	1	2L, ID	CS	C+V	3
O'Melia	2	P	1	1D	CS	C	1
Larsen	3	P	1	1L	CS	C	1
Lorenzen	2	P	1	1L	CS	C+V	1
Thomann 1	8	P,N,C	1	2L	CS	C+V	1
Thomann 2	10	P,N,C	1	2L	CS	C	1
Thomann 3	15	P,N,C	67	2L	CS	–	1
Chen&Orlob	15	P,N,C	sev.	2L	CS	C	min. 2
Patten	33	P,N,C	1	1L	CS	C	1
Di Toro	7	P,N	7	1L	CS	C+V	1
Biermann	14	P,N,Si	1	1L	NC	C	1
Canale	25	P,N,Si	1	2L	CS	C	1
Jørgensen	17–20	P,N,C,	1	1-2L	NC	C+V	26
Cleaner	40	P,N,C,Si	sev. sev.	L	CS	C	many
Nyholm, Lavsoe	7	P,N	1–3	1-2L	NC	C+V	25
Aster/ Melodia	10	P,N,Si	1	2L	CS	C+V	1
Baikal	>16	P,N	10	3L	CS	C+V	1
Chemsee	>14	P,N,C,S	1	profile	CS	C+V	many
Minlake	9	P,N	1	1	CS	C+V	>10
Salmo	17	P,N	1	2L	CS	C+V	16

Notes:
*CS, constant stoichiometric; NC, independent nutrient cycle.
**C, calibrated; V, validated.

Table 7.6 indicates the characteristic features of the models, the number of case studies to which it has been applied (with some modification from case study-to-case study, as site-specific properties should be reflected in the selected modification, unless the model is very simple), and whether the model has been calibrated and validated.

It is not possible to review all complex models in detail. Therefore, one model among the more complex models has been selected and presented in detail here. Eutrophication models are illustrative examples because they demonstrate quite clearly the ideas behind biogeochemical models. The calibration and validation of the selected model and its use to develop scenarios will be discussed. The results demonstrate what can be achieved by using ecological models, provided that sufficient effort is expended to obtain good data and good ecological background knowledge about the modelled ecosystem.

The conceptual diagrams of the nutrient cycles are presented in Figures 2.1 and 7.3. This model was developed for Lake Glumsø — a case study that has the following advantages:

1. The lake is shallow (mean depth 1.8 m) and no formation of a thermocline takes place. The case study is thus relatively simple.
2. The lake is small (volume 420,000 m^3) and well mixed, which implies it is unnecessary for a model to consider hydrodynamics and it can instead focus on ecological processes.
3. Retention time is short (<6 months), which means that any change due to a management action can be observed fairly rapidly.
4. A radical change in nutrient input occurred in April 1981, and subsequent water quality changes were observed (Jørgensen, 1986).
5. It is unique, in that a prediction of the water quality was published before any changes actually took place (Jørgensen et al. 1978). It has since been possible to validate this prediction.
6. The lake was intensely studied from 1973 to1984. The model is therefore based on comprehensive data.

The success of this model has led to its application to at least 25 other case studies — of course with the necessary modifications.

The Lake Glumsø model is probably one of the most well-examined eutrophication models. The results represent what is obtainable in relation to validation under almost unchanged loading, to accuracy in predictions, and to general applicability. Therefore, these results are emphasized in the following presentation.

The ecology of Lake Glumsø was investigated before the model was developed (Jørgensen et al. 1973). The phases in modelling development presented in Chapter 2 were followed very carefully to obtain a model with the predictive power needed for use as a management instrument.

Figures 2.1 and 7.3 are the conceptual diagrams of the N- and P-flows of the model. Many of the equations can be found in other eutrophication models. It seems of little value to present all of the model's equations, and the following sections are devoted to the most characteristic features of the model to illustrate typical modelling considerations. They are:

1. Independent cycling of N, P, and C, which is a result of the two-step process description of phytoplankton growth.
2. A more detailed description of the water-sediment interactions is extremely important for many lakes where a significant amount of the nutrient is stored in the sediment.
3. The equation applied for the description of the grazing of phytoplankton by zooplankton, which takes into account a threshold concentration of phytoplankton and a carrying capacity of the lake.

The two steps describing the phytoplankton growth are (see also Figure 7.4):

1. Uptake of nutrients according to Monod's kinetics
2. Growth determined by the internal substrate concentration

In other words, independent nutrient cycles of phosphorus, nitrogen, and carbon are considered. Phytoplankton biomass, as well as carbon, phosphorus, and nitrogen in algal cells must be included as state variables, all expressed in the units g/m^3. This is more complex than the constant stoichiometric approach, which is applied in most eutrophication models (see Table 7.6). The most frequent equation applied for this approach is:

$$\text{Growth of phytoplankton} = \mu_{max} \min (NS/(k_n + NS), (PS/(PS + k_p)) \quad (7.17)$$

where μ_{max} is the maximum growth rate and k_n and k_p are Michaelis-Menten half saturation constants. It presumes that phosphorus and

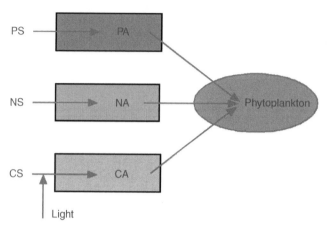

FIGURE 7.4 The two-steps model of phytoplankton growth. The first step is uptake of nutrients PS, NS, and CS, followed by a growth of phytoplankton, which is dependent on the nutrient concentrations in the phytoplankton cells, PA, NA, and CA. The carbon uptake is dependent on the light, while the uptake of phosphorus and nitrogen can take place even in darkness. It is a more physiologically correct description of phytoplankton growth than the equations (7.17)

nitrogen (and maybe also silica and carbon) are taken up in a given ratio. Jørgensen (1976a) showed that it was impossible to obtain an accurate time at which the maximum phytoplankton concentration and production occurred using the simpler noncausal Monod's kinetic for phytoplankton growth. The proportions of nitrogen and phosphorus in both zooplankton and fish should be included in the model to ensure element conservation. The two-step phytoplankton growth (see Figure 7.4) is described using a growth rate coefficient μ_{max}, which is limited by four factors:

1. A temperature factor:

$$FT1 = \exp(A(T - T_{opt})) \, (T_{max}T)/(T_{max} - T_{opt})A(T_{max} - T_{opt}) \tag{7.18}$$

where A, T_{opt}, and T_{max} are species dependent constants. T is temperature.

2. A factor for intracellular nitrogen, NC:

$$FN3 = I - NC_{min}/NC \tag{7.19}$$

3. A parallel factor for intracellular phosphorus:

$$FP3 = I - PC_{min}/PC; \text{ and similarly} \tag{7.20}$$

4. A factor for intracellular carbon:

$$FC3 = 1 - CC_{min}/CC \tag{7.21}$$

The phytoplankton growth is limited only by the minimum of the last three factors. It means that we have:

$$dPhyt/dt = \mu_{max}FT1 \cdot \min(FP3, FN3, FC3) \tag{7.22}$$

NC, PC, and CC are determined by nutrient uptake rates:

$$UC = UC_{max}FC1 \cdot FC2 \cdot FRAD \tag{7.23}$$

$$UN = UN_{max}FN1 \cdot FN2 \tag{7.24}$$

$$UP = UP_{max} \cdot FP1 \cdot FP2 \tag{7.25}$$

where UC_{max}, UN_{max}, and UP_{max} are species-dependent constants (maximum uptake rates); generally, UC_{max} will be greater the smaller the size of the considered phytoplankton. FCI, FNI, and FPI are expressions that give the limitations in uptake:

$$FC1 = (FC_{max} - FCA)/(FCA_{max} - FCA_{min}) \tag{7.26}$$

$$FN1 = (FN_{max} - FNA)/(FNA_{max} - FNA_{min}) \tag{7.27}$$

$$FP1 = (FP_{max} - FPA)/(FPA_{max} - FPA_{min}) \tag{7.28}$$

where FCA_{max}, FCA_{min}, FNA_{max}, FNA_{min}, FPA_{max}, and FPA_{min} are constants indicating the maximum and minimum contents, respectively, of nutrients in phytoplankton. FCA, FNA, and FPA are determined as CC/PHYT, NC/PHYT, and PC/PHYT. FC2, FN2, and FP2 give the limitations in uptake caused by the nutrient level in the lake water:

$$FC2 = C/(KC + C) \tag{7.29}$$

$$FN2 = NS/(NS + KN) \tag{7.30}$$

$$FP2 = PS/(PS + KP) \tag{7.31}$$

C, NS, and PS are the concentrations of soluble inorganic forms in the water of carbon, nitrogen, and phosphorus. These expressions are in accordance with the Michaelis-Menten formulation. KC, KN, and KP are half-saturation constants. FRAD is a complex expression, covering the influence of solar radiation. This influence is integrated over depth and the self-shading effect is included.

The intracellular nitrogen, phosphorus, and carbon can now be determined by differential equations:

$$dNC/dt = UN \cdot PHYT - (SA + GZ/F + Q/V)NC \tag{7.32}$$

$$dPC/dt = UP \cdot PHYT - (SA + GZ/F + Q/V)PC \tag{7.33}$$

$$dCC/dt = UC \cdot PHYT - (SA + RESP + GZ/F + Q/V)CC \tag{7.34}$$

where PHYT is the phytoplankton concentration, GZ, is the grazing rate corresponding to gross zooplankton growth, F is a yield factor (approximately 2/3, i.e., zooplankton utilizes 66.7% of the food), Q is the outflow rate, SA is the settling rate (day^{-1}), and V is the volume. RC is the respiration rate, found as:

$$RC = RC_{max}(CC/CC_{max})^{2/3} \qquad (7.35)$$

A more detailed sediment submodel is another characteristic feature of the presented model. As the sediment accumulates nutrients, it is important to describe quantitatively the processes determining the mass flows from sediment to water, particularly in shallow lakes, where the sediment may contain the major part of nutrients. To what extent will accumulated compounds in the sediment be redissolved in the lake water? The exchange processes between mud and water of phosphorus and nitrogen have been extensively studied, as these processes are important for the eutrophication of lakes. Several of the very early developed models did not consider the importance of these sediment water interactions and ignored the exchange of nutrients between mud and water. As pointed out by Jørgensen, Kamp-Nielsen, and Jacobsen (1975), this will inevitably produce a poor result. Ahlgren (1973) applied a constant flow of nutrients between sediment and water, and Dahl-Madsen and Strange-Nielsen (1974) used a simple first-order kinetic to describe the exchange rate.

A more comprehensive submodel (Figure 7.5) for the exchange of phosphorus has been developed by Jørgensen et al. (1975). The settled

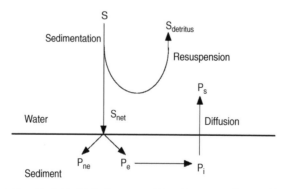

FIGURE 7.5 Sedimentation, S, divided into $S_{detritus}$ and S_{net}. P_{ne}, nonexchangeable phosphorus in unstabilized sediment; P_e, exchangeable phosphorus in unstabilized sediment; P_i, phosphorus in interstitial water; and P_s, dissolved phosphorus in water.

material, S, is divided into $S_{detritus}$ and S_{net}, the first is mineralized by microbiological activity in the water body, and the latter is material actually transported to the sediment. S_{net} can also be divided into two flows:

$$S_{net} = S_{net,s} + S_{net,e} \tag{7.36}$$

where $S_{net,s}$ = flow to the stable nonexchangeable sediment, and $S_{net,e}$ = mass flow to the exchangeable unstable sediment.

Correspondingly, P_{ne} and P_e — nonexchangeable and exchangeable phosphorus concentrations — both based on the total dry matter in the sediment, can also be distinguished. An analysis of the phosphorus profile in the sediment (Figure 7.6) produces the ratio, (f), of the exchangeable to the total settled phosphorus:

$$f = (S_{net} - S_{net,s})/S_{net} = S_{net,e}/S_{net} \tag{7.37}$$

$$\text{and } dP_e/dt = a f S_{net,e} - K_5 * P_e K_6{}^{(T-20)} \tag{7.38}$$

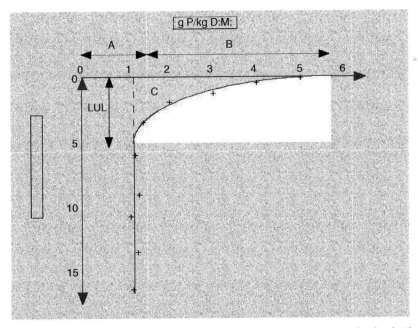

FIGURE 7.6 Analysis of core from Lake Esrom. mg P/g dry matter is plotted against the depth. The area C represents exchangeable phosphorus, $f = (B.A^{-1})$, and LUL is the unstabilized layer.

where a = factor for converting water concentration units to concentration units in the sediment (mg P kg^{-1} DM). $S_{net,e}$ is found from sediment profile studies. The increases of the stabilized sediment are found by numerous methods — the application of lead isotopes, for example, is a fast and reliable method. Exchangeable phosphorus is mineralized similarly to detritus in a water body, and a first-order reaction as indicated gives a reasonably good description of the conversion of P_e into interstitial phosphorus, P_i: $K_5 * P_e K_6^{(T-20)}$, where K_5 = a rate coefficient, K_6 = a temperature coefficient, and T = temperature.

Finally, the interstitial phosphorus, P_i, is transported by diffusion from the pore water to the lake water. This process, which has been studied by Kamp-Nielsen (1975), can be described following the empirical equation (valid at 7°C):

$$\text{Release of P} = 1.21\,(P_i - P_s) - 1.7 \ (\text{mg P m}^{-2}\ 24\text{h}^{-1}) \tag{7.39}$$

where P_s is the dissolved phosphorus in the lake water.

It thus turns out that:

$$d\,P_i/dt = K_5 * P_e K_6^{(T-20)} - (1.21(P_i - P_s) - 1.7) \times T/280 \tag{7.40}$$

T is the absolute temperature as the release rate was found to be proportional to T. Notice that the phosphorus released from the sediment is diluted in the lake water corresponding to the ratio between the active sediment layer and the depth of the lake — see also the four state variable eutrophication model presented in Section 7.3.

This submodel was validated in three case studies (Jørgensen et al., 1975) examining sediment cores in the laboratory. Kamp-Nielsen (1975) added an adsorption term to these equations.

A similar submodel for the nitrogen release has been set up by Jacobsen and Jørgensen (1975). The nitrogen release from sediment is expressed as a function of the nitrogen concentration in the sediment and the temperature, considering both aerobic and anaerobic conditions.

The grazing on phytoplankton by zooplankton, Z, and the predation on zooplankton by fish, F, are both expressed by a modified Monod expression:

$$\mu Z = \mu Z_{max}(\text{PHYT} - \text{GL})/(\text{PHYT} - \text{KA}) \tag{7.41}$$

$$\mu F = \mu F_{max}(\text{ZOO} - \text{KS})/(\text{ZOO} - \text{KZ}) \tag{7.42}$$

where GL, KA, KS, and KZ are constants. These expressions are according to Steele (1974). GL and KS express the very low concentrations at which grazing and predation do not take place. The time to find the food and the energy spent on searching after food is simply too high at low concentration.

The following points in the model were changed from 1979 to 1983 and this gave a better validation:

1. FC3, FN3, and FP3 were changed to:

$$(CC - CC_{min})/(CC_{max} - CC_{min}) \qquad (7.43)$$

   and similarly for FN3 and FP3.

2. The T_{opt} in the temperature factor was changed to the actual temperature in the lake water during the summer months to allow for temperature adaptation.

3. The temperature dependence of phytoplankton respiration was changed to an exponential expression.

4. RC was changed to:

$$RC = RC_{max} \, CC/CC_{max} \qquad (7.44)$$

   The exponent 2/3 in Eq. (7.35) is valid for individual cells as the surface is approximately proportional to the weight or volume of the cells, but since phytoplankton concentration is used here, application of the exponent 2/3 is irrelevant.

5. As previously mentioned, only part of the settled phosphorus is exchangeable. In the Lake Glumsø study it was found that 15% of the settled phosphorus was nonexchangeable to account for the observed phosphorus profile in the sediment. In the new version, exchangeable and nonexchangeable nitrogen were also distinguished. These changes gave a better correspondence between the modelled and the observed nitrogen balance.

6. A carrying capacity of zooplankton was introduced to give a better simulation of zooplankton and phytoplankton. Carrying capacities are often observed in ecosystems (see Eq. 2.4), but their necessity in this case may be because of a simulation of the grazing process that is too simple. Phytoplankton might not be grazed by all zooplankton species present, and some species might use detritus

as a food source. The zooplankton growth rate, mZ, is computed in accordance with these modifications as:

$$\mu Z = \mu Z_{max} \cdot FPH \cdot FT2 \cdot F2CK \tag{7.45}$$

where FPH = (PHYT − GL)/(PHYT − KA) — see the expression Eq. (7.41) — FT2 is a temperature regulation expression, and F2CK accounts for carrying capacity:

$$F2CK = 1 - ZOO/CK \tag{7.46}$$
$$\text{where } CK = 26 \text{ mg/L} \tag{7.47}$$

was chosen in this case.

An intensive measuring period was applied to improve parameter estimation as described in Chapter 2. The results of this effort can be summarized as follows:

(A) The previously applied expression for the influence of temperature on phytoplankton growth — a simple Arrhenius expression $1.05^{(T-20)}$ — produced unacceptable parameters with standard deviations that were too high. A better expression, Eq. (7.18), was introduced as a result of the intensive measuring period.

(B) It was possible to improve the parameter estimation, which gives, for some of the parameters, more realistic values. Whether this would give an improved validation when observations from a period with drastic changes in the nutrients loading are available could not be stated.

(C) Two zooplankton state variables based on phytoplankton grazing and detritus feeding were tested but did not produce any advantages.

(D) The other expressions applied for process descriptions were confirmed.

It is necessary to validate models against an independent set of measurements. No general method of validation is available, but almost the same method suggested by WMO (1975) for validation of hydrological models was applied for this model. Table 7.7 provides results of the validation improved as described previously. The following numerical validation criteria were applied:

Table 7.7   Numerical Validation of the Described Model

Validation Criteria	State Variable	Value
Y	All	0.31
R	Ptotal (P4)	0.26
R	Psoluble (PS)	0.16
R	Ntotal (N4)	0.02
R	Nsoluble (NS)	0.14
R	Phytoplankton	
	(CA)	0.10
R	Zooplankton (Z)	0.27
R	Production	0.03
A	Ptotal (P4)	0.12
A	Psoluble (PS)	0.18
A	Ntotal (N4)	0.07
A	Nsoluble (NS)	0.03
A	Phytoplankton (CA)	0.15
A	Zooplankton (Z)	0.00
A	Production	0.08
TE	Ptotal (P4)	105 days
TE	Psoluble (PS)	60 days
TE	Ntotal (N4)	15 days
TE	Nsoluble (NS)	15 days
TE	Phytoplankton, (CA)	0 days*
		120 days**
TE	Zooplankton (Z)	60 days
TE	Production	0 days

Notes:

*Based on measuring suspended matter 1–60 μm.

**Based on chlorophyll.

1. Y, coefficient of variation of the residuals of errors for the state variables for the validation period, defined as:

$$Y = \frac{\sqrt{(-y_c - y_m)^2}}{n\,Y_{a,m}} \qquad (7.48)$$

where y_c = calculated values of the state variables, y_m = measured values of the state variables, n = number of comparisons, and $Y_{a,m}$ = average of measured values over the validation period.

(2) R, the relative error of mean values:

$$R = (Y_{a,c} - Y_{a,m})/Y_{a,m} \qquad (7.49)$$

where $Y_{a,c}$ is the average of measured values over the validation period.

(3) A, the relative error of maximum values:

$$A = (Y_{max,c} - Y_{max,m})/Y_{max,m} \qquad (7.50)$$

where $Y_{max,c}$ is the maximum value of the calculated state variable in the validation period, and $Y_{max,m}$ is the maximum value of the measured state variable in the validation period. A for the phytoplankton concentration or the production (dPhyt/dt) are often considered the most important validation criteria, as they describe the "worst-case" situation. This is also often reflected in validations of prognoses.

(4) TE, timing error:

$$TE = \text{Date of } Y_{max,c} - \text{date of } Y_{max,m} \qquad (7.51)$$

Y, R, and A produce the errors in relative terms. By multiplying by 100, the errors are obtained as a percentage. The standard deviation, Y, for all measured state variables, is 31%. It is the standard deviation for one comparison of model value and measured value. As the standard deviation for a comparison of n sets of model values and measured values is $\sqrt{n}$ times smaller and n is in the order of 225, the overall average picture of the lake is given with a standard deviation of about 2%, which is acceptable. Y is generally 5 times larger for hydrodynamics models (WMO, 1975).

The relative errors of mean values, R, are 3% for production, 10% for phytoplankton, and 2% for nitrogen — all acceptable values. The relative error for total phosphorus is 26% and for zooplankton 27%, which must be considered too high. The relative errors of the maximum values, A, are from 0 to 18%, which is acceptable. The ability of the model to predict maximum production and maximum phytoplankton concentration has special interest for a eutrophication model; the relative errors of 8 and 15%, respectively, are fully acceptable.

The ability to predict the time when maximum values occur is expressed by using TE. Production and phytoplankton (use for suspended matter 1–60 m) have good agreement between model values and measured values. TE for total and soluble nitrogen is also acceptable, while the zooplankton and phosphorus values are on the high side. All in all the validation has demonstrated that the model should have value as a predictive tool, although the dynamics of phosphorus and zooplankton could be improved.

The changes in the model made between 1979 and 1983 included the six points mentioned earlier and improved the validation further, as Y was reduced from 31 to 16%.

As mentioned in the introduction to this model, it has been applied with modifications to 25 other case studies. The changes in the model were all based on ecological observations. Table 7.8 reviews the modifications needed in the 25 case studies to get a workable model. By calibration carried out according to Chapter 2, it was found that the most crucial parameters were all in the range of values found in the literature. Note that the parameters were all found by:

1. Using literature values as initial guesses (see Jørgensen, S.E., Nors Nielsen and L.A. Jørgensen, 1991 and Jørgensen, L.A. Nors Nielsen and S.E. Jørgensen, 2000)
2. Using frequent measuring periods to get good first estimations of parameters
3. A first rough calibration of the model to improve parameter estimations
4. Use of an automatic calibration procedure to allow a finer calibration of 6–8 of the most important (most sensitive to the phytoplankton concentration) parameters with ranges partly based on the frequent measurements. This procedure was repeated at least twice and only when the same parameter values were found was the calibration considered satisfactory.

The presented model and other models of similar complexity are widely applied as environmental management tools. They represent what can be achieved by the use of ecological models, provided all steps of the procedure shown in Section 2.3 are carefully included in the model development. Eutrophication models represent the type of ecological model that has received most attention and effort during the last

Table 7.8   Survey of Eutrophication Studies Based Upon the Application of a Modified Glumsø Model

Ecosystem	Modification	Level*
Glumsø, version A	Basis version	7
Glumsø, version B	Nonexchangeable nitrogen	7
Ringkøbing Firth	Boxes, nitrogen fixation	5
Lake Victoria	Boxes, thermocline, other food chain	4
Lake Kyoga	Other food chain	4
Lake Mobuto Sese Seko	Boxes, thermocline, other food chain	4
Lake Fure	Boxes, nitrogen fixation, thermocline	7
Lake Esrom	Boxes, Si-cycle, thermocline	4
Lake Gyrstinge	Level fluctuations, sediment exposed to air	4–5
Lake Lyngby	Basis version	6
Lake Bergunda	Nitrogen fixation	2
Broia Reservoir	Macrophytes, 2 boxes	2
Lake Great Kattinge	Resuspension	5
Lake Svogerslev	Resuspension	5
Lake Bue	Resuspension	5
Lake Kornerup	Resuspension	5
Lake Søbygaard	SDM	7
Lake Balaton	Adsorption to suspended matter	2
Roskilde Fjord	Complex hydrodynamics	4
Lagoon of Venice	Ulva/Zostera competition	6
Lake Annone	SDM	6
Lake Balaton	SDM	6
Lake Mogan, Ankara	Only P cycle, competition submerged vegetation/ phytoplankton + SDM	6
Stadsgraven, Copenhagen	4–6 interconnected basins	5 (level 6: 93)
Internal lakes of Copenhagen	5–6 interconnected basins	5

SDM, Structurally Dynamic Model.
for*, see p. 207

35 years. The results reflect what could be obtained for all ecosystem models, if sufficient effort is used in their examination and development.

Level 1: Conceptual diagram selected
Level 2: Verification carried out
Level 3: Calibration using intensive measurements
Level 4: Calibration of entire model
Level 5: Validation — object function and regression coefficient are found
Level 6: Validation of a prognosis for significant changed loading or development of structurally dynamic models (SDMs)
Level 7: Validation of a prognosis and development of SDMs

As the validation was acceptable, the model was applied to predict the production, phytoplankton concentration, and transparency under conditions in which the phosphorus input to the lake was reduced 90%. Such a reduction was easy to achieve by a well-controlled chemical precipitation. Before the reduction of the phosphorus input, the lake was very eutrophied, which can be seen by the following typical observations:

Total P g/m^3: 1.1
Phytoplankton concentration peak value: (mg chl. a/m^3) 850
Production (g C/(m^2 year)) 1050
Minimum transparency at spring bloom (m) 0.18

Fortunately, the water residence time of Lake Glumsø is only 6 months, so it was possible to validate the simulation properly within a few years. A comparison of the prediction and the actual observations after 90% reduction of the phosphorus input is shown in Table 7.9. The standard deviations indicated in the table are for the prediction based on the validation results shown earlier and for the measurement based on a general determination of the standard deviations for measurement on 10%, relatively.

The prediction validation is fully acceptable except for the daily production (g C/(m^2 24h)) at spring bloom during the second year. In the beginning of the second year, the phytoplankton species shifted from Scenedesmus to various species of diatoms. It is always more difficult

Table 7.9 Validation of the Prognosis for Glumsø Lake

Comparison of:	Prediction			Measurements	
	Time	Value	St. Dev.	Value	St. Dev.
Min. transparency	1 year	0.20 m	0.03	0.20 m	0.02
Min. transparency	2 year	0.30 m	0.05	0.25 m	0.025
Min. transparency	3 year	0.45m	0.07	0.50 m	0.05
Max. production	2 year	6.0 gC/24hm^2	0.3	11.0 gC/24hm^2	1.1
Max production	3 year	5.0 gC/24hm^2	0.3	6.2 gC/24hm^2	0.6
Max. chl.a.	1 year	750 mg/m^3	112	800 mg/m^3	80
Max. chl.a	2 year	520 mg/m^3	78	550 mg/m^3	55
Max. chl.a	3 year	320 mg/m^3	48	380 mg/m^3	38
Annual production	2 year	720 gC/y m^2	15*	750 gC/y m^3	19*
Annual production	3 year	650 gC/y m^3	13*	670 gC/y m^3	17*

Notes:
*A standard deviation of 8% is used for the prognosis divided by $\sqrt{15}$ and 10% divided by $\sqrt{15}$ for the measurements, because the determination of the annual production is based on 15 measurements and 15 prognosis values.

to predict accurately a rate, such as production, than a state variable, such as phytoplankton.

The shift implied that the well-determined parameters for phytoplankton were no longer valid, which may explain the discrepancy between the prediction and the observations, particularly the second year after reducing the phosphorus loading. However, the shift has clearly demonstrated the need for a structurally dynamic modelling approach, as discussed in Chapter 10 (see also Jørgensen et al., 2004).

7.6. Model of Subsurface Wetland

The model presented in this section very clearly illustrates the basic ideas behind biogeochemical models. It was developed as a result of a Danida project promoting the cooperation between Copenhagen and Dar es Salaam University in Tanzania. Later, the UNEP-IETC developed software based on this model that could be used by developing countries to design subsurface wetlands. The software was called Sub-wet. Fleming College, the center for alternative waste water treatment, wanted to use the software for wetland design in cold climates, so they supported further development of the software to be applied both for warm and cold climates. This version of the software is denoted Sub-wet 2.0 and it is available for download from the home page of UNEP-IETC (2009). See also Jørgensen, Chon, and Recknagel (2009). The model presentation in the next section follows the Sub-wet 2.0 manual used as the basis for the model information.

The scope of the model is to design and manage a subsurface wetland based on defined removal efficiencies of organic matter (expressed in terms of BOD_5), nitrate, ammonium, organic nitrogen, and phosphorus. Thus, it is necessary to know the:

1. Water flow
2. Concentrations of the previously mentioned constituents in the water
3. Required removal efficiencies for these constituents (i.e., their concentrations in the treated water).

The modelled subsurface wetland consists of a constructed or a natural wetland area. The constructed wetland has gravel soil, ensuring a good water flow through the wetland. The core design parameter is the area and the volume of the wetland (denoted V and A).

The conceptual diagram of the model is presented in Figures 7.7–7.9. Figure 7.7 illustrates the organic matter sub-model, while Figure 7.8 illustrates the nitrogen submodel with three different nitrogen compounds (organic nitrogen, ammonium and nitrate). Figure 7.9 illustrates the phosphorus submodel. The model state variables are: BOD_5, nitrate (NIT), ammonium (AMM), total phosphorus (TP), and organic nitrogen (ORN) in 5 successive boxes, denoted A, B, C, D, and E. Totally, the model has 25 state variables, all using the units mg/L or g/m^3:

BOD_5-A, BOD_5-B, BOD_5-C, BOD_5-D. BOD_5-E (mg O_2/L)
NIT-A, NIT-B, NIT-C, NIT-D, NIT-E (mg N/L)
AMM-A, AMM-B, AMM-C, AMM-D, AMM-E (mg N/L)
TPO-A, TPO-B, TPO-C, TPO-D, TPO-E (mg P/L)
ORN-A, ORN-B, ORN-C, ORN-D, ORN-E (mg N/L)

The model variables are expressed by three letters (e.g., NIT for nitrate), followed by IN, OUT, or A,B,C,D,E, with the parameters using two letters.

The model has the following forcing functions, which the user must specify for a given model run:

Volume of wetland (m^3; possible range 10–10,000,000)
Flow of water (QIN, expressed as m^3/24 h; possible range 1–1,000,000)
Porosity (as fraction of POR; no unit; range 0––1; default value 0.46)
Input concentration of BOD$_5$ (BOD-IN; mg O$_2$/L; range 0–1000)
Input concentration of ammonium (AMM-IN; mg N/L; range 0–100)
Input concentration of nitrate (NIT-IN; mg N/L; range 0–100)
Input concentration of total phosphorus (TPO-IN; mg P/L; range 0–50)
Concentration in of organic nitrogen (ORN-IN; mg N/L (range 0–200)
Fraction of BOD$_5$ as suspended matter (POM; no unit; range 0–1)
Fraction of organic-N matter as suspended matter (PON; no unit)
Fraction of phosphorus as suspended matter (POP; no unit)
Average oxygen concentration in Box A (AOX; mg/L; range 0–20)
Average oxygen concentration in Box B (BOX; mg/L; range 0–20)
Average oxygen concentration in Box C (COX; mg/L; range 0–20)
Average oxygen concentration in Box D (DOX; mg/L; range 0–20)
Average oxygen concentration in Box E (EOX; mg/L; range 0–20)
Default value for AOX, BOX, COX, DOX, EOX = 0.4 mg/L
Average Temperature (TEMP; as function of time; daily average temperature is listed for the number of days to be simulated with the model)
The length of model simulations must be indicated as number of days.

The following forcing functions are calculated, and included in the forcing function table, along with the forcing functions:

Retention time, RTT (= VOL*POR/Q; 24h)
Retention time per box, RTB (= RTT/5; 24h)
Box volume, BOV (= VOL*POR/5)

7.6.1. Process Equations

A continuous transfer takes place from one state variable to another in the model simulations. This section identifies the processes that take place in the subsurface wetland. It is noted that the same processes take place in each box, although the concentrations are different in each box. Thus, the equations are repeated in the model program with an indication of the concentrations in the five different model boxes — A,B,C,D, and E. All of the processes are expressed by four letters, followed by A, B, C, D, or E, corresponding to the five boxes. It is reiterated that the model expressions are the same for each box, although the applied concentrations of the modelled materials differ for each box. Exponent is expressed by the notation (^).

The following equations are repeated in each box with an indication of the letters of the box (Figures 7.7–7.9):

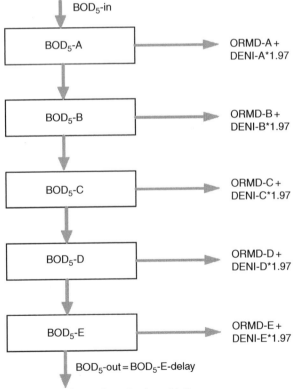

ORMD = decomposition of organic matter by oxidation
DENI* 1.97 = decomposition of organic matter by denitrification

FIGURE 7.7 The BOD_5 submodel.

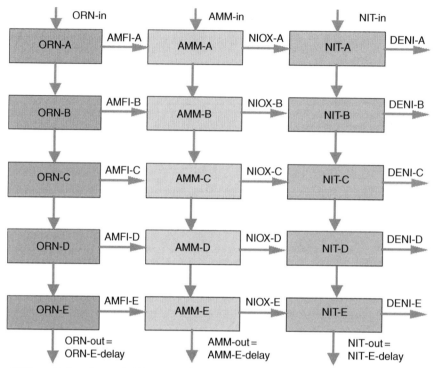

AMFI = oxidation of organic N to ammonium.
NIOX = nitrification (ammonium-> nitrate.)
DENI = denitrification (nitrate -> dinitrogen<)

FIGURE 7.8 The nitrogen submodel illustrating the three nitrogen compounds (organic–N, ammonium, and nitrate).

Ammonification = AMFI = ORN*AC* TA ^ (TEMP-20)
Nitrification = NIOX = AMM*NC*INOX*TN ^ (TEMP-20)/
(AMM+MA)
Oxidation of BOD_5 = ORMD = BOD_5*OC *INOO*TO ^ (TEMP-20)
Denitrification = DENI = NIT*DC*TD ^ (TEMP-20)/(NIT + MN)
INOX-A = AOX/(AOX + KO), and so on for boxes B, C, D, and E, using
the notations BOX, COX, DOX, and EOX; however, KO is the same
parameter for all five boxes
INOO-A = AOX/(AOX + OO), and so on for boxes B, C, D, and E,
using the notations BOX, COX, DOX, and EOX; however, OO is the
same parameter for all five boxes.
Plant uptake of ammonium = PUAM = AMM*PA

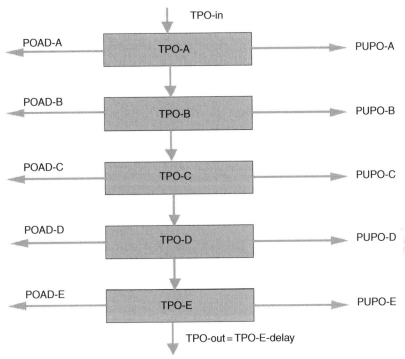

TPO = toal phosphorus PUPO = plant uptake of phopshorus
POAD = adsorption of phosphorus to the gravel.

FIGURE 7.9 The phosphorus submodel.

Plant uptake of nitrate = PUNI = NIT*PN
Plant uptake of phosphorus = PUPO-A = TPO-A*PP*(1-POP) for box
A, while PUPO-B = TPO-B*PP; PUPO-C = TPO-C*PP; PUPO-D =
TPO-D*PP; and PUPO-E = TPO-E*PP (note that the multiplication by
(1 − POP) only applied to box A)
Adsorption of phosphorus = POAD-A = TPO-A*(1-POP)*(POR) − AF*
(1-POR), if POAD>0; otherwise POAD = 0 for box A, while the
following equation is applied for the other boxes: POAD = TPO*POR −
AF*(1-POR).

The model uses delay values (i.e., the concentrations of the five constituents in the five boxes during one box-retention time (RTB) earlier).
For example:

AMM-A-delay = AMM-A at time t − RTB, when t>RTB; if t<RTB,
AMM-A-delay is 0.

These equations are repeated for all five constituents in all five boxes, and the delay concentrations are indicated with "-delay".

Further, the simulated results also are used to determine the removal efficiencies, which are also shown on graphs. They are found as function of time, as follows:

Efficiency of BOD_5-removal (%) = 100*(BOD_5-in − BOD_5-out)/BOD_5-in

Efficiency of Nitrate removal (%) = 100*(NIT-in-NIT-out)/NIT-in

Efficiency of Ammonium removal (%) = 100*(AMM-in-AMM-out)/AMM-in

Efficiency of Organic-N removal (%) = 100*(ORN-in − ORN-out) /ORN-in

Efficiency of Nitrogen removal (%) = 100* ((NIT-in + AMM-in + ORN-in) − (NIT-out +AMM-out + ORN-out))/(NIT-in + AMM-in + ORN-in)

Efficiency of Phosphorus removal (%) = 100*(TPO-in – TPO-out)/TPO-in

7.6.2. Parameters

The model parameters are as follows: (The values correspond to the parameters valid for warm climate conditions. Other parameters are recommended for use in cold climate; see the software.

AC = 0.05–0.8 [default value 0.5 (1/24h)]
NC = 0.1–1.5 [default value 0.8 (1/24h)]
OC = 0.05–0.8 [default value 0.5 (1/24h)]
DC = 0.25–5 [default value 2.2 (1/24h)]
TA = 1.02–1.06 [default value 1.04 (no unit)]
TN = 1.02–1.07 [default value 1.047 (no unit)]
TO = 1.02–1.06 [default value 1.04 (no unit)]
TD = 1.05–1.12 [default value 1.09 (no unit)]
KO = 0.1–2 [default value 1.3 (mg/L)]
OO = 0.1–2 [default value 1.3 (mg/L)]
MA = 0.05–2 [default value 1 (mg/L)]
MN = 0.01–1 [default value 0.1 (mg/L)]
PA = 0.00–1 [default value 0.01 (1/24h)]

PN = 0.00–1 [default value 0.01 (1/24h)]
PP = 0.00–1 [default value 0.003 (1/24h)]
AF = 0–100 [default value 1.0]

7.6.3. Differential Equations

The 25 differential equations in the model are as follows:

BOV*d BOD-5-A/dt = QIN*BOD-IN − QIN*(1-POM)*BOD-A-delay − BOV*ORMD-A-DENi-A*1.97

BOV*d BOD-5-B/dt = QIN*(1-POM)*BOD-A-delay − BOV*ORMD-B − QIN*BOD-B-delay − DENI-B*1.97

BOV*d BOD-5-C/dt = QIN*BOD-B-delay − BOV*ORMD-C − QIN*BOD-C-delay − DENI-C*1.97

B OV*d BOD-5-D/dt = QIN*BOD-C-delay − BOV*ORMD-D − QIN*BOD-D-delay − DENI-D*1.97

BOV*d BOD-5-E/dt = QIN*BOD-D-delay − BOV*ORMD-E − QIN*BOD-E-delay − DENI-E*1.97 (QIN*BOD-E-delay indicates BOD$_5$-OUT, which is eventually shown on a graph, together with measured values of the BOD$_5$-OUT, while BOD$_5$-A, B, C, D, and E are shown in a table as function of time)

BOV*dNIT-A/dt = QIN*NIT-IN − QIN*NIT-A-delay − BOV*DENI-A + BOV*NIOX-A − BOV*PUNI-A;

BOV*dNIT-B/dt = QIN*NIT-A-delay − BOV*DENI-B + BOV*NIOX-B − BOV*PUNI-B − QIN*NIT-B-delay

BOV*dNIT-C/dt = QIN*NIT-B-delay − BOV*DENI-C + BOV*NIOX-C − BOV*PUNI-C − QIN*NIT-C-delay

BOV*dNIT-D/dt = QIN*NIT-C-delay − BOV*DENI-D + BOV*NIOX-D − BOV*PUNI-D − QIN*NIT-D-delay

BOV*dNIT-E/dt = QIN*NIT-D-delay − BOV*DENI-E + BOV*NIOX-E − BOV*PUNI-E − QIN*NIT-E-delay (QIN*NIT-E-delay indicates NIT-OUT, which is eventually shown on a graph, together with measured values of NIT-OUT, while NIT-A, NIT-B, NIT-C, NIT-D, and NIT-E are all shown in a table as a function of time)

BOV*dAMM-A/dt = QIN*AMM-IN − QIN*AMM-A-delay − BOV*NIOX-A + BOV*AMFI-A − BOV*PUAM-A

BOV*dAMM-B/dt = QIN*AMM-A-delay + BOV*AMFI-B − BOV*NIOX-B − BOV*PUAM-B − QIN*AMM-B-delay

BOV*dAMM-C/dt = QIN*AMM-B-delay + BOV*AMFI-C – BOV*NIOX-C – BOV*PUAM-C – QIN*AMM-C-delay

BOV*dAMM-D/dt = QIN*AMM-C-delay + BOV*AMFI-D – BOV*NIOX-D – BOV*PUAM-D – QIN*AMM-D-delay

BOV*dAMM-E/dt = QIN*NIT-D-delay + BOV*AMFI-E – BOV*NIOX-E – BOV*PUNI-E – QIN*AMM-E-delay (QIN*AMM-E-delay indicates AMM-OUT, which is eventually shown on a graph, together with measured values of AMM-OUT, while AMM-A, AMM-B, AMM-C, AMM-D, and AMM-E are all shown in a table as a function of time);

BOV*dORN-A/dt = QIN*ORN-IN – QIN*(1-PON)*ORN-A-delay – BOV*AMFI-A

BOV*dORN-B/dt = QIN*ORN-A-delay – BOV*AMFI-B – QIN*ORN-B-delay

BOV*dORN-C/dt = QIN*ORN-B-delay – BOV*AMFI-C – QIN*ORN-C-delay

BOV*dORN-D/dt = QIN*ORN-C-delay – BOV*AMFI-D – QIN*ORN-D-delay

BOV*dORN-E/dt = QIN*ORN-D-delay – BOV*AMFI-E – QIN*ORN-E-delay (IN*ORN-E-delay indicates ORN-OUT, which is eventually shown on a graph, together with measured values of ORN-OUT, while ORN-A, ORN-B, ORN-C, ORN-D, and ORN-E are all shown in a table as a function of time)

BOV*dTPO-A/dt = QIN-TPO-IN – QIN*(1-POP)*TPO-A-delay – BOV*PUPO-A-BOV*POAD-A

BOV*dTPO-B/dt = QIN*(1-POP)*TPO-A-delay – BOV*PUPO-B - BOV*POAD-B – QIN*TPO-B-delay

BOV*dTPO-C/dt = QIN*TPO-B-delay – BOV*PUPO-C – BOV*POAD-C – QIN*TPO-C-delay

BOV*dTPO-D/dt = QIN*TPO-C-delay – BOV*PUPO-D – BOV*POAD-D – QIN*TPO-D-delay

BOV*dTPO-E/dt = QIN*TPO-D-delay – BOV*PUPO-E – BOV*POAD-E – QIN*TPO-E-delay (QIN*TPO-E-delay indicates TPO-OUT, which is eventually shown on a graph, together with measured values of TPO-OUT, while TPO-A, TPO-B, TPO-C, TPO-D. and TPO-E are all shown in a table as a function of time).

7.6.4. Model Results

As previously mentioned, the simulated values of BOD_5-out, nitrate-out (NIT-out), ammonium-out (AMM-out), total phosphorus-out (TPO-out), and organic nitrogen-out (ORN-out) are shown in the form of tables and graphs. If the measured values are available, then they are shown on the same graphs to allow for a direct comparison.

The simulated results of the removal efficiencies also are shown on graphs, including:

Efficiency of BOD_5-removal (%)
Efficiency of nitrate removal (%)
Efficiency of ammonium removal (%)
Efficiency of organic-N removal (%)
Efficiency of nitrogen removal (%)
Efficiency of phosphorus removal (%)

It also may be useful to include the predicted concentrations of the five constituents in the five boxes as a means of illustrating where the removal processes are most effective in the wetland and where they are less effective. It may be possible to apply such information to improve the overall removal efficiencies by imposed changes in the composition of the waste water, or by changes directly in the wetland (e.g., addition of oxygen). The predicted concentrations in the boxes obtained with the model simulations are listed in a table for each day in the simulation period as follows: BOD_5-A, BOD_5-B, BOD_5-C, BOD_5-D, BOD_5-E, NIT-A, NIT-B, NIT-C, NIT-D, NIT-E, AMM-A, AMM-B, AMM-C, AMM-D, AMM-E, TPO-A, TPO-B, TPO-C, TPO-D, TPO-E, ORN-A, ORN-B, ORN-C, ORN-D, and ORN-E.

7.6.5. Practical Information About Forcing Functions and Parameters

Usually, the model is applied to design a wetland that has not yet been constructed. Thus, it is recommended that the following information is obtained to indicate the forcing functions:

1. Measure the temperature every 6 hours, for example, in a shallow aquatic ecosystem (i.e., before the wetland is constructed, and directly in the wetland after it has been constructed) to get an initial indication of the temperature variations. Alternatively, use some other local temperature measurements. Find the daily average temperature to be listed on the forcing function screen image. Although the temperature expression is not linear, using the average daily temperature for the temperature expression applied herein is only a minor error.

2. Use the default values indicated when no information about the forcing functions is available.

3. For the uptake rate coefficients of nutrients (nitrate, ammonium, and phosphorus; PA, PN, and PP), use knowledge about the phosphorus (P) and nitrogen (N) content and the growth rate of plants that will be applied in the wetland.

7.6.6. Use of the Model for Wetland Design

If the model is used to design a wetland that has not yet been constructed, then the concentrations in the wetland, the water flow through the wetland, and the expected concentrations after wetland treatment should be known. The model is then used to predict the results for simulating different volumes. The simulated results that meet the criteria for the treated water are used, and the corresponding volume is used for the design. It is recommended that a volume 10–15% greater than that predicted with this method be used to take into account the uncertainties of the model. Further, a depth of 0.6–1.00 m is used and a maximum flow rate of about 1.25 m/h applied in order to determine the width and length of the wetland.

7.7. Global Warming Model

Global climate change due to the emission of greenhouse gases such as carbon dioxide, methane, and nitrogen oxides is probably the environmental problem that has attracted most recent attention. It is therefore not surprising that many models have been developed to predict the temperature change as a result of the emission of greenhouse gases. Some of the global warming models are extremely complex and detailed. They consider not only the change of the average global temperature but also how the temperature is distributed geographically. These models are usually developed by climatologists and require an enormous computer capacity. It is, however, sufficient to reveal a relationship between the emission of greenhouse gases and the global average temperature, which can be done by much smaller models. A global warming model that can predict the increase of the global average temperature is presented in this section to illustrate an ecological modelling approach for this central environmental problem. The model is developed using of STELLA and the diagram is shown in Figure 7.10. The model equations using the STELLA format are shown Table 7.10. The model has the following *state variables:*

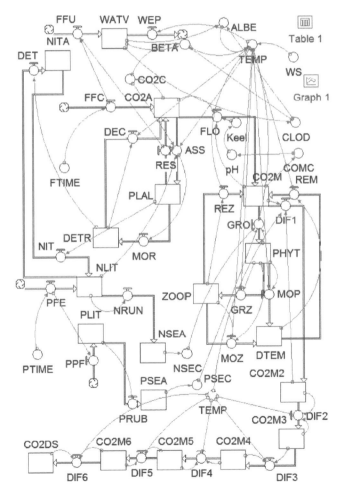

FIGURE 7.10 The global warming model presented in Section 7.6.

Table 7.10   The Global Warming Model: Equations in the STELLA Format

```
CO2A(t) = CO2A(t - dt) + (FFC + DEC + RES - ASS - FLO) * dt
INIT CO2A = 775
INFLOWS:
FFC = 6*FTIME*1.02^TIME
DEC = DETR*66*1.05^(TEMP-15.2)/1500
RES = 0.5*ASS
OUTFLOWS:
ASS = (120*13*CO2C)/(7*(CO2C+300))*(NLIT/865)*1.05^
(TEMP-15.2)
FLO = 0.15*CO2M*(((CO2A-775)/CO2A)/(Keel))
CO2DS(t) = CO2DS(t - dt) + (DIF6) * dt
INIT CO2DS = 92094
INFLOWS:
DIF6 = 3987*((CO2M6/125-CO2DS/2800)/(2800))*(TEMP/
15.2)^0.5
CO2M(t) = CO2M(t - dt) + (FLO + REZ + REM - GRO - DIF1) * dt
INIT CO2M = 2466.8
INFLOWS:
FLO = 0.15*CO2M*(((CO2A-775)/CO2A)/(Keel))
REZ = ZOOP*10*1.05^(TEMP-15.2)
REM = 0.02*DTEM*1.05^(TEMP-15.2)
OUTFLOWS:
GRO = PHYT*(1-PHYT/20)*1.05^(TEMP-15.2)*15.3*MAX(NSEC/
(NSEC+2*10^(-7)),PSEC/(PSEC+3.5*10^(-8)))
DIF1 = (3987*(CO2M/75-CO2M2/125)/75)*(TEMP/15.2)^0.5
CO2M2(t) = CO2M2(t - dt) + (DIF1 - DIF2) * dt
INIT CO2M2 = 4110
INFLOWS:
DIF1 = (3987*(CO2M/75-CO2M2/125)/75)*(TEMP/15.2)^0.5
OUTFLOWS:
DIF2 = (3987*(CO2M2-CO2M3)/(125*125))*(TEMP/15.2)^0.5
CO2M3(t) = CO2M3(t - dt) + (DIF2 - DIF3) * dt
INIT CO2M3 = 4110
INFLOWS:
DIF2 = (3987*(CO2M2-CO2M3)/(125*125))*(TEMP/15.2)^0.5
OUTFLOWS:
DIF3 = 3987*((CO2M3-CO2M4)/(125*125))*(TEMP/15.2)^0.5
CO2M4(t) = CO2M4(t - dt) + (DIF3 - DIF4) * dt
INIT CO2M4 = 4110
INFLOWS:
DIF3 = 3987*((CO2M3-CO2M4)/(125*125))*(TEMP/15.2)^0.5
OUTFLOWS:
DIF4 = 3987*((CO2M4-CO2M5)/(125*125))*(TEMP/15.2)^0.5
CO2M5(t) = CO2M5(t - dt) + (DIF4 - DIF5) * dt
INIT CO2M5 = 4110
INFLOWS:
DIF4 = 3987*((CO2M4-CO2M5)/(125*125))*(TEMP/15.2)^0.5
OUTFLOWS:
DIF5 = 3987*((CO2M5-CO2M6)/(125*125))*(TEMP/15.2)^0.5
CO2M6(t) = CO2M6(t - dt) + (DIF5 - DIF6) * dt
```

Table 7.10 The Global Warming Model: Equations in the STELLA
Format—cont'd

```
INIT CO2M6 = 4110
INFLOWS:
DIF5 = 3987*((CO2M5-CO2M6)/(125*125))*(TEMP/15.2)^0.5
OUTFLOWS:
DIF6 = 3987*((CO2M6/125-CO2DS/2800)/(2800))*(TEMP/
15.2)^0.5
DETR(t) = DETR(t - dt) + (MOR - DEC) * dt
INIT DETR = 1500
INFLOWS:
MOR = 0.49*ASS
OUTFLOWS:
DEC = DETR*66*1.05^(TEMP-15.2)/1500
DTEM(t) = DTEM(t - dt) + (MOP + MOZ - REM) * dt
INIT DTEM = 3000
INFLOWS:
MOP = PHYT*4.5*1.05^(TEMP-15.2)
MOZ = (4.8*ZOOP+0.2222*ZOOP*PHYT)*1.05^(TEMP-15.2)
OUTFLOWS:
REM = 0.02*DTEM*1.05^(TEMP-15.2)
NITA(t) = NITA(t - dt) + (DET - NIT) * dt
INIT NITA = 3800000
INFLOWS:
DET = 0.000063*DETR
OUTFLOWS:
NIT = PLAL*0.00016
NLIT(t) = NLIT(t - dt) + (PFE + NIT - DET - NRUN) * dt
INIT NLIT = 852
INFLOWS:
PFE = 0.03*PTIME^TIME
NIT = PLAL*0.00016
OUTFLOWS:
DET = 0.000063*DETR
NRUN = NLIT*0.000214
NSEA(t) = NSEA(t - dt) + (NRUN) * dt
INIT NSEA = 1904
INFLOWS:
NRUN = NLIT*0.000214
PHYT(t) = PHYT(t - dt) + (GRO - GRZ - MOP) * dt
INIT PHYT = 5
INFLOWS:
GRO = PHYT*(1-PHYT/20)*1.05^(TEMP-15.2)*15.3*MAX(NSEC/
(NSEC+2*10^(-7)),PSEC/(PSEC+3.5*10^(-8)))
OUTFLOWS:
GRZ =
PHYT*16*TEMP*1.05^(TEMP-15.2)
MOP = PHYT*4.5*1.05^(TEMP-15.2)
PLAL(t) = PLAL(t - dt) + (ASS - MOR - RES) * dt
INIT PLAL = 560
```

Continued

Table 7.10 The Global Warming Model: Equations in the STELLA Format—cont'd

```
INFLOWS:
ASS = (120*13*CO2C)/(7*(CO2C+300))*(NLIT/865)*1.05^
(TEMP-15.2)
OUTFLOWS:
MOR = 0.49*ASS
RES = 0.5*ASS
PLIT(t) = PLIT(t - dt) + (PPF - PRUB) * dt
INIT PLIT = 60
INFLOWS:
PPF = PFE*0.1
OUTFLOWS:
PRUB = PLIT*0.00005
PSEA(t) = PSEA(t - dt) + (PRUB) * dt
INIT PSEA = 129.5
INFLOWS:
PRUB = PLIT*0.00005
WATV(t) = WATV(t - dt) + (FFU - WEP) * dt
INIT WATV = 67580
INFLOWS:
FFU = 0.63*FFC+RES
OUTFLOWS:
WEP = 0.0547*(TEMP-15.2)
ZOOP(t) = ZOOP(t - dt) + (GRZ - REZ - MOZ) * dt
INIT ZOOP = 1
INFLOWS:
GRZ =
PHYT*16*TEMP*1.05^(TEMP-15.2)
OUTFLOWS:
REZ = ZOOP*10*1.05^(TEMP-15.2)
MOZ = (4.8*ZOOP+0.2222*ZOOP*PHYT)*1.05^(TEMP-15.2)
ALBE = 0.301+2.1*10^(-6)*(WATV-67580)+1.0*(CLOD-0.5)
BETA = 0.399+118*10^(-6)*(CO2C-350)+0.563*(CLOD-0.5)+
2.73*10^(-6)*(WATV-67580)
CLOD = WATV*0.5/(67580)
CO2C = CO2A*29/(12*5.35)
COMC = CO2M*3800/(75*1.36*10^3)
FTIME = 1
Keel = 10+(8.4-pH)*0.7
NSEC = NSEA/1.36*10^9
pH = 8.4 -1.2* COMC/919
PSEC = PSEA/1.36*10^9
PTIME = 1.05
TEMP = (WS*(1-ALBE)/(28840*(1-BETA)))^0.25-273.3
WS = 1.73*10^14
```

Water in the atmosphere, WATV
Nitrogen in the atmosphere, NITA
Nitrogen in the lithosphere, NLIT
Nitrogen in the hydrosphere, NSEA
Phosphorus in the lithosphere, PLIT
Phosphorus in the hydrosphere, PSEA
Carbon as carbon dioxide in the atmosphere, CO2A
Carbon in plant biomass in the lithosphere, PLAL
Carbon in detritus in the lithosphere, DETR
Carbon in the upper layer of the sea, CO2M
Carbon in phytoplankton in the sea, PHYT
Carbon in the zooplankton in the sea, ZOOP
Carbon in detritus in the sea, DTEM
Carbon in 6 deeper layers of the sea, CO2M2, CO2M3, ... CO2M6, CO2DS

The forcing functions are:

Use of fossil fuel, FFC
Use of nitrogen fertilizer, PFE
Use of phosphorus fertilizer, PPF

Figure 7.11 shows the model results corresponding to a 2% increase in the use of fossil fuels since time = 0, which represents 1990 as a reference year. The starting average global temperature is 15.71°C and the model simulation shows the temperature increasing to 20.16°C, corresponding to an increase of the global average temperature of about 4.5°C. The temperature in 2010 is 16.32 or 0.62 centigrade higher than in 1990. This value corresponds very closely to the recorded increase of the global average temperature during the last 20 years. The carbon dioxide concentration has increased from about 350 parts per million to about 390 parts per million in 2010, which also matches the measured carbon dioxide concentration increase in the atmosphere. The model simulation projects that carbon dioxide concentration is expected to be about 720 parts per million in 2100. If the temperature increase is to be held to only two degrees Celsius during this century as recommended by the climate panel of The United Nations, then it is necessary to phase out fossil fuel use during the next 30 to 50 years. A prediction based on a

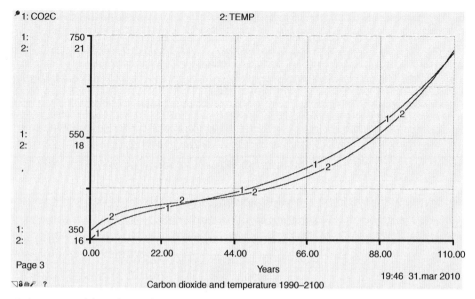

FIGURE 7.11 Model simulation showing atmospheric carbon dioxide concentration and the average global temperature from 1990 to 2100.

continuous 2% annual increase of fossil fuel use during the next ten years followed by a linear decrease to no fossil fuel use by 2070 shows a temperature increase to 18.1°C, This scenario results in a 1.78°C increase compared with today and 2.4°C higher than 1990.

The model has the following features and includes the following *processes*:

1. The global cycling of N, P, and C are included as the nitrogen and phosphorus cycles interact with the carbon cycle
2. The carbon dioxide diffusion in the sea is described by a multilayer model
3. The ability of the ocean to take up carbon dioxide is a function of pH, which is dependent on the carbon dioxide concentration relative to the concentrations of hydrogen carbonate and carbonate
4. The ability of the oceans to take up carbon dioxide is a function of the temperature
5. Increased photosynthesis by increased carbon dioxide concentration according to a Michaelis-Menten expression; $CO2C / (CO''C + 300)$; see ASS Table 7.10

6. Water content in the atmosphere changed when the temperature changed and atmospheric water is also a greenhouse gas

7. The cloudiness changed when the water content in the atmopshere changed, which also changes the albedo;

8. Deforestration, 100,000 km^2 per year according to the Food and Agriculture Organization (FAO), is included

9. The change of albedo due to decreased ice coverage is estimated

10. The change in primary production in the sea, in wetlands, and in forests due to changed temperature and due to the increased carbon dioxide concentration (see also point 5) is considered

11. Permafrost melting for tundra regions is considered; it will decrease the albedo and increase the primary production

Problems

1. Two alternatives exist for improving the visual quality of Lake X: (1) Increase the dilution (flushing) rate and (2) decrease the concentration of nutrients in the inflow by waste water treatment. The present detention time is 8 months and the average inflow of phosphorus, which is considered the most limiting nutrient, is 120 mg L^{-1}. The lake can be considered a completely mixed reactor. Which alternative would you choose and why?

2. The average flow velocity of a stream is 0.7 m/s and the average depth is 1.5 m. Estimate the rate of oxygen transfer from the atmosphere to the water at 12, 15, and 20°C.

3. A stream has the following characteristics during a low flow period: flow rate 70 m^3 s^{-1} and 0.4 m s^{-1}, temperature 24°C, depth 2 m, dissolved oxygen 85%, and BOD_5 2 mg/L at point X. How many kg of BOD_5 can be discharged into the stream at point X, if a minimum of 5 mg/L is to be maintained in the stream? Average rate constants can be assumed. Nitrification is negligible.

4. A steam receives wastewater at a rate of 7 m^3 s^{-1}. The wastewater has BOD_5 12 mg/L and the ammonium concentration is 23 mg/L. The stream has a flow rate of 60 m^3 s^{-1} and 0.5 m s^{-1}, temperature 18°C, depth 2 m, dissolved oxygen 95%. Which minimum oxygen concentration will be recorded in the stream at which

distance from the discharge point? Use the constant presented in the text.

5. Estimate the difference in the estimation of the reaeration coefficient using all of the expressions in the text.

6. BOD_5 at room temperature $20°C$ is found to be 14 mg/L in a sample. What is BOD_7 at $18°C$?

7. Determine the BOD_5 and the oxygen concentration in a completely mixed lake with an inflow of 40 L/s, a depth of 3 m and an area of 15 ha. The average wind speed is approximately 4.5 m/s, the oxygen concentration in the inflow is 8 mg/L and contains no BOD. 120 kg of BOD is discharged to the lake by waste water per day. The lake has a sandy bottom. The photosynthesis corresponds to 3 mg oxygen/(l day).

8. Set up a STELLA program for Lorenzen's model.

9. Explain why the relationship between summer chlorophyll and annual average phosphorus concentration is so different for the various investigations of the relationship.

10. Find the transparency for a lake with an annual average phosphorus concentration of 1 mg/L and a depth of 2 m using Table 7.5. Use Eq. (7.13) to find the chl. a. concentration and Figure 7.3. Explain the discrepancy.

11. Explain why any new lake model development inevitably requires an examination of possible model modifications.

12. Why is validation of a model compulsory?

13. How will you describe the generality of eutrophication models?

14. Explain why it is expected that a structural dynamic model will be able to offer a better validation.

APPENDIX 1

TABLE 1 Dissolved oxygen (ppm, mg/l) in fresh, brackish and sea water at different temperatures and at different chlorinities (%). Values are at saturation

Temp	0%	0.2%	0.4%	0.6%	0.8%	1.0%	1.2%	1.4%	1.6%	1.8%	2.0%
1	14.24	13.87	13.54	13.22	12.91	12.58	12.29	11.99	11.70	11.42	11.15
2	13.74	13.50	13.18	12.88	12.56	12.26	11.98	11.69	11.40	11.13	10.86
3	13.45	13.14	12.84	12.55	12.25	11.96	11.68	11.39	11.12	10.85	10.59
4	13.09	12.79	12.51	12.22	11.93	11.65	11.38	11.10	10.83	10.59	10.34
5	12.75	12.45	12.17	11.91	11.63	11.36	11.09	10.83	10.57	10.33	10.10
6	12.44	12.15	11.86	11.60	11.33	11.07	10.82	10.56	10.32	10.09	9.86
7	12.13	11.85	11.58	11.32	11.06	10.82	10.56	10.32	10.07	9.84	9.63
8	11.85	11.56	11.29	11.05	10.80	10.56	10.32	10.07	9.84	9.61	9.40
9	11.56	11.29	11.02	10.77	10.54	10.30	10.08	9.84	9.61	9.40	9.20
10	11.29	11.03	10.77	10.53	10.30	10.07	9.84	9.61	9.40	9.20	9.00
11	11.05	10.77	10.53	10.29	10.06	9.84	9.63	9.41	9.20	9.00	8.80
12	10.80	10.53	10.29	10.06	9.84	9.63	9.41	9.21	9.00	8.80	8.61
13	10.56	10.30	10.07	9.84	9.63	9.41	9.21	9.01	8.81	8.61	8.42
14	10.33	10.07	9.86	9.63	9.41	9.21	9.01	8.81	8.62	8.44	8.25
15	10.10	9.86	9.64	9.43	9.23	9.03	8.83	8.64	8.44	8.27	8.09
16	9.89	9.66	9.44	9.24	9.03	8.84	8.64	8.47	8.28	8.11	7.94
17	9.67	9.46	9.26	9.05	8.85	8.65	8.47	8.30	8.11	7.94	7.78
18	9.47	9.27	9.07	8.87	8.67	8.48	8.31	8.14	7.97	7.79	7.64
19	9.28	9.08	8.88	8.68	8.50	8.31	8.15	7.98	7.80	7.65	7.49
20	9.11	8.90	8.70	8.51	8.32	8.15	7.99	7.84	7.66	7.51	7.36
21	8.93	8.72	8.54	8.35	8.17	7.99	7.84	7.69	7.52	7.38	7.23
22	8.75	8.55	8.38	8.19	8.02	7.85	7.69	7.54	7.39	7.25	7.11
23	8.60	8.40	8.22	8.04	7.87	7.71	7.55	7.41	7.26	7.12	6.99
24	8.44	8.25	8.07	7.89	7.72	7.56	7.42	7.28	7.13	6.99	6.86
25	8.27	8.09	7.92	7.75	7.58	7.44	7.29	7.15	7.01	6.88	6.85
26	8.12	7.94	7.78	7.62	7.45	7.31	7.16	7.03	6.89	6.86	6.63
27	7.98	7.79	7.64	7.49	7.32	7.18	7.03	6.91	6.78	6.65	6.52
28	7.84	7.65	7.51	7.36	7.19	7.06	6.92	6.79	6.66	6.53	6.40
29	7.69	7.52	7.38	7.23	7.08	6.95	6.82	6.68	6.55	6.42	6.29
30	7.56	7.39	7.25	7.12	6.96	6.83	6.70	6.58	6.45	6.32	6.19

8
Ecotoxicological Models

CHAPTER OUTLINE

8.1. Classification and Application of Ecotoxicological Models

Ecotoxicological models are increasingly applied to assess the environmental risk of chemical emissions to the environment. We distinguish between fate models and effect models. Fate models provide the concentration of a chemical in one or more environmental compartments; for instance, the concentration of a chemical compound in a fish or in a lake. Effect models translate a concentration or body burden in a biological compartment to an effect either on an organism, a population, a community, an ecosystem, a landscape (consisting of two or more ecosystems), or the entire ecosphere.

The results of a fate model can be used to find the ratio (RQ), between the computed concentration, predicted environmental concentration (PEC), and the nonobserved-effect concentration (NOEC), which is determined through literature values or laboratory experiments. Further detail about the procedure for environmental risk assessment (ERA) and how to account for the uncertainty of the assessment will be presented in the next section.

The effect models presume that we know the concentration of a chemical in a focal compartment, either by a model or by analytical determinations. The effect models translate the found concentrations into an effect on either the growth of an organism, the development of a population or the community, the changes of an ecosystem or a landscape, or on the entire ecosphere.

It is also possible to merge fate models with effect models, combining the two approaches. We could call such models fate-transport-effect-models (FTE-models).

Many fate models, fewer effect models, and only a few FTE-models have been applied to solve ecotoxicological problems and perform ERAs. However, the development is toward a wider application of effect and FTE-models.

A. Fate models may be divided into three classes:
 I. Models that map the fate and transport of a chemical in a region or a country. These models are sometimes called Mackay-type models after Don Mackay, who first developed them. A detailed discussion of the application of these models can be found in Mackay et al., (1991, 1992) and SETAC (1995). This type of fate model is rarely calibrated and validated, although indicating the standard deviation of the results has been attempted (see SETAC, 1995).
 II. Models that consider a specific case of toxic substance pollution; for instance, a discharge of a chemical to a coastal zone from a chemical plant or a sewage treatment plant. This type of fate model must always be calibrated and validated.
 III. Models that focus on a chemical used locally. It implies that an evaluation of the risk requires the determination of a typical concentration (which is much higher than the regional concentration that would be obtained from model type I) in a typical locality. A typical example is the application of pesticides, where the model has to look into a typical application on an agriculture field close to a stream and with a ground water mirror close to the surface. This model type can be considered a hybrid

of model types I and II. The conceptual diagram and the equations of the type III model are similar to model type II, but the interpretation of the model results are similar to model type I. This model type should always be calibrated and validated by data obtained for a typical case study, but the prognosis is most commonly applied for development of "a worst-case situation" or "an average situation," which may be different from the case study applied for the calibration and validation.

Examples of all three model types are presented in this chapter. Chapter 6, on steady-state models, has already presented an ecotoxicological model type II. Only examples of dynamic models are included in this chapter.

B. Effect models may be classified according to the hierarchical level of concern:

  I. Organism models. The core of the model is the influence of a toxic substance on an organism, for example, a relationship between the growth parameters and the concentration of a toxic substance.
  II. Population models. The population models presented in Chapter 5, including individual-based models (IBMS), may include relationships between toxic substance concentrations and the model parameters.
  III. Ecosystem models. The influences of a toxic substance on several parameters are included. The result of these chemical impacts is an ecosystem with a different structure and composition.
  IV. Landscape models. As ecosystems are open systems, the effects of chemicals may change several interrelated ecosystems. Landscape models can be used in these cases.
  V. Global models. The impacts of chemicals are the core of this model. A typical global model represents the ozone layer and its decomposition due to the discharge of chemicals (i.e., freon).

FTE-models can be any combination of fate and effect models, although the combinations of AII and AIII fate models with BII and BIII effect models will be practical for ecotoxicological management.

The applied effect models are mainly type I and II, although the effects on ecosystem levels may be of particular importance due to their frequent irreversibility. Ecosystems may, in some cases, change their composition and structure significantly due to a discharge of toxic substances. In such cases, it is recommended to apply structurally dynamic models (SDMs), which are also called variable parameter models (see Chapter 10).

Ecotoxicological models are applied for registration of chemicals, to solve site-specific pollution problems, or to follow ecosystem recovery after pollution abatement or remediation has taken place.

Type AI and AIII models are widely used for registration of chemicals. About 100,000 chemicals are registered, but only about 20,000 chemicals are used at a scale that may threaten the environment with high probability. It is the long-term goal to perform an ERA for all 20,000 chemicals in use if ER continue the present rate of evaluation prior to 1984, when an ecotoxicological evaluation of all new chemicals became compulsory in the European Union (EU). Among the 20,000 chemicals, 2500 have been selected as high volume chemicals that are of most concern. Among the 2500 chemicals, 140 have been selected by the EU to be examined in detail including an ERA, which requires the application of models. These are called highly expected regulatory output chemicals (HERO-chemicals). A proper ecotoxicological evaluation of the chemicals in use prior to 1984 is important; it will take 100 years before we have a proper ecotoxicological evaluation of the 2500 high volume chemicals and 800 years before we have evaluated all chemicals in use. Unfortunately, by this time there will be many new chemicals.

About 300–400 new chemicals are registered per year. These chemicals have to be evaluated properly, although it may be possible in some cases for the chemical manufacturers to postpone the evaluation and the final decision a few years.

AII fate models and BII, BIII, and, in a few cases, BIV effect models are applied, sometimes in combination as an FTE model to solve site-specific pollution problems caused by toxic substances or to make predictions on the recovery of ecosystems after the impacts have been removed. These applications are mainly carried out by environmental protection agencies and rarely by chemical manufacturers.

In conclusion, there is an urgent need for good ecotoxicological models as well as for wider experience in the applicability of these models. The application of ecotoxicological models up to now has been minor compared to the environmental management possibilities that these models offer.

Section 8.2 reviews the performance of an ERA. Section 8.3 presents the characteristics and structure of ecotoxicological models. Section 8.4 gives an overview of some of the most illustrative, ecotoxicological models published during the last 20 years. The description of the chemical, physical, and biological processes will generally be according to the equations presented in Chapter 2. Section 8.5 is devoted to parameter estimations methods, which are important to ecotoxicological models.

The following sections are used to present ecotoxicological models of case studies. Section 8.6 presents a very simple ecotoxicological model of chromium pollution in Fåborg Fjord, Denmark. This case study clearly illustrates that a simple model can give an acceptable and sufficiently accurate answer to an environmental management question, provided the modeller knows the ecosystem and can select the processes of importance for the management question in focus. The case study in Section 8.7 covers an ecotoxicological model for relating contamination of agricultural products by cadmium with the heavy metal pollution of soil due to the content of cadmium in fertilizers, dry deposition, and sludge. Section 8.8 presents the development of class 1 fate model by use of equilibrium calculation and fugacity. It contains two illustrative examples to show how to develop this type of models, which is mostly applied for contamination of a region by a toxic substance.

8.2. Environmental Risk Assessment

8.2.1. Overview of Environmental Risk Assessment

A brief introduction to the concepts of ERA is given in this section to introduce readers to the concepts and ideas behind the application of ecotoxicological models to assess an environmental risk.

Treatment of industrial wastewater, solid waste, and smoke is very expensive. Consequently, the industries attempt to change their products and production methods in a more environmentally friendly direction to reduce the treatment costs. Therefore, industries need to know how much the different chemicals, components, and processes are polluting our environment. In other words: What is the environmental risk of using a specific material or chemical compared with other alternatives? If industries can reduce their pollution just by switching to another chemical or process, then they will reduce their environmental costs and improve their green image. An assessment of the environmental risk associated with the use of a specific chemical and a specific process enables industries to make the right selection of materials, chemicals, and processes to benefit the economy of the enterprise and the quality of the environment.

Similarly, society needs to know the environmental risks of all chemicals to phase out the most environmentally threatening chemicals and set standards for the use of all other chemicals. The standards should ensure there is no serious risk in using the chemicals, provided that the standards are followed carefully. Modern abatement of pollution includes ERA, which is defined as the process of assigning magnitudes and probabilities to the adverse effects of human activities. This process involves identification of hazards such as the release of toxic chemicals to the environment by quantifying the relationship between an activity associated with an emission to the environment and its effects. The entire ecological hierarchy is considered in this context including the effects on the cellular (biochemical) level, the organism level, the population level, the ecosystem level, and the entire ecosphere.

The application of ERA is rooted in the recognition that:

1. The elimination cost of all environmental effects is impossibly high.
2. Practical environmental management decisions must always be made on the basis of incomplete information.

We use about 100,000 chemicals in amounts that might threaten the environment, but we know only about 1% of what is necessary to make a proper and complete ERA of these chemicals. Section 8.5 is a short introduction to available estimation methods to apply if information about properties of chemical compounds is unavailable in the literature. A list of relevant properties and how they impact the environment is also given.

ERA is in the same family as environmental impact assessment (EIA), which attempts to assess the impact of a human activity. EIA is predictive, comparative, and concerned with all possible effects on the environment, including secondary and tertiary (indirect) effects, whereas ERA attempts to assess the probability of a given (defined) adverse effect as a result of human activity.

Both ERA and EIA use models to find the expected environmental concentration (EEC), which is translated into impacts for EIA and to risks of specific effects for ERA. Development of ecotoxicological models for assessing environmental risks is detailed in the following section. An overview of ecotoxicological models is given in Jørgensen et al. (1995).

Legislation and regulation of domestic and industrial chemicals for the protection of the environment have been implemented in Europe and North America for decades. Both regions distinguish between existing chemicals and introduction of new substances. For existing chemicals, the EU requires a risk assessment to humans and the environment according to a priority setting. An informal priority setting (IPS) is used for selecting chemicals among the 100,000 listed in "The European Inventory of Existing Commercial Chemical Substances." The purpose of the IPS is to select chemicals for detailed risk assessment from among the EEC high production volume compounds, that is, >1000 t/y (about 2500 chemicals). Data necessary for the IPS and an initial hazard assessment are called Hedset and cover issues such as environmental exposure, environmental effects, exposure to humans, and human health effects.

At the UNCED meeting on the Environment and Sustainable Development in Rio de Janeiro in 1992, it was decided to create an Intergovernmental Forum on Chemical Safety (IGFCS, Chapter 19 of Agenda 21). Its primary task is to stimulate and coordinate global harmonization in the field of chemical safety and covers the following principal themes: assessment of chemical risks, global harmonization of classification and labeling, information exchange, risk reduction programs, and capacity building in chemical management.

Uncertainty plays an important role in risk assessment (Suter, 1993). Risk is the probability that a specified harmful effect will occur or, in the case of a graded effect, the relationship between the magnitude of the effect and its probability of occurrence.

Risk assessment has traditionally emphasized risks to human health over the concerns of ecological effects. However, some chemicals such as chlorine, ammonia, and certain pesticides — which have no risk or only a small amount of risk to human health — cause severe effects on ecosystems such as aquatic organisms. An up-to-date risk assessment is comprised of considerations of the entire ecological hierarchy, which is the ecologist's worldview in terms of levels of organization. Organisms interact directly with the environment, so they can be exposed to toxic chemicals. The species-sensitivity distribution is therefore more ecologically credible (Calow, 1998). A reproducing population is the smallest meaningful level ecologically. However, populations do not exist in a vacuum; they require a community of other organisms of which the population is a part. The community occupies a physical environment with which it forms an ecosystem.

Moreover, both the various adverse effects and the ecological hierarchy have different scales in time and space, which must be included in a proper ERA (Figure 8.1). For example, oil spills occur at a spatial scale similar to those of populations, but they are briefer than population processes. Therefore, a risk assessment of an oil spill requires the consideration of reproduction and recolonization on a longer time scale to determine the magnitude of the population response and its significance to natural population variance.

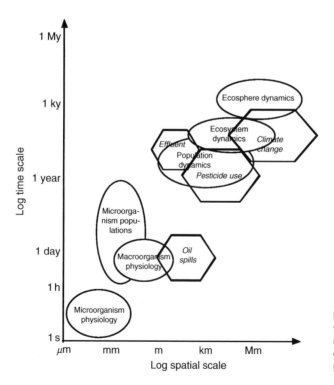

FIGURE 8. 1 The spatial and time scale for various hazards (hexagons, italic) and for the various levels of the ecological hierarchy (circles, non-italic).

8.2.2. Uncertainties in Risk Assessment

Uncertainties in risk assessment are taken into account by application of safety factors. Uncertainties have three basic causes:

1. Inherent randomness of the world (stochasticity)
2. Errors in execution of assessment
3. Imperfect or incomplete knowledge

Inherent randomness refers to uncertainty that can be described and estimated but not reduced because it is characteristic of the system. Meteorological factors such as rainfall, temperature, and wind are effectively stochastic at levels of interest for risk assessment. Many biological processes such as colonization, reproduction, and mortality also need to be described stochastically.

Human errors are inevitably attributes of all human activities. This type of uncertainty includes incorrect measurements, data recording errors, computational errors, and so on.

Uncertainty is addressed using an assessment (safety) factor from 10 to 1000. The choice of assessment factor depends on the quantity and quality of toxicity data (Table 8.1). The assessment or safety factor is used in step 3 of the ERA procedure presented in the following section. Relationships other than the uncertainties originating from randomness, errors, and lack of knowledge may be considered when the assessment factors are selected (e.g., cost-benefit). This implies that the assessment factors for drugs and pesticides may be given a lower value due to their possible benefits.

Table 8.1 Selection of Assessment Factors to Derive Predicted No Effect Concentration

Data Quantity and Quality	Assessment Factor
At least one short-term LC_{50} from each of the three trophic levels of the base set (fish, zooplankton, and algae)	1000
One long-term NOEC, either for fish or daphnia	100
Two long-term NOECs from species representing two trophic levels	50
Long-term NOECs from at least three species (normally fish, daphnia, and algae) representing three trophic levels	10
Field data or model ecosystems	Case by case

PNEC, Predicted No Effect Concentration. Note: See also step 3 of the procedure presented below.

Lack of knowledge results in an undefined uncertainty that cannot be described or quantified. It is a result of practical constraints on our ability to describe, count, measure, or quantify accurately everything that pertains to a risk estimate. Clear examples are the inability to test all toxicological responses of all species exposed to a pollutant and the simplifications needed in the model used to predict the EEC.

The most important feature distinguishing risk assessment from impact assessment is the emphasis in risk assessment on characterizing and quantifying uncertainty. Therefore, it is of particular interest in risk assessment to estimate the analyzable uncertainties, such as natural stochasticity, parameter errors, and model errors. Statistical methods may provide direct estimates of uncertainties, and they are widely used in model development.

The use of statistics to quantify uncertainty is complicated in practice by the need to consider errors in both the dependent and independent variables and to combine errors when multiple extrapolations should be made. Monte Carlo analysis is often used to overcome these difficulties (Bartell et al. 1992).

Model errors include inappropriate selection or aggregation of variables, incorrect functional forms, and incorrect boundaries. The uncertainty associated with model errors is usually assessed by field measurements utilized for calibration and validation of the model (see Chapter 2). The modelling uncertainty for ecotoxicological models is no different from what was previously discussed in Chapter 2.

8.2.3. Step-by-Step Guide for Ecological Risk Assessment

Chemical risk assessment is divided into nine steps shown in Figure 8.2. The nine steps correspond to questions that the risk assessment attempts to answer when quantifying the risk associated with the use of a chemical.

Step 1: Which hazards are associated with the application of the chemical? This involves gathering data on the types of hazards such as possible environmental damage and human health effects. The health effects include congenital, neurological, mutagenic, endocrine disruption (e.g., estrogen), and carcinogenic effects. It may also include characterization of the behavior of the chemical within the body (interactions with

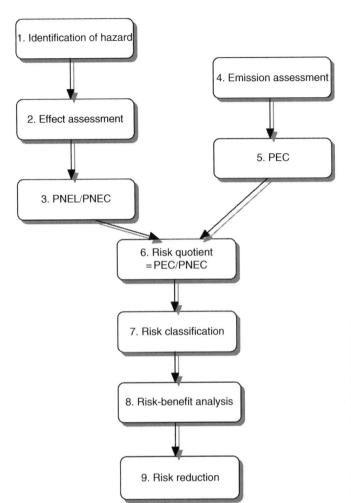

FIGURE 8.2 The presented procedure in nine steps to assess the risk of chemical compounds. Steps 1–3 require extensive use of ecotoxicological handbooks and ecotoxicological estimation methods to assess the toxicological properties of the chemical compounds considered, while step 5 requires the selection of a proper ecotoxicological model.

organs, cells, or genetic material). The possible environmental damage including lethal effects and sub-lethal effects on growth and reproduction of various populations is considered in this step. As an attempt to quantify the potential danger posed by chemicals, a variety of toxicity tests have been devised. Some of the recommended tests involve experiments with subsets of natural systems, such as microcosms, or with entire ecosystems. The majority of testing new chemicals for possible effects has, however, been confined to studies in the laboratory on a limited number of test species. Results from these laboratory assays provide useful information for quantification of the relative toxicity of different chemicals. They are used to forecast effects in natural systems, although their justification has been seriously questioned (Cairns et al. 1987).

Step 2: What is the relation between dose and responses of the type defined in step 1? It implies knowledge of NEC and LD_x values (dose that is lethal to x% of the organisms considered), LC_y values (concentration lethal to y% of the organisms considered), and EC_z values (concentration giving the indicated effect to z% of the considered organisms) where x, y, and z express a probability of harm. The answer can be found by laboratory examination or we may use estimation methods. Based upon these answers, a most probable level of no effect (NEL) is assessed. Data needed for steps 1 and 2 are obtained directly from scientific libraries, but are increasingly found via online data searches in bibliographic and factual databases. Data gaps should be filled with estimated data. It is very difficult to completely know about a chemical's effect on all levels from cells to ecosystem as some effects are associated with very small concentrations (the estrogen effect). Therefore it is far from sufficient to know NEC, LD_x-, LC_y-, and EC_z-values.

Step 3: Which uncertainty (safety) factors reflect the amount of uncertainty that must be taken into account when experimental laboratory data or empirical estimation methods are extrapolated to real situations? Usually, safety factors of 10–1000 are used. The choice was discussed earlier and is usually in accordance with Table 8.1. If good knowledge about the chemical is available, then a safety factor of 10 may be applied. If, on the other hand, it is estimated that the available information has a very high uncertainty, then a safety factor of 10,000 may be recommended. Most frequently, safety factors of 50–100 are applied. NEL times the safety factor is the predicted noneffect level (PNEL). The complexity of ERA is often simplified by deriving the predicted no-effect concentration (PNEC) for different environmental components (water, soil, air, biotas, and sediment).

Step 4: What are the sources and quantities of emissions? The answer requires thorough knowledge of the production and use of the considered chemical compounds, including an assessment of how much of the chemical is wasted in the environment by production and use. The chemical may also be a waste product, which makes it very difficult to determine the amounts involved; for instance, the very toxic dioxins are waste products from incineration of organic waste.

Step 5: What is (are) the actual exposure concentration(s)? The answer to this question is the PEC. Exposure can be assessed by measuring environmental concentrations. It may also be predicted by a model when the emissions are known. The use of models is necessary in most cases either because we are considering a new chemical, or because the assessment of environmental concentrations requires a very large number of measurements to determine the variations in concentrations. Furthermore, it provides an additional certainty to compare model results with measurements, which implies that it is always recommended both to develop a model and to make at least a few measurements of concentrations in the ecosystem components when and where it is expected that the highest concentration will occur. Most models demand an input of parameters, describing the properties of the chemicals and the organisms, which also requires extensive application of handbooks and a wide range of estimation methods. The development of an environmental, ecotoxicological model requires extensive knowledge of the physical-chemical-biological properties of the chemical compound(s) considered. The selection of a proper model is discussed in this chapter and in Chapter 2.

Step 6: What is the ratio PEC/PNEC? This ratio is often called the risk quotient. It should not be considered an absolute assessment of risk but rather a relative ranking of risks. The ratio is usually found for a wide range of ecosystems such as aquatic and terrestrials well as ground water. Steps 1–6 shown in Figure 8.3 agree with Figure 8.2 and the information given in the previous six steps.

Step 7: How will you classify the risk? Risk valuation decides on risk reductions (step 9). Two risk levels are defined: (1) the upper limit, that is, the maximum permissible level (MPL); and (2) the lower limit, that is, the negligible level, NL. It may also be

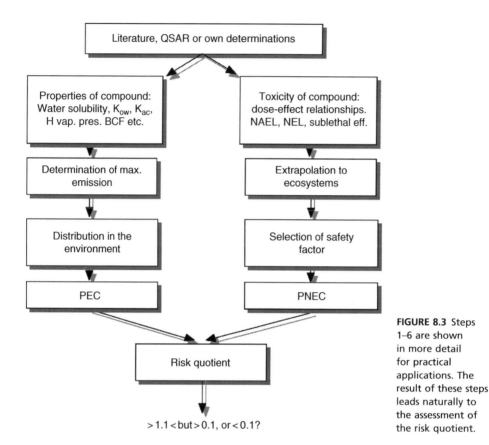

FIGURE 8.3 Steps 1–6 are shown in more detail for practical applications. The result of these steps leads naturally to the assessment of the risk quotient.

defined as a percentage of MPL, for instance, 1% or 10% of MPL. The two risk limits create three zones: a black, unacceptable, high risk zone >MPL; a gray, medium risk level; and a white, low risk level <NL. The risk of chemicals in the gray and black zones must be reduced. If the risk of the chemicals in the black zone cannot be reduced sufficiently, then phasing out the use of these chemicals should be considered.

Step 8: What is the relation between risk and benefit? This analysis involves examination of socioeconomic, political, and technical factors, which are beyond the scope of this volume. The cost-benefit analysis is difficult because the costs and benefits are often of a different order and dimension.

Step 9: How can the risk be reduced to an acceptable level? The answer to this question requires deep technical, economic, and legislative investigation. Assessment of alternatives is often an important aspect in risk reduction.

Steps 1, 2, 3, and 5 require knowledge of the properties of the focal chemical compounds, which again implies an extensive literature search and/or selection of the best feasible estimation procedure. In addition to "Beilstein," (http://www.reaxys.com/info/) it is recommended to have on hand the following handbooks of environmental properties of chemicals and methods for estimation of these properties in case literature values are not available:

Jørgensen, S. E., Nielsen, S. N., and Jørgensen, L. A. (1991). *Handbook of Ecological Parameters and Ecotoxicology*, Elsevier, Amsterdam. Year 2000 published as a CD called Ecotox. It contains three times the amount of parameters then the 1991 book edition. See also Chapter 2 for further details about Ecotox.
Howard, P. H. et al. (1991). *Handbook of Environmental Degradation Rates*. Lewis Publishers, New York.
Verschueren, K. (2007). Several editions have been published, the latest in 2007. *Handbook of Environmental Data on Organic Chemicals*. Van Nostrand Reinhold, New York.
Mackay, D., Shiu, W. Y., and Ma, K. C. (1991, 1992). *Illustrated Handbook of Physical-Chemical Properties and Environmental Fate for Organic Chemicals*. Volume I. Mono-aromatic Hydrocarbons. Chloro-benzenes and PCBs, 1991. Volume II. Polynuclear Aromatic Hydrocarbons, Polychlorinated Dioxins, and Dibenzofurans, 1992. Volume III. Volatile Organic Chemicals, 1992. Lewis Publishers, New York.
Jørgensen, S. E., Halling-Sørensen, B., and Mahler, H. (1997). *Handbook of Estimation Methods in Environmental Chemistry and Ecotoxicology*. Lewis Publishers, Boca Raton, FL.

Steps 1–3 are sometimes denoted as effect assessment or effect analysis, and steps 4–5 are exposure assessment or effect analysis. Steps 1–6 may be called risk identification, while ERA encompasses all 9 steps presented in Figure 8.2. Step 9 is very demanding, as several possible steps in reduction of the risk should be considered, including treatment methods, cleaner technology, and substitutes to replace the examined chemical.

8.2.4. Risk Assessment of medicinal and veterinarian chemicals

In North America, Japan, and the EU, medicinal products are considered similar to other chemical products because there is no difference between a medicinal product and other chemical products. In the EU, technical directives for human medicinal products do not include any reference to ecotoxicology and the assessment of their potential risk. However, a detailed technical draft guideline issued in 1994 indicated that the applied approach for veterinary medicine would also apply to human medicinal products. Presumably, ERA will be applied to all medicinal products in the near future when sufficient experience with veterinary medicinal products has been achieved. Veterinary medicinal products, on the other hand, are released in larger amounts to the environment as manure. It is also possible to perform an ERA where the human population is in focus. Ten steps corresponding to Figure 8.3 are shown in Figure 8.4, which is not significantly different from Figure 8.3. The principles for the two types of ERA are the same. Figure 8.4 uses the nonadverse effect level (NAEL) and nonobserved adverse effect level (NOAEL) to replace the PNEC, and the PEC is replaced by the tolerable daily intake (TDI).

This type of ERA is of particular interest to veterinary medicine that may contaminate food products for human consumption. For instance, the use of antibiotics in pig feed has attracted a lot of attention, as they may be found as residue in pig meat or may contaminate the environment though the application of manure as natural fertilizer.

Selection of a proper ecotoxicological model is the first step in the development of an environmental exposure model, as required in step 5. It will be discussed in more detail in the next section.

8.3. Characteristics and Structure of Ecotoxicological Models

Toxic substance models are most often biogeochemical models because they attempt to describe the mass flows of the considered toxic substances. But there are effect models of the population dynamics that

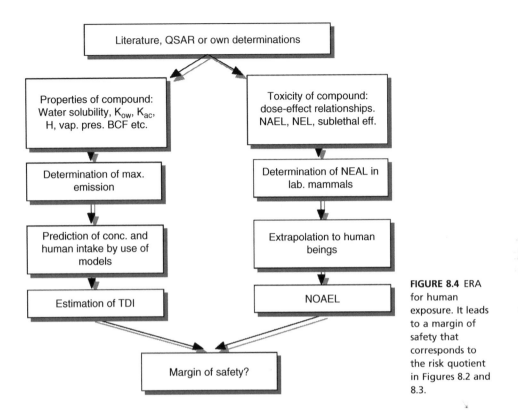

FIGURE 8.4 ERA for human exposure. It leads to a margin of safety that corresponds to the risk quotient in Figures 8.2 and 8.3.

include the influence of toxic substances on the birth rate and/or the mortality, and therefore should be considered toxic substance models.

Toxic substance models differ from other ecological models in that:

1. The need for parameters to cover all possible toxic substance models is great, and general estimation methods are widely used. Section 8.5 is devoted to this question. It has also been discussed in Section 2.8.
2. The safety margin should be high; for instance, expressed as the ratio between the predicted concentration and the concentration that gives undesired effects. This is discussed in Section 8.2, where RQ = PEC/NOEC is applied after an assessment factor (a safety margin) has been applied. The selection of the assessment factor is presented in Section 8.2.
3. They require possible inclusion of an effect component, which relates the output concentration to its effect. It is easy to include an

effect component in the model; it is, however, often a problem to find a well-examined relationship to base it on.

4. Toxic substance models need to be simple due to points 1 and 2, and our limited knowledge of process details, parameters, sub-lethal effects, and antagonistic and synergistic effects is limited.

It may be an advantage to outline the approach before developing a toxic substance model according to the procedure presented in Section 2.3.

1. Obtain the best knowledge about the possible processes of the toxic substances in the ecosystem.
2. Attempt to get parameters from the literature and/or from your own experiments (*in situ* or in the laboratory).
3. Estimate all parameters by the methods presented in Sections 2.10 and 8.5.
4. Compare the results from 2 and 3 and attempt to explain discrepancies.
5. Estimate which processes and state variables are feasible and relevant to include in the model. When in doubt, at this stage it is better to include too many processes and state variables rather than too few.
6. Use a sensitivity analysis to evaluate the significance of the individual processes and state variables. This often may lead to further simplification.

To summarize, ecotoxicological models differ from ecological models by:

1. Often being simpler conceptually
2. Requiring more parameters
3. Using a wider range of parameter estimation methods
4. Including of an effect component

Ecotoxicological models may be divided into five classes according to their structure. These classes illustrate the possibilities of simplification, which are urgently needed:

1. *Food chain or food web dynamic models*
   This class of models considers the flow of toxic substances through the food chain or food web. It can also be described as an ecosystem

model focusing on the transfer of a toxic substance to ecological and nonecological components. Such models are relatively complex and contain many state variables. The models contain many parameters that often have to be estimated by one of the methods presented in Section 8.5. This model type is typically used when many organisms are affected by a toxic substance or the entire structure of the ecosystem is threatened by the presence of a toxic substance. Because of the complexity of these models, they have not been widely used. They are similar to the more complex eutrophication models that consider the nutrient flow through the food chain or even through the food web. Sometimes they are even constructed as submodels of a eutrophication model (Thomann et al., 1974). Figure 8.5 shows a conceptual diagram of an ecotoxicological food chain model for lead. There is a flow of lead from atmospheric fallout and wastewater to an aquatic ecosystem where it is concentrated through the food chain by "bioaccumulation." A simplification is hardly possible for this model type because it is the model's purpose to describe and quantify the bioaccumulation through the food chain.

2. *Static models of toxic substance mass flows*
   If the seasonal changes are minor, or of minor importance, then a static model of the mass flows will often be sufficient to describe the situation and show the expected changes if the input of toxic substances is reduced or increased. This model type is based upon a mass balance as seen from the example in Figure 8.6. It will often contain more trophic levels, but the modeller is frequently concerned with the flow of the toxic substance through the food chain. If there are some seasonal changes, then this type, which usually is simpler than food chain or food web dynamic models, can still be advantageous if the modeller is concerned with the worst case or the average case and not with the changes.

3. *Dynamic models of a toxic substance in one trophic level*
   It is often only the toxic substance concentration in one trophic level that is studied. This includes the abiotic environment (sometimes called the zeroth trophic level), — soil, water, or air. Figure 8.7 illustrates an example with a model of copper contamination in an aquatic ecosystem. The main concern is the copper concentration in

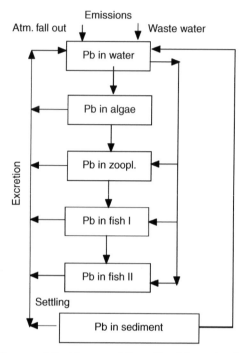

FIGURE 8.5 Conceptual diagram of the bioaccumulation of lead through a food chain in an aquatic ecosystem.

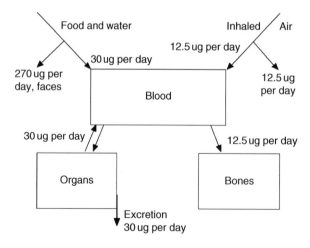

FIGURE 8.6 A static model of the lead uptake by an average Dane in 1980 before lead in gasoline was banned.

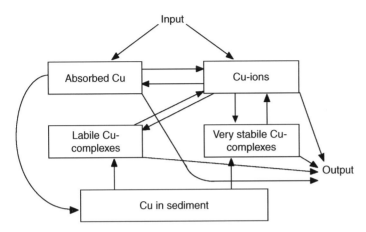

FIGURE 8.7 Conceptual diagram of a simple copper model.

the water, as it may reach a toxic level for the phytoplankton. Zooplankton and fish are much less sensitive to copper contamination, so the alarm rings first at the concentration level that is harmful to phytoplankton. However, only the ionic form is toxic so it is necessary to model the partition of copper in ionic form, complex bound form, and adsorbed form. The exchange between copper in the water phase and in the sediment is also included because the sediment can accumulate relatively large amounts of heavy metals. The amount released from the sediment may be significant under certain circumstances, such as low pH. Figure 8.8 shows an example where the main concern is the DDT (dichlorodiphenyltrichloroethane) concentration in fish. There may be such a high concentration of DDT that, according to the World Health Organization (WHO) standards, the fish are not recommended for human consumption. This model can be simplified to just the fish instead of the entire food chain. Some physical-chemical reactions in the water phase are still important and they are included as shown on the conceptual diagram in Figure 8.8. As seen from these examples, simplifications are often feasible when the problem is well defined, including which component is most sensitive to toxic matter and which processes are most important for concentration changes.

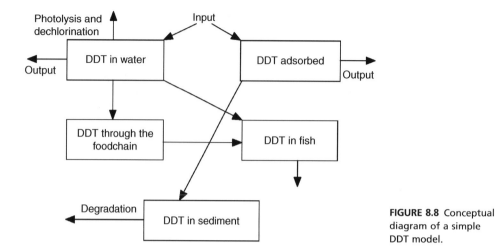

FIGURE 8.8 Conceptual diagram of a simple DDT model.

Figure 8.9 shows the processes of interest for modelling the concentration of a toxic component at one trophic level. The inputs are uptake from the medium (water or air) and from digested food = total food − nondigested food. The outputs are mortality (transfer to detritus), excretion, and predation from the next level in the food chain.

4. *Ecotoxicological models in population dynamics*
Population models are biodemographic models and the number of individuals or species are state variables. Simple population models consider only one population. Population growth is a result of the difference between natality and mortality:

$$dN/dt = B*N - M*N = r*N, \tag{8.1}$$

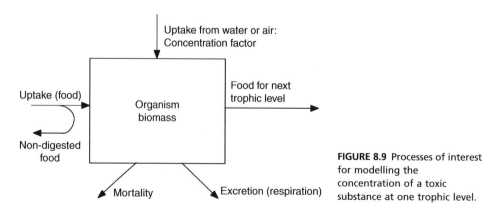

FIGURE 8.9 Processes of interest for modelling the concentration of a toxic substance at one trophic level.

where N is the number of individuals; B is the natality, that is, the number of new individuals per unit of time and per unit of population; M is the mortality, that is, the number of organisms that die per unit of time and per unit of population; and r is the increase in the number of organisms per unit of time and per unit of population and is equal to B − M. B, N, and r are not necessarily constants as in the exponential growth equation, but are dependent on N, the carrying capacity, and other factors. The concentration of a toxic substance in the environment or in the organisms may influence the natality and the mortality, and if the relation between a toxic substance concentration and these population dynamic parameters is included in the model, it becomes an ecotoxicological model of population dynamics. Population dynamic models may include two or more trophic levels, and ecotoxicological models include the influence of the toxic substance concentration on natality, mortality, and interactions between these populations. In other words, an ecotoxicological model of population dynamics is a general model of population dynamics with the inclusion of relations between toxic substance concentrations and some important model parameters.

5. *Ecotoxicological models with effect components*
Although class 4 models already may include relations between concentrations of toxic substances and their effects, these are limited to population dynamic parameters, not to a final assessment of the overall effect. In comparison, class 5 models include more comprehensive relations between toxic substance concentrations and effects. These models may include lethal and/or sub-lethal effects as well as effects on biochemical reactions or on the enzyme system. These effects may be considered on various levels of the biological hierarchy from the cells to the ecosystems.
In many problems, it may be necessary to go into more detail about the effect to answer the following questions:

• Does the toxic substance accumulate in the organism?
• What is the long-term concentration in the organism when uptake rate, excretion rate, and biochemical decomposition rate are considered?
• What is the chronic effect of this concentration?

- Does the toxic substance accumulate in one or more organs?
- What is the transfer between various parts of the organism?
- Will decomposition products eventually cause additional effects?

A detailed answer to these questions may require a model of the processes that take place in the organism, and a translation of concentrations in various parts of the organism into effects. This implies that the intake = (uptake by the organism)*(efficiency of uptake) is known. Intake may either be from water or air, which also may be expressed (at steady state) by concentration factors, such as the ratio between the concentration in the organism and in the air or water.

But, if all the previously mentioned processes were taken into consideration for just a few organisms, the model would easily become too complex, contain too many parameters to calibrate, and require more detailed knowledge than it is possible to provide. Often we do not even have all the relations needed for a detailed model, as toxicology and ecotoxicology are not completely well understood. Therefore, most models in this class do not consider too many details of the partition of the toxic substances in organisms and their corresponding effects, but instead are limited to the simple accumulation in the organisms and their effects. Usually, accumulation is rather easy to model and the following simple equation is often sufficiently accurate:

$$dC/dt = (ef * Cf * F + em * Cm * V)/W - Ex * C = (INT)/W - Ex * C \qquad (8.2)$$

where C is the concentration of the toxic substance in the organism; ef and em are the efficiencies for the uptake from the food and medium, respectively, (water or air); Cf and Cm are the concentration of the toxic substance in the food and medium, respectively; F is the amount of food uptake per day; V is the volume of water or air taken up per day; W is the body weight either as dry or wet matter; and Ex is the excretion coefficient (1/day). From Eq. (8.2), INT covers the total intake of toxic substance per day.

This equation has a numerical solution, and the corresponding plot is shown in Figure 8.10:

$$C/C(max) = (INT * (1 - \exp(Ex * t)))/(W * Ex) \qquad (8.3)$$

where C(max) is the steady-state value of C:

$$C(max) = INT/(W * Ex) \qquad (8.4)$$

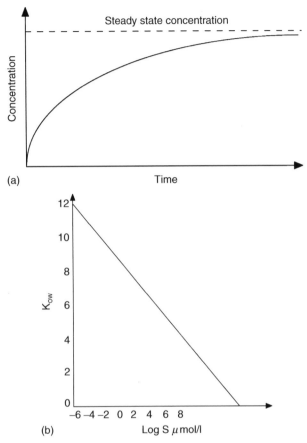

FIGURE 8.10 (a) Concentration of a toxic substance in an organism versus time. (b) Relationship between water solubility (unit: μmol/L) and octanol-water distribution coefficient.

Synergistic and antagonistic effects have not been discussed so far. They are rarely considered in this type of model for the simple reason that we do not have much knowledge about these effects. If we have to model combined effects of two or more toxic substances, then we can only assume additive effects unless we can provide empirical relationships for the combined effect.

A complete solution of an ecotoxicological problem requires four submodels, of which the fate model may be considered the first model in the chain (Figure 8.11). In Figure 8.11, the four components are (Morgan, 1984):

1. A fate or exposure model that should be as simple as possible and as complex as needed

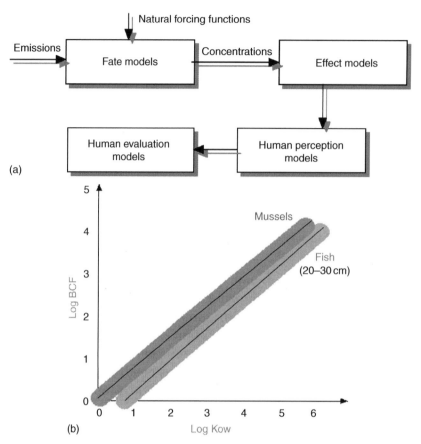

FIGURE 8.11 (a) Four submodels of a total ecotoxicological model. (b) Two applicable relationships for octanol-water distribution coefficient and the biological concentration factor for fish and mussels.

2. An effect model that translates the concentration into an effect; see class type 5 and the different levels of effects presented in Section 8.1

3. A model for human perception

4. A model for human evaluation

The first two submodels are "objective," predictive models corresponding to the structural model types 1–5 described previously, or the classes described from an application point of view as described in Section 8.1. They are based upon physical, chemical, and biological processes. They are very similar to other environmental models and are founded upon mass transfer and mass balances and physical, chemical. and biological processes.

Submodels 3 and 4 are different from the generally applied environmental management models and are briefly discussed in the following section. A risk assessment component, associated with the fate model, comprises human perception and evaluation processes (Figure 8.11). These submodels are explicitly value laden, but must build on objective information concerning concentrations and effects. They are often considered in the ERA procedure when deciding on the assessment factor.

Factors that may be important to consider in this context include:

1. Magnitude and time constant of exposure
2. Spatial and temporal distribution of concentration
3. Environmental conditions determining the process rates and effects
4. Translation of concentrations into magnitude and duration of effects
5. Spatial and temporal distribution of effects
6. Reversibility of effects

The uncertainties relating to the information on which the model is based and the uncertainties related to the development of the model are crucial in risk assessment. In addition to the discussion of the assessment factor in Section 8.2 and partly in Section 8.3 where the focus was on the effects on the trophic levels, the uncertainty of risk assessment may be described by the following five categories:

1. Direct knowledge and statistical evidence on the important components (state variables, processes, and interrelations of the variables) of the model are available.
2. Knowledge and statistical evidence on the important submodels are available, but the aggregation of the submodels is less certain.
3. Adequate knowledge of the model components for the considered system is not available, but good data are available for the same processes from a similar system, and it is estimated that these data may be applied directly or with minor modifications to the model development.
4. Some, but insufficient, knowledge is available from other systems. Attempts are made to use these data without the necessary transferability. Attempts are made to eliminate gaps in knowledge by

using additional experimental data as far as it is possible within the limited resources available for the project.

5. The model is to a large extent, or at least partly, based on the subjective judgment of experts.

The acknowledgement of the uncertainty is of great importance and may be taken into consideration either qualitatively or quantitatively. Another problem is where to take the uncertainty into account. Should the economy or the environment benefit from the uncertainty? The ERA procedure presented in Section 8.2 has definitely facilitated the possibility of considering the environment more than the economy.

Until 10–15 years ago, researchers had developed very little understanding of the processes by which people actually perceive the exposures and effects of toxic chemicals. These processes are just as important for the risk assessment as the exposures and effects processes themselves. The characteristics of risks and effects are important for the perceptions of people. These characteristics may be summarized in the following list.

Characteristics of risk:

Voluntary or involuntary?
Are the levels known to the exposed people or to science?
Is it novel or familiar?
Is it common or dreaded (e.g., does it involve cancer)?
Are mishaps controllable?
Are future generations threatened?
What scale: global, regional, or local?
Function of time? How (for instance, whether increasing or decreasing)?
Can it easily be reduced?

Characteristics of effects:

Immediate or delayed?
On many or a few people?
Global, regional, or local?
Involve death?
Are effects of mishaps controllable?

Observable immediately?

How are they a function of time?

A factor analysis was performed by Slovic et al. (1982), which showed, among other results, an unsurprising correlation between people's perception of dreaded and unknown risks. Broadly speaking, there are two methods of selecting the risks we will deal with.

The first is described as the "rational actor model," involving people that look systematically at all risks they face and make choices about which they will live with and at what levels. For decision making, this approach uses single, consistent, objective functions and a set of decision rules.

The second method is called the "political/cultural model." It involves interactions between culture, social institutions, and political processes for the identification of risks and determination of those people will live with and at what level.

Both methods are unrealistic, as they are both completely impractical in their pure form. Therefore, we must select a strategy for risk abatement founded on a workable alternative based on the philosophy behind both methods.

Several risk management systems are available, but no attempt is made here to evaluate them. However, some recommendations should be given for the development of risk management systems:

1. Consider as many of the previously listed characteristics as possible and include the human perceptions of these characteristics in the model.
2. Do not focus too narrowly on certain types of risks. This may lead to suboptimal solutions. Attempt to approach the problem as broadly as possible.
3. Choose strategies that are pluralistic and adaptive.
4. Benefit-cost analysis is an important element of the risk management model, but it is far from the only important element and the uncertainty in evaluation of benefit and cost should not be forgotten. The variant of this analysis applicable to environmental risk management may be formulated as follows:

$$\text{net social benefit} = \text{social benefits of the project} \\ - \text{"environmental"costs of the project} \quad (8.5)$$

5. Use multi-attribute utility functions, but remember that people have trouble thinking about more than two or three and, at most, four attributes in each outcome.

The application of the estimation methods, presented in Section 8.5, renders it feasible to construct ecotoxicolgical models, even with limited parameter knowledge. The estimation methods have a high uncertainty, but a great safety factor (assessment factor) helps in accepting this uncertainty. On the other hand, our knowledge about the effects of toxic substances is very limited, particularly at the ecosystem, organism, and organ level. Therefore, models with effect components are only able to give a rough picture of what is currently known in this area.

8.4. An Overview: The Application of Models in Ecotoxicology

A number of toxic substance models have been published in the last 35 years and several models are now available in ecotoxicology. During the last ten years many of the models developed from 1975 to 2000 have been applied in environmental management, while only a few new models have been developed. This is probably because the spectrum of available toxic substance models is sufficient to cover the relevant ecotoxicological problems. Most models reflect the proposition that good knowledge of the problem and ecosystem can be used to make reasonable simplifications. Ecotoxicological modelling has been approached from two sides: population dynamics and biogeochemical flow analysis. As the second approach mostly focuses on environmental management, it has been natural to also approach the toxic substance problems from this angle. The most difficult part of modelling the effect and distribution of toxic substances is to obtain the relevant knowledge about the behavior of the toxic substances in the environment and to use this knowledge to make the feasible simplifications. The modeller of ecotoxicological problems is challenged to select the appropriate and balanced complexity, and there are many examples of rather simple ecotoxicological models that can solve the focal problem.

It can be seen from the overview in Table 8.2 that many ecotoxicological models have been developed during 1970s and 1980s. Before

Table 8.2   Examples of Toxic Substance Models

Toxic Substance Model Class	Model Characteristics	Reference
Cadmium (1)	Food chain similar to a eutrophication model	Thomann et al. (1974)
Mercury (1)	6 state variables: water, sediment, suspended matter, invertebrates, plant, and fish	Miller (1979)
Vinyl chloride (3)	Chemical processes in water	Gillett et al. (1974)
Methyl parathion (1)	Chemical processes in water and benzothiophene microbial degradation, adsorption, 2-4 trophic levels	Lassiter (1978)
Methyl mercury (4)	A single trophic level: food intake, excretion, metabolism growth	Fagerstrøm & Aasell (1973)
Heavy metals (3)	Concentration factor, excretion, bioaccumulation	Aoyama et al. (1978)
Pesticides in fish DDT & methoxy-chlor (5)	Ingestion, concentration factor, adsorption on body, defecation, excretion, chemical decomposition, natural mortality	Leung, (1978)
Zinc in algae (3)	Concentration factor, secretion hydrodynamical distribution	Seip (1978)
Copper in sea (5)	Complex formation, adsorption sub-lethal effect of ionic copper	Orlob et al. (1980)
Radionuclides in sediment (3)	Photolysis, hydrolysis, oxidation, biolysis, volatilization, and resuspension	Onishi and Wise (1982)
Metals (2)	A thermodynamic equilibrium model	Felmy et al. (1984)
Sulfur deposition (3)	Box model to calculate deposition of sulfur	McMahon et al. (1976)
Radionuclides (3)	Distribution of radionuclides from a nuclear accident release	ApSimon et al. (1980)
Sulfur transport (3)	Long-range transmission of sulfur pollutants	Prahm and Christensen (1976)
Lead (5)	Hydrodynamics, precipitation, toxic effects of free ionic lead on algae, invertebrates, and fish	Lam and Simons (1976)
Radionuclides (3)	Hydrodynamics, decay, uptake, and release by various aquatic surfaces	Gromiec & Gloyna (1973)
Radionuclides (2)	Radionuclides in grass, grains, vegetables, milks, eggs, beef, and poultry are state var.	Kirschner & Whicker (1984)
SO2, NOx, and heavy metals	Threshold model for accumulation effect of fire pollutants on spruce in forests	Kohlmaier et al. (1984)
Toxic environmental chemicals (5)	Hazard ranking and assessment from physic-chemical data and a limited number of laboratory tests	Bro-Rasmussen & Christiansen (1984)
Heavy metals (3)	Adsorption, chemical reactions, ion exchange	Several authors
Polycyclic aromatic hydrocarbons (3)	Transport, degradation, bioaccumulation	Bartell et al. (1984)
Persistent toxic organic substances (3)	Groundwater movement, transport, and accumulation of pollutants in groundwater	Uchrin (1984)
Cadmium, PCB (2)	Hydraulic overflow rate (settling), sediment interactions, steady-state food chain submodel	Thomann (1984)

Continued

Table 8.2   Examples of Toxic Substance Models—cont'd

Toxic Substance Model Class	Model Characteristics	Reference
Mirex (3)	Water-sediment exchange processes, adsorption, volatilization, bioaccumulation	Halfon (1984)
Toxins (aromatic hydrocarbons, Cd)	Hydrodynamics, deposition, resuspension, volatilization, photooxidation, decomposition, adsorption, complex formation, (humic acid)	Harris et al. (1984)
Heavy metals (2)	Hydraulic submodel, adsorption	Nyholm et al. (1984)
Oil Slicks	Transport and spreading, influence of surface tension, gravity, and weathering processes	Nihoul (1984)
Acid rain (soil) (3)	Aerodynamic, deposition	Kauppi et al. (1984)
Persistent organic chemicals (5)	Fate, exposure, and human uptake	Mackay (1991)
Chemicals, general (5)	Fate, exposure, ecotoxicity for surface water and soil	Matthies et al. (1987)
Toxicants, general (4)	Effect on populations of toxicants	de Luna and Hallam (1987)
Chemical hazard (5)	Basin-wide ecological fate	Morioka and Chikami (1986)
Pesticides (4)	Effects on insect populations	Longstaff (1989)
Insecticides (2)	Resistance	Schaalje et al. (1988)
Mirex and Lindane (1)	Fate in Lake Ontario	Halfon (1986)
Acid rain (5)	Effects on forest soils	Kauppi et al. (1986)
Acid rain (5)	Cation depletion of soil	Jørgensen et al. (1995)
pH, Calcium, and aluminum (4)	Survival of fish populations	Breck et al. (1988)
Photochemical smog (5)	Fate and risk	Wratt et al. (1992)
Nitrate (3)	Leaching to groundwater	Wuttke et al. (1991)
Oil spill (5)	Fate	Jørgensen et al. (1995)
Toxicants (4)	Effects on populations	Gard (1990)
Pesticides (3)	Loss rates	Jørgensen et al. (1995)
TCDD (3)	Photodegradation	Jørgensen et al. (1995)
Toxicants (4)	Effects general on populations	Gard (1990)
Pesticides and surfactants (3)	Fate in rice fields	Jørgensen et al. (1997)
Toxicants (3)	Migration of dissolved toxicants	Monte (1998)
Growth promoters (3)	Fate, agriculture	Jørgensen et al. (1998)
Toxicity (3)	Effect on eutrophication	Legovic (1997)
Pesticides (3)	Mineralization	Fomsgaard (1997)
Mecoprop (3)	Mineralization in soil	Fomsgaard and Kristensen (1999)

1975, toxic substances were hardly associated with environmental modelling because the problems seemed straightforward. The many pollution problems associated with toxic substances could easily be solved simply by eliminating the source of the toxic substance. During the 1970s, it was acknowledged that the environmental problems of toxic substances were very complex due to the interaction of many sources and many simultaneously, interacting processes and components. Several accidental releases of toxic substances into the environment have reinforced the need for models. The list given in Table 8.2 presents a comprehensive survey of the available ecotoxicological models, but the list should not be considered complete or even nearly complete as the table is not a result of a complete literature review. This table is meant to illustrate the spectrum of available models, to demonstrate that all five types of models have been developed, and to help the reader to find a reference to a specific toxic substance modelling problem.

8.5. Estimation of Ecotoxicological Parameters

Slightly more than 100,000 chemicals are produced in such an amount that they threaten or may threaten the environment. They cover a wide range of applications: household chemicals, detergents, cosmetics, medicines, dye stuffs, pesticides, intermediate chemicals, auxiliary chemicals in other industries, additives to a wide range of products, chemicals for water treatment, and so on. They are viewed as mostly indispensable in modern society, resulting in increased production of chemicals about 40-fold during the last four decades. A proportion of these chemicals reaches the environment through their production, transport, application, or disposal. In addition, the production or use of chemicals may cause unforeseen waste or byproducts, for example, chloro-compounds from the use of chlorine as a disinfectant. Because we would like to have the benefits of using the chemicals and not accept the harm they may cause, several urgent questions have been raised that have already been discussed in this chapter. These questions cannot be answered without models, and we cannot develop models without knowing the most important parameters, at least within some ranges. The Organization for Economic Cooperation and Development (OECD) has reviewed the common properties that we should know for all chemicals. These include the boiling point and melting point, which are necessary to know the chemical form (solid, liquid, or gas) found in the environment. We also must know the distribution of the chemicals in the five spheres: hydrosphere, atmosphere, lithosphere, biosphere, and technosphere (anthroposphere). This requires knowing the solubility in water; the partition coefficient water/lipids; Henry's constant; the vapor pressure; the rate of degradation by hydrolysis, photolysis, chemical oxidation, and microbiological processes; and the adsorption equilibrium between water and soil — all as a function of the temperature. We need to discover the interactions between living organisms and the chemicals, which implies that we should know the biological concentration factor (BCF), the magnification through the food chain, the uptake rate, and the excretion rate by the organisms and where in the organisms the chemicals will be concentrated. We must also know the effects on a wide range of different organisms. This means we should be able to find the LC_{50} and LD_{50} values, the MAC and NEC values (MAC = maximum allowable concentration and NEC = non-effect concentration), the relationship between the various possible sub-lethal effects and concentrations, the influence of the chemical on fecundity, and the carcinogenic and teratogenic properties. We should also know the effect on the ecosystem level. How do the chemicals affect populations and their development and interactions, that is, the entire network of the ecosystem?

Table 8.3 presents an overview of the most relevant physical-chemical properties of organic compounds and their interpretation with respect to the behavior in the environment, which should be reflected in the model.

ERAs also require information about chemicals' properties regarding their interactions with living organisms. It might not be necessary to know the properties with the high accuracy provided by measurements in a laboratory, but it would be beneficial to know the properties with sufficient accuracy to make it possible to utilize the models for management and risk assessments. Therefore, estimation methods have been developed as an urgently needed alternative to measurements. These are based on the structure of the chemical compounds (the so-called QSAR and SAR methods), but it may also be possible to use allometric principles to transfer rates of interaction processes and concentration factors between a chemical and one or a few organisms to other

Table 8.3 Overview of the Most Relevant Environmental Properties of Organic Compounds and Their Interpretation

Property	Interpretation
Water solubility	High water solubility corresponds to high mobility
K_{ow}	High K_{ow} means that the compound is lipophilic. It implies that it has a high tendency to bioaccumulate and be sorbed to soil, sludge, and sediment. BCF and K_{oc} are correlated with K_{ow}.
Biodegradability	This is a measure of how fast the compound is decomposed to simpler molecules. A high biodegradation rate implies that the compound will not accumulate in the environment, while a low biodegradation rate may create environmental problems related to the increasing concentration in the environment and the possibilities of a synergistic effect with other compounds.
Volatilization, vapor pressure	High rate of volatilization (high vapor pressure) implies that the compound will cause an air pollution problem
Henry's constant, He	He determines the distribution between the atmosphere and the hydrosphere.
pK	If the compound is an acid or a base, pH determines whether the acid or the corresponding base is present. As the two forms have different properties, pH becomes important for the properties of the compounds.

organisms. This section focuses on these methods and attempts to give a brief overview on how these methods can be applied and what approximate accuracy they can offer. A more detailed overview of the methods can be found in Jørgensen et al. (1997).

It may be interesting here to discuss the obvious question: Why is it sufficient to estimate a property of a chemical in an ecotoxicological context with 20% or 50% or higher uncertainty? Ecotoxicological assessment usually produces an uncertainty of the same order of magnitude, which means that the indicated uncertainty may be sufficient from the modelling viewpoint. But can a result with such an uncertainty be used? The answer is often yes, because in most cases we want to assure that we are (very) far from a harmful or very harmful level. We use (see also Section 8.2) a safety factor of 10–1000 (most often 50–100). When we are concerned with very harmful effects, such as the collapse of an ecosystem or a health risk for a large human population, we will inevitably select a safety factor that is very high. In addition, our lack of knowledge about synergistic effects and the presence of many compounds in the environment at the same time force us to apply a very high safety factor. In such a context, we usually go for a concentration in the environment that is magnitudes lower than those corresponding to a slightly harmful effect or considerably lower than the NEC. It is analogous to civil engineers constructing bridges. They make very sophisticated calculations (develop models) that account for wind, snow, temperature changes, and so on, and afterwards they multiply the results by a safety factor of 2 to 3 to ensure that the bridge will not collapse. They use safety factors because the consequences of a bridge collapse are unacceptable.

The collapse of an ecosystem or a health risk to a large human population is also completely unacceptable. So, we should use safety factors in ecotoxicological modelling to account for the uncertainty. Due to the complexity of the system, the simultaneous presence of many compounds, and our present knowledge or rather lack of knowledge, we should use 10–100, or even 1000, as a safety factor. If we use safety factors that are too high, then the risk is only that the environment will be less contaminated at maybe a higher cost. Besides, there are no alternatives to the use of safety factors. We can increase our ecotoxicological knowledge step by step, but it will take decades before it may be reflected in considerably lower safety factors. A measuring program of all processes and components is impossible due to the high complexity of the ecosystems. This does not imply that we should not use the information of measured properties that are available. Measured data are usually more accurate than estimated data. Furthermore, the use of measured data within the network of estimation methods improves the accuracy of estimation methods. Several handbooks on ecotoxicological parameters are fortunately available. The most important references were listed in Section 8.2. Estimation methods for the physical-chemical properties of chemical compounds were already applied 40 to 60 years ago, as they were urgently needed in chemical engineering. They are based on contributions to a focal property by molecular groups and the molecular weight: the boiling point, the melting point, and the vapor pressure as function of the temperature. These are examples of properties that are frequently estimated in chemical engineering by these methods. In addition, a number of auxiliary properties results from these estimation methods, such as the critical data and the molecular volume. These properties may not have a direct application as ecotoxicological parameters in ERA, but they are used as intermediate parameters as a basis for estimating other parameters.

The water solubility; the partition coefficient octanol-water, K_{ow}; and Henry's constant are crucial parameters in our network of estimation methods, because many other parameters are well correlated with these two parameters. These three properties can be found for a number of compounds or estimated with reasonably high accuracy using knowledge of the chemical structure — the number of various elements, the number of rings, and the number of functional groups. In addition, there is a good relationship between water solubility and K_{ow} (Figure 8.10). Recently, many good estimation methods for these three core properties have been developed.

During the last 20 years, several correlation equations have been developed based upon a relationship between the water solubility, K_{ow}, or Henry's constant on the one hand, and physical, chemical, biological, and ecotoxicological parameters for chemical compounds on the other. The most important of these parameters are the adsorption isotherms soil–water; the rate of the chemical degradation processes such as hydrolysis, photolysis, and chemical oxidation; the BCF, the ecological magnification factor (EMF); the uptake rate; excretion rate; and a number of ecotoxicological parameters. Both the ratio of concentrations in the sorbed phase and in water at equilibrium, K_a, and BCF, — defined as the ratio of the concentration in an organism and in the medium (water for aquatic organisms) at steady state presuming that both the medium and the food are contaminated — may often be estimated with relatively good accuracy from expressions like K_a, K_{oc}, or BCF $= a \log K_{ow} + b$. K_{oc} is the ratio between the concentration in soil consisting of 100% organic carbon and in water at equilibrium between the two phases. Numerous expressions with different a and b values have been published (Jørgensen et al., 1991, 1997, 2000; Jørgensen, 2000). Some of these relationships are shown in Table 8.4 and Figure 8.11.

The biodegradation in waste treatment plants is often of particular interest, in which case the %BOD may be used. It is defined as the 5-day BOD as a percentage of the theoretical BOD. It may also be indicated as the BOD_5-fraction; for instance, a BOD_5-fraction of 0.7 means that BOD_5 corresponds to 70% of the theoretical BOD. It is also possible to find an indication of BOD_5 percentage removal in an activated sludge plant.

Table 8.4 Regression Equations for Estimation of the BCF

Indicator	Relationship	Correlation Coefficient	Range (Indicator)
K_{ow}	$\log BCF = -0.973 + 0.767 \log K_{ow}$	0.76	$2.0*10\text{-}2 - 2.0*106$
K_{ow}	$\log BCF = 0.7504 + 1.1587 \log K_{ow}$	0.98	$7.0 - 1.6*104$
K_{ow}	$\log BCF = 0.7285 + 0.6335 \log K_{ow}$	0.79	$1.6* - 1.4*104$
K_{ow}	$\log BCF = 0.124 + 0.542 \log K_{ow}$	0.95	$4.4 - 4.2*107$
K_{ow}	$\log BCF = -1.495 + 0.935 \log K_{ow}$	0.87	$1.6 - 3.7*106$
K_{ow}	$\log BCF = -0.70 + 0.85 \log K_{ow}$	0.95	$1.0 - 1.0*107$
K_{ow}	$\log BCF = 0.124 + 0.542 \log K_{ow}$	0.90	$1.0 - 5.0*107$
S (µg/L)	$\log BCF = 3.9950 - 0.3891 \log S$	0.92	$1.2 - 3.7*107$
S (µg/L)	$\log BCF = 4.4806 - 0.4732 \log S$	0.97	$1.3 - 4.0*107$
S (µmol/L)	$\log BCF = 3.41 - 0.508 \log S$	0.96	$2.0*10\text{-}2 - 5.0*103$

BCF - Biological Concentration Factor

Biodegradation is, in some cases, very dependent on the concentration of microorganisms. Therefore, it may be beneficial to indicate it as a rate coefficient relative to the biomass of the active microorganisms in the units mg/(g dry wt 24h).

In the microbiological decomposition of xenobiotic compounds, an acclimatization period from a few days to one to two months must pass before the optimum biodegradation rate can be achieved. The two types of biodegradation are primary and ultimate. Primary biodegradation is any biologically induced transformation that changes the molecular integrity. Ultimate biodegradation is the biologically mediated conversion of an organic compound to inorganic compound and products associated with complete and normal metabolic decomposition.

The biodegradation rate is expressed by a wide range of units:

1. First-order rate constant - (1/24h)
2. Half-life time - (days or hours)
3. mg per g sludge per 24h - (mg/(g 24h))
4. mg per g bacteria per 24 h - (mg/(g 24h))
5. ml of substrate per bacterial cell per 24h - (ml/(24h cells))
6. mg COD per g biomass per 24 h (mg/(g 24h))
7. ml of substrate per gram of volatile solids inclusive microorganisms - (ml/(g 24h))

8. $BODx/BOD_8$, that is, the biological oxygen demand in x days compared with complete degradation (-), named the BODx-coefficient.
9. BODx/COD, that is, the biological oxygen demand in x days compared with complete degradation, expressed by means of COD(-)

The biodegradation rate in water or soil is difficult to estimate because the number of microorganisms varies several orders of magnitude from one type of aquatic ecosystem to the next and from one type of soil to the next.

Models enlisting artificial intelligence have been used as a promising tool to estimate this important parameter. However, a (very) rough, first estimation can be made on the basis of the molecular structure and the biodegradability. The following rules can be used to set up these estimations:

1. Polymer compounds are generally less biodegradable than monomer compounds. 1 point for a molecular weight > 500 and equal to 1000, 2 points for a molecular weight > 1000.
2. Aliphatic compounds are more biodegradable than aromatic compounds. 1 point for each aromatic ring.
3. Substitutions, especially with halogens and nitro groups, will decrease the biodegradability. 0.5 points for each substitution, although 1 point if it is a halogen or a nitro group.
4. Introduction of a double or triple bond generally means an increase in the biodegradability (double bonds in aromatic rings are not included in this rule). 1 point for each double or triple bond.
5. Oxygen and nitrogen bridges (-O- and -N- (or =)) in a molecule will decrease the biodegradability. 1 point for each oxygen or nitrogen bridge.
6. Branches (secondary or tertiary compounds) are generally less biodegradable than the corresponding primary compounds. 0.5 point for each branch.

Sum the total number of points and use the following classification:

= 1.5 points: The compound is readily biodegraded. More than 90% will be biodegraded in a biological treatment plant.
2.0– 3.0 points: The compound is biodegradable. Probably about 10%– 90% will be removed in a biological treatment plant. BOD_5 is 0.1–0.9 of the theoretical oxygen demand.

3.5– 4.5 points: The compound is slowly biodegradable. Less than 10% will be removed in a biological treatment plant. $BOD_{10} = 0.1$ of the theoretical oxygen demand.

5.0–5.5 points: The compound is very slowly biodegradable. It will hardly be removed in a biological treatment plant and a 90% biodegradation in water or soil will take 6 months.

= 6.0 points: The compound is refractory. The half-life time in soil or water is counted in years.

Several useful methods for estimating biological properties are based upon the similarity of chemical structures. If we know the properties of one compound, then they may be used to find the properties of similar compounds. For example, if we know the properties of phenol, which is named the parent compound, then they may be used to more accurately estimate the properties of monochloro-phenol, dichloro-phenol, trichloro-phenol, and so on, as well as for the corresponding cresol compounds. Estimation approaches based on chemical similarity generally produce a more accurate estimation, but they are also more cumbersome to apply as they cannot be used because each estimation has a different starting point; namely the parent compound with its own particular properties.

Allometric estimation methods presume (Peters, 1983) there is a relationship between the value of a biological parameter and the size of the affected organism. These estimation methods were presented in Section 2.9, as they are closely related to the energy balances of organisms. The toxicological parameters LC_{50}, LD_{50}, MAC, EC, and NEC can be estimated from a wide spectrum of physical and chemical parameters, although these estimation equations generally are more inaccurate than the estimation methods for physical, chemical, and biological parameters. Both molecular connectivity and chemical similarity usually offer better accuracy for estimating toxicological parameters.

The various estimation methods may be classified into two groups:

A. General estimation methods based on an equation of general validity for all types of compounds: Some of the constants may be dependent on the type of chemical compound or calculated by adding contributions (increments) based on chemical groups and bonds.

B. Estimation methods valid for a specific class of chemical compounds such as aromatic amines, phenols, aliphatic hydrocarbons, and so on. The property of at least one key compound is known. Based upon the structural differences between the key compounds and all other compounds of the considered type — for instance, two chlorine atoms have substituted hydrogen in phenol to get 2,3-dichloro-phenol — and the correlation between the structural differences and the differences in the considered property, the properties for all compounds of the considered class can be found. These methods are therefore based on chemical similarity.

Methods of class B are generally more accurate than methods of class A, but they are more difficult to use because it is necessary to find the right correlation for each chemical type. Furthermore, the requested properties should be known for at least one key component, which sometimes may be difficult when a series of properties are needed. If estimation of the properties for a series of compounds belonging to the same chemical class is required, then it is tempting to use a suitable collection of class B methods.

.Methods of class A form a network that facilitates linking the estimation methods together in a computer software system such as EEP (see www.ecologicalmodel.net), which contains many estimation methods. The relationship between the two properties is based on the average result obtained from a number of different equations found in the literature. There is, however, a price for using such "easy-to-go" software. The accuracy of the estimations is not as good as with the more sophisticated methods based upon similarity in chemical structure, but in many contexts, particularly modelling, the results found by EEP can offer sufficient accuracy. It is always useful to come up with a first intermediate guess.

With this software it is also possible to start the estimations from the properties of the chemical compound already known. The accuracy of the estimation from using the software can be improved considerably by having knowledge about a few key parameters such as the boiling point and Henry's constant. Because it is possible to get software that estimates Henry's constant and K_{ow} with higher accuracy than EEP, a combination of separate estimations of these two parameters prior to using EEP are recommended. Another possibility would be to estimate a couple of key properties using chemical similarity methods and then use these estimations as known values in EEP. These methods for improving estimation accuracy will be discussed in the next section. The network of EEP as an example of these estimation networks is illustrated in Figure 8.12. EEP is a network of class A methods, so the accuracy of its estimations is not as high as those obtained by the more specific class B methods. However, using EEP makes it possible to estimate the most pertinent properties directly and relatively from the structural formula. The latest version of EEP contains an estimation of the biodegradation based on a further development of the system previously presented.

EEP is based on average values of results obtained by simultaneous use of several estimation methods for most of the parameters. It implies increased accuracy of the estimation, mainly because it gives a

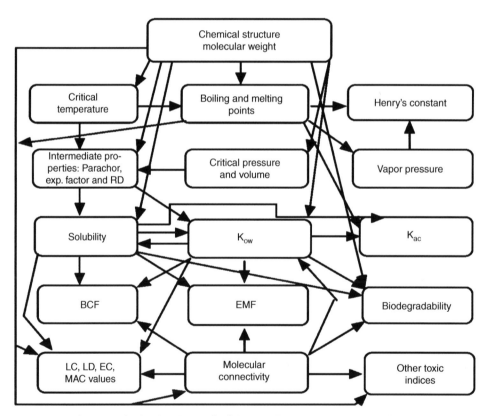

FIGURE 8.12 The network of estimation methods in EEP. The arrow represents a relationship between two or more properties. BCF - Biological Concentration Factor; EEP - Software denoted Estimation of Environmental Parameters; MAC - maximum allowable concentration; EMF – ecological magnification factor

reasonable accuracy for a wider range of compounds. If several methods are used in parallel, then a simple average of the parallel results have been used in some cases, while a weighted average is used in cases where it is beneficial for the overall accuracy of the program. When parallel estimation methods give the highest accuracy for different classes of compounds, use of weighting factors seems to offer a clear advantage. It is generally recommended to apply as many estimation methods as possible for a given case study to increase the overall accuracy. If the estimation by EEP can be supported by other recommended estimation methods, then it is strongly recommended to use those methods.

8.6. Ecotoxicological Case Study I: Modelling the Distribution of Chromium in a Danish Fjord

This case study requires an FTE-model combining a fate model type AII (a specific ecosystem is considered) with an effect model type BI (focus on the organism level). The structure of the model is a class 2 (see Section 8.3), as it focuses on a steady-state situation, although the spatial distribution is also considered. Only one trophic level is considered. It is an illustrative case study because:

1. The case study shows what can be achieved by a simple model.
2. It is possible to validate the results set up eight years previously. Model validation is necessary for development of reliable models, which was the case here. Since there are only a few cases of validated predictions, it was considered significant to include this case study.
3. The model development clearly shows how important it is to know the system and its processes if the right model with the right simplifications is to be selected.

A tanning plant discharged wastewater with a high concentration of chromium(III) into the fjord for decades. In 1958, production was expanded significantly and there was a pronounced increase in the chromium concentration of the sediment (Mogensen & Jørgensen, 1979). For further details see Mogensen, 1978.

It was the goal of this investigation to set up a model for the distribution of chromium in the fjord based on analysis of chromium in phytoplankton, zooplankton, fish, benthic fauna, water (dissolved as well as suspended), and sediment. Already, during the first phase of the investigation, it was clear that the phytoplankton, zooplankton, and fish were not contaminated by chromium, while the sediment and the benthic fauna clearly showed a raised concentration of chromium. This was easy to explain: the chromium(III) precipitates as hydroxide by contact with the seawater that has a pH of 8.1 compared with 6.5–7.0 for the wastewater.

The overall analysis showed that the important processes include:

1. Settling of the precipitated chromium(III) hydroxide and other insoluble chromium compounds
2. Diffusion of the chromium, mainly as suspended matter, throughout the fjord caused mainly by tides; this implies that an eddy diffusion coefficient has to be found
3. Bioaccumulation from sediment to benthic fauna

Processes 1 and 2 can be combined in one submodel, while process 3 requires a separate submodel.

The distribution model is based on the equations of advection and diffusion processes, which have been expanded to include settling:

$$\partial C/\partial t = D * \partial^2 C/\partial X^2 - Q * \partial C/\partial X - K * (C - C_0)/h \tag{8.6}$$

where C is the concentration of total chromium in water in mg/L; C_0 is the solubility of chromium(III) in seawater at pH = 8.1 in mg/L; Q is the inflow to the fjord = outflow by advection (m^3/24h); D is the eddy diffusion coefficient considering the tide (m^2/24h); X is the distance from the discharge point in m; K is the settling rate in m/24h; and h is the mean depth in m.

For a tidal fjord such as Faaborg Fjord with only insignificant advection, Q may be set to 0. Since the tanning plant has discharged a near constant amount of chromium(III) during the last 20 years, we can consider the stationary situation:

$$\partial C/\partial t = 0 \tag{8.7}$$

Equation (8.6) therefore takes the form:

$$D * \partial^2 C/\partial X^2 = K * (C - C_0)/h \tag{8.8}$$

This second-order differential equation has an analytical solution. Cu = the total discharge of chromium in g per 24 h is known. This information is used together with F = cross-sectional area (m^2) to state the boundary conditions. The following expression is obtained as an analytical solution:

$$C - C_0 = (Cu/F) * \sqrt{(h/D * K)} * \exp[-\sqrt{(K/h * D)} * X] + IK \qquad (8.9)$$

F is known only approximately in this equation due to the nonuniform geometry of the fjord. The total annual discharge of chromium is 22,400 kg. Both the consumption of chromium by the tanning factory and the analytical determinations of the wastewater discharged by the factory confirm this number. The depth, h, is about 8 m on average and IK is an integration constant.

Equation (8.9) may be transformed to:

$$Y = K * (C - C_0) = (Cu/F) * \sqrt{(h * K/D)} * \exp[-\sqrt{K/h * D}] * X + K * IK \qquad (8.10)$$

Y is the amount of chromium (g) settled per 24h and per m^2. Eq. (10) gives Y as a function of X.

Y is, however, known from the sediment analysis. A typical chromium profile for a sediment core is shown in Figure 8.13. We know that the increase in the chromium concentration took place about 25 years before the model was built, so it is possible to find the sediment rate in mm or cm per year: 75mm/25 y = 3 mm/y. Because we know the concentration of chromium in the sediment, we can calculate the amount of chromium settled per year, or 24 h, and per m^2, and this is Y. The Y values found by this method are plotted versus X in Figure 8.14.

A nonlinear regression analysis was used to fit the data to an equation of the following form:

$$Y = a * \exp(-bX + c) \qquad (8.11)$$

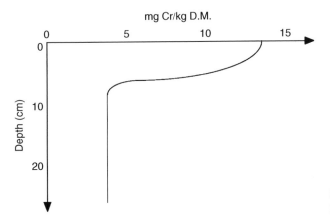

FIGURE 8.13 Typical chromium profile of sediment core.

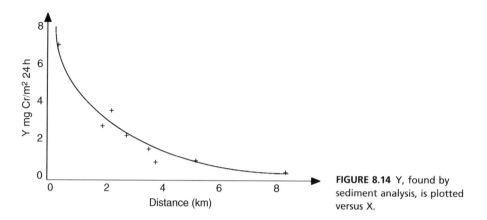

FIGURE 8.14 Y, found by sediment analysis, is plotted versus X.

a, b, and c are constants, which are found by the regression analysis.

Table 8.5 shows Y = f(X). Table 8.6 lists the estimations of a, b, and c found by the statistical analysis. Table 8.7 illustrates the result of the statistical analysis, which shows that the model found with the values of a,

Table 8.5   Y versus X

Station Number	g Cr/m² year	(Y) mg Cr/m² day	(X) Distance from discharge point (m)
1	2.55	7.0	500
2	2.39	6.5	500
3	1.47	4.0	1500
4	0.35	1.0	2750
5	0.78	2.1	2750
6	0.14	0.38	5250
7	0.03	0.082	8500
8	0.20	0.55	3250
9	0.06	0.16	3500
10	0.58	1.6	2000

Table 8.6   Estimations of a, b, and c

	Estimate	Asymptotic St. error
a	0.009909	0.00084
b	0.000723	0.00015
c	+0.000081	0.00045

Table 8.7 Statistical Analysis

	Degree of freedom	Sum of squares	Mean square
Model	3	0.00011337	0.00003779
Residual	6	0.00000233	0.00000033
Total	9		
	F = 114.5		

b, and c from Table 8.6 have a very high probability. The F value found is 114.5, while an F value with a probability of 0.9995 is only 30.4.

Table 8.8 translates the constants a, b, and c into parameters of the model. D is found on the basis of an average value for K, 1.6 m/24h. This value is found from the definition of Y. Y is known as shown in Eq. (8.11). Furthermore, C_o (the solubility of chromium(III) hydroxide) is known from the solubility constant and pH = 8.1 to be 0.2 mg/m^3, and as C is measured for all stations, K may be found from:

$$K = Y/(C - C_O) \tag{8.12}$$

The settling rates found by this method are shown in Table 8.9.

As seen from Table 8.9, the settling rate is approximately the same at three of the five stations. Stations 6 and 7 are given a lower value. It should be expected that the settling rate decreases with increasing distance from the discharge point. Yet, it should not be forgotten that the determination of the chromium concentration in the water is not very accurate, as the concentration is low. K should be compared with

Table 8.8 Parameters

From the regression analysis we have:

$$\frac{Cu * (hK)^{1/2}}{FD} = 0.00990 = a$$

and

$$\frac{(K)^{1/2}}{h * D} = 0.000723 = b$$

which gives

$$Cu * h/F = a/b = 13.7$$

F = 35,800 m^2, which seems a reasonable average value of the cross-sectional area. From analysis of C at stations 2, 5, 6, 7, and 8, we get an estimation of K since

$$Y = \frac{mgCr}{m^2 day} = K(C - C_O)(C_O \text{ is found to be } 0.2 mg/m^3)$$

Table 8.9 Settling rates

Station	mg Cr/m²day	C − C₀ (mg m⁻³)	K (m day⁻¹)
2	6.5	2.5	2.6
5	2.1	0.9	2.3
6	0.4	0.6	0.7
7	0.1	0.2	0.5
8	0.6	0.3	2.0

settling rates of phytoplankton and detritus, which are in the range of 0.1 to 0.5 m/24h. It is expected that the settling rate for chromium(III) hydroxide is higher than the settling for phytoplankton and detritus, which is confirmed by the results in Table 8.9.

The value for the diffusion coefficient found from the settling rate corresponds to 4.4 m^2/s — a reasonable value compared with other D values from similar situations (estuaries). The value for F is based on a width slightly more than the width of the inner fjord, but as a weighted average for the inner and outer fjord, it seems a reasonable value.

Integration from 0 to infinity over a half circle area results in 22 t of chromium(III); that is, almost all the chromium discharged may be explained by the model assuming that the distribution takes place over a half circle area.

All in all, it may be concluded that the distribution model produces acceptable results, particularly of the high sediment chromium concentration. The use of sediment analysis, as demonstrated, is recommended for developing a distribution model for a component that settles readily.

The second submodel focuses on the chromium contamination of the benthic fauna. It may be shown (Jørgensen, 1979) that under steady-state conditions the relation between the concentration of a contaminant in the nth link in the food chain and the corresponding concentration in the $(n − 1)^{th}$ link can be expressed using the following equation:

$$C_n = (MY(n) * C_{n-1} * YT(n))/(MY(n) * YF(n) - RESP(n) + EXC(n)) = K' * C_{n-1}, \quad (8.13)$$

where MY(n) = the maximum growth rate for n^{th} link of the food chain (1/day), C_n = the chromium concentration in the nth link of the food chain (mg/kg), and C_{n-1} = the chromium concentration in the $(n − 1)^{th}$

link of the food chain (mg/kg). YT(n) = the utility factor of chromium in the food for the nth link of the food chain (-), and YF(N) = the utility factor of the food in the nth link of the food chain (-). RESP(n) = the respiration rate of the nth link of the food chain (1/day) and EXC(n) = the excretion rate of chromium for the nth link of the food chain (1/day).

For some species present in Faaborg Fjord these parameter values can be found in the literature (Jørgensen et al., 1991, 2000). The mussel *Mytilus edulis* was found at almost all the stations and the following parameters are valid: YT(n) and YF(n) are found for other species:

MY(n) = 0.03 1/day
YT(n) = 0.07
YF(n) = 0.66
RESP(n) = 0.001 1/day
EXC(n) = 0.04 1/day

Using these values gives K' = 0.036 for *M. edulis*. In other words, the concentration of chromium in *M. edulis* should be expected to be 0.036 times the concentration in the sediment.

Twenty-one mussels from Faaborg Fjord were analyzed and by statistical analysis it was found that the relation between the concentration in the sediment and in the mussels is linear:

$$C_n = C_{n-1} * K' \tag{8.14}$$

where K' was found to be 0.015 ± 0.002. The discrepancy from the theoretical value is fully acceptable, when it is considered that the parameters are found in the literature and they may not be exactly the same values for all environments for all conditions. In general, biological parameters can only be considered approximate values. The relatively low standard deviation of the observed K' value confirms, however, the relation used. It is recommended that the highest K' value = 0.036 is used when the model is used for environmental management, because that way the uncertainty of the K' value is "used to the benefit of the environment."

The model was used as a management tool and the acceptable level of the chromium concentration in the sediment of the most polluted area was assessed to be 70 mg/kg dry matter. That corresponds to a chromium concentration of 70*0.036 = 2.5 mg/kg dry biomass in mussels, or about 2.5 times the concentration found in uncontaminated areas of the open sea. This was considered the NOEC and accepted by the environmental authorities of the district (council).

The distribution model is now used to assess the total allowable discharge of chromium (kg/y) to see if the chromium concentration in the sediment should be reduced to 70 mg/kg dry matter in the most polluted areas (stations 1 and 2). It was found that the total discharge of chromium should be reduced to 2000 kg or less per year to achieve a reduction of about 92%.

Consequently, the environmental authorities required the tanning plant to reduce its chromium discharge to ≤2000 kg per year. The tanning plant has complied with the standards since 1980.

A few samples of sediment (4) and mussels (7) taken from 1987 to 1988 have been analyzed and used to validate the model results (Table 8.10). Settled chromium in $mg/m^2/day$ was found on the basis of the previously determined sedimentation rate. The model validation was fully acceptable as the deviation between the prediction and observed average values for chromium in mussels is approximately 12%.

8.7. Ecotoxicological Case Study II: Contamination of Agricultural Products by Cadmium and Lead

Agricultural products are contaminated by lead and cadmium originating from air pollution, the application of sludge from municipal wastewater plants as a soil conditioner, and from the use of fertilizers.

The uptake of heavy metals from municipal sludge by plants has previously been modelled (see Jørgensen, 1976b). Depending on the soil composition, it is possible to find a distribution coefficient for various

Table 8.10 Validation of the Prognosis

| Item | mg per kg dry matter | | |
	Observed Value	Range	Predicted Value
Cr in sediment	65	57–81	70
Cr in mussels	2.2	1.4–4.5	2.5
mg Cr/m^2 day	0.59	0.44–0.830.67	

heavy metal ions, that is, the fraction of the heavy metal dissolved in the soil–water relative to the total amount. The distribution coefficient was found by examining the dissolved heavy metals relative to the total amount for several different types of soil. The correlation between pH, the concentration of humic substances, clay, and sand in the soil, as well as the distribution coefficient were also determined. The uptake of heavy metals was considered a first-order reaction of the dissolved heavy metal.

This model does not consider:

1. Direct uptake from atmospheric fallout onto the plants
2. Other contamination sources such as fertilizers and the long-term release of heavy metal bound to the soil and the unharvested parts of the plants

The objective of the model is to include these sources in a model for lead and cadmium contamination of plants. This model is a fate type A3 (see Section 8.1). Published data on lead and cadmium contamination in agriculture are used to calibrate and validate the model, which is intended to be used for a more applicable risk assessment for the use of fertilizers and sludge that contains cadmium and lead as contaminants. The structure of the model is type 3 (see Section 8.3).

The basis for the model is the lead and cadmium balance for average Danish agricultural land. Figures 8.15 and 8.16 illustrate the balances, modified from Andreasen (1985) and Knudsen and Kristensen (1987), to account for the changes of the mass balances year 1999. The atmospheric fallout of lead has gradually been reduced due to reduction of lead concentration in gasoline, while the most important source of cadmium contamination is fertilizer. The latter can only be reduced by using less contaminated sludge and phosphorus ore for the production of phosphorus fertilizer. The amounts of lead and cadmium coming from domestic animals and plant residues after harvest are significant contributions.

8.7.1. The Model

Figure 8.17 shows a conceptual diagram of the Cd-model. STELLA software was used to construct a model with four state variables: Cd-bound, Cd-soil, Cd-detritus, and Cd-plant. An attempt was made to use one or two state variables for cadmium in the soil, but to develop acceptable agreement between data and model output, three state variables were

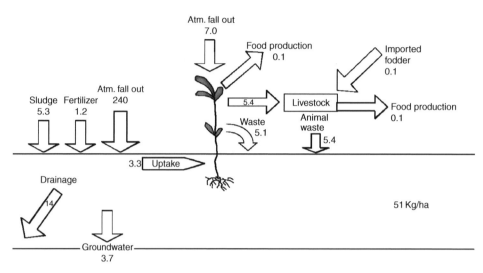

FIGURE 8.15 Lead balance of average Danish agriculture land. All rates are g Pb/ha y.

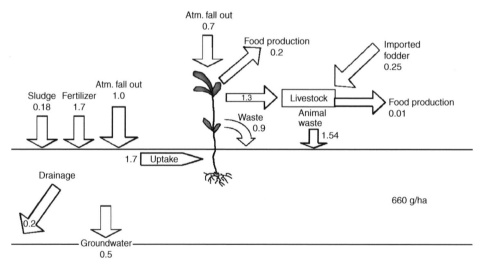

FIGURE 8.16 Cadmium balance of average Danish agriculture land. All rates are g Cd/hay.

needed. This can be explained by the presence of several soil components that bind the heavy metal differently (Christensen, 1981, 1984; EPA Denmark, 1979; Hansen & Tjell, 1981; Jensen & Tjell, 1981; Chubin & Street, 1981). Cd-bound covers the cadmium bound to minerals and refractory material, Cd-soil covers the cadmium bound by

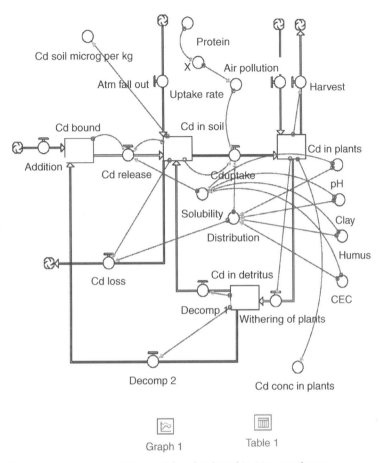

FIGURE 8.17 Conceptual diagram of the model as developed in STELLA software.

adsorption and ion exchange, and Cd-detritus is the cadmium bound to organic material with a wide range of biodegradability. The forcing functions are airpoll, Cd-air, Cd-input, yield, and loss.

The atmospheric fallout is known, and the allocation of this source to the soil (airpoll) and the plants (Cd-air) follows Hansen and Tjell (1981) and Jensen and Tjell (1981). Cd-input covers the heavy metal in the fertilizer, which comes as a pulse at day 1 and afterward with a frequency of every 180 days (Table 8.11). The yield corresponds to the harvested part of the plants, which is also expressed as a pulse function at day 180, and afterward with an occurrence every 360 days. In this table it is 40% of the plant biomass (Table 8.11).

Table 8.11   Model Equations

```
Cd-detritus = Cd-detritus + dt * ( Cd-waste -
  mineralization - minquick )
INIT(Cd-detritus) = 0.27
Cd-plant = Cd-plant + dt * ( Cduptake - yield - Cd-waste +
  Cd-air )
INIT(Cd-plant) = 0.0002
Cd-soil = Cd-soil + dt * ( -Cduptake - loss + transfer +
  minquick + airpoll )
INIT(Cd-soil) = 0.08
Cdtotal = Cdtotal + dt * ( Cd-input - transfer +
  mineralization )
INIT(Cdtotal) = 0.19
airpoll =0.0000014
Cd-air = 0.0000028+STEP(-0.0000028,180)+STEP
  (+0.0000028,360)+STEP(-0.0000028,540)+STEP
  (+0.0000028,720)+STEP(-0.0000028,900)
Cd-input = PULSE(0.0014,1,180)
Cduptake = distributioncoeff*Cd-soil*uptake rate
Cd-waste = PULSE(0.6*Cd-plant,180,360)+PULSE(0.6*Cd-
  plant,181,360)
CEC = 33
clay = 34.4
distributioncoeff =0.0001*(80.01-6.135*pH-
  0.2603*clay-0.5189*humus-0.93*CEC)
humus = 2.1
loss = 0.01*Cd-soil*distributioncoeff
mineralization = 0.012*Cd-detritus
minquick = IF TIME_180 THEN 0.01*Cd-detritus ELSE
  0.0001*Cd-detritus
pH = 7.5
plantvalue = 3000*Cd-plant/14
protein = 47
solubility = 10o(+6.273-
  1.505*pH+0.00212*humus+0.002414*CEC)*112.4*350
transfer = IF Cd-soil<solubility THEN 0.00001*Cdtotal
  ELSE 0.000001*Cdtotal
uptake rate = x +STEP(-x,180)+STEP(x,360)+STEP(-x,540)+
  STEP(x,720)+STEP(-x,900)
x = 0.002157*(-0.3771+0.04544*protein)
yield = PULSE(0.4*Cd-plant,180,360)+PULSE(0.4*Cd-
  plant,181,360)
```

The loss covers transfer to the soil and groundwater below the root zone. It is expressed as a first-order reaction with a rate coefficient dependent on the distribution coefficient found from the soil composition and pH, according to the correlation found by Jørgensen (1976b). The rate constant is dependent on the hydraulic conductivity of the soil. Here in Table 8.11, the constant 0.01 reflects the dependence of the hydraulic conductivity.

The transfer from Cd-bound to Cd-soil indicates the slow release of cadmium due to a slow decomposition of the refractory material to which cadmium is bound. The cadmium uptake by plants is expressed as a first-order reaction, where the rate is dependent on the distribution coefficient, as only dissolved cadmium can be taken up. It is also dependent on the plant species. It will be shown that the uptake is a step function where grass is 0.0005 during the growing season and zero after the harvest until the next growing season starts. Cd-waste covers the transfer of plant residues to detritus after harvest. It is a pulse function, which is 60% of the plant biomass, as the remaining 40% has been harvested.

Cd-detritus covers a wide range of biodegradable matter and the mineralization is accounted for in the model by two mineralization processes: one for Cd-soil and one for Cd-total.

8.7.2. Model Results

Data from Jensen and Tjell (1981) and Hansen and Tjell (1981) were used for model calibration and validation. This phase of the modelling procedure revealed that three state variables for heavy metal in soil were needed to get acceptable results. It was particularly difficult to obtain the right values for heavy metal concentrations the second and third year after municipal sludge had been used as a soil conditioner. This use of models may be called experimental mathematics or modelling, where simulations with different models are used to deduce which model structure should be preferred. The results of experimental mathematics must be explained by examining the processes involved and can be referred to the references Jensen and Tjell.

The results of the validation demonstrate good agreement between observations and model prediction (Figure 8.18), especially considering the low model complexity. Wider use of the model requires more data from experiments with many plant species to test the model applicability. It can

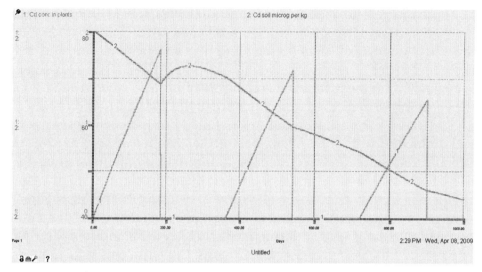

FIGURE 8.18 Cadmium concentration in plants and soil in μg/kg dry matter. The harvest takes place at day 180, 540, and 900. The cadmium concentration according to observations was found at the three harvests to be, respectively, 1.7, 1.1, and 0.8μg/kg dry matter. The cadmium in soil is reduced over the simulation period from about 80 μg/kg dry matter in soil to about 45 μg/kg dry matter in soil.

be concluded from these results that the model structure must account for at least three state variables for the heavy metal in soil to cover the ability of different soil components to bind the heavy metal by various processes.

The problem modelled is very complex and many processes are involved. On the other hand, an ecotoxicological management model should be somewhat simple and not involve too many parameters. The model can obviously be improved, but it gives at least a first rough picture of the important factors in the contamination of agricultural crops. It is not possible to get accurate results with toxic substance models, but as we want to use somewhat large safety factors, the need for high accuracy is not pressing.

8.8. Fugacity Fate Models

This A1 type of fate model, seen in Section 8.1, is applied mainly to compare two or more chemicals in order to select the least environmentally harmful one or to point out particularly hazardous chemicals. This model type, originally developed by Mackay (1991), has a wide application in environmental chemistry with many different models developed by different authors (SETAC, 1995).

These models are based on the concept of fugacity, $f = c/Z$, where c is the concentration in the considered phase and Z is the fugacity capacity (measured in mol/m^3 Pa or moles/L atm). Fugacity is defined as the escaping tendency, and has the units of pressure (atmosphere or Pa) and is identical to the partial pressure of ideal gases. By equilibrium between two phases, the fugacity of the two phases is equal. If the two Zs are known, then it is possible to calculate the concentrations in the two phases. If there is no equilibrium, then the rate of transfer from one phase to the other is proportional to the difference in fugacity.

If the equation for ideal gases can be applied, we have $pV = nRT$, where n is the number of moles, R the gas constant = 8.314 Pa m^3/mole K, and T is the absolute temperature. This leads to $p = cRT$, and:

$$c = p/RT = f/(RT) \qquad (8.15)$$

By acceptable approximation (application of the equation for ideal gases and the activity is equal to the concentration) the fugacity capacity in air is:

$$Z_a = 1/RT \qquad (8.16)$$

At equilibrium between water and air, the fugacity is the same in the two phases, as already mentioned:

$$c_a Z_a = c_w Z_w \qquad (8.17)$$

where w is used as index for water.

Based upon Henry's law: $p = k_H*y$, where, k_H is Henry's constant and as used above, $p = c_a RT$ and $y = c_w/(c_w + [H_2O])$, we can find the distribution between air and water. The concentration of water in water is with good approximation $1000/18 >> c_w$, which means that we get $p = c_a RT = k_H$ $y = k_H c_w/(c_w + [H_2O])$, $= k_H c_w 18/1000$. Equation (8.17) yields:

$$c_a/c_w = Z_a/Z_w = 18/1000RT \qquad (8.18)$$

It implies, that $Z_w = 1000/18k_H$.

Similarly, the distribution between water and soil (index s) can be applied to find the fugacity capacity of soil:

$$c_s/c_w = Z_s/Z_w = K_{ac} \qquad (8.19)$$

Z_s is found as $Z_w* K_{ac} = 1000 K_{ac}/18k_H$. In a parallel manner Z_o, the fugacity capacity for octanol can be found as $1000 k_H K_{ow}/18$ and the fugacity capacity for biota, Z_b as $1000 k_H BCF/18$. Table 8.12 presents

Table 8.12   Fugacity Capacity in moles/L atm

Phase	In mol/L atm.
Atmosphere	1/RT (R=0.0820)
Hydrosphere	$1000/k_H$ 18
Litosphere (soil)	1000 Koc/18 k_H
Octanol	1000 Kow/18 k_H
Biota	1000 BCF/18 k_H

Note: If the unit moles/m³ Pa is required divide by 101.325.

an overview of the found fugacity capacities in mole/L atm. R = 0.0820 atm L/(moles K), when these units are applied. If m^3 is used as a volume unit and Pa as a unit for pressure, then we get 1 atm = 101 325 Pa and 1 = 1/1000 m^3. It implies that R has the units J/mole K corresponding to the value 0.082 × 101 325/1000 = 8.3J/(moles K). Figure 8.19 shows a conceptual diagram of the most simple fugacity model.

Multimedia models are applied on four levels. An equilibrium distribution (level 1) is found from the known fugacity capacities and equal fugacities in all spheres. If advection and chemical reactions must be included in one or more phases, but the equilibrium is still valid, then we have level 2. The fugacities are still the same in all phases. Level 3 presumes steady state but no equilibrium between the phases. Transfer between the phases is therefore taking place. The transfer rate is proportional to the fugacity difference between the two phases. Level 4 is a dynamic version of level three, which implies that all concentrations and possibly also the emissions change over time.

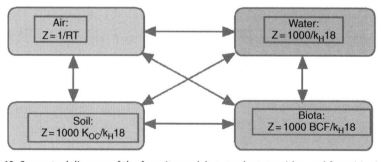

FIGURE 8.19 Conceptual diagram of the fugacity model at steady state with equal fugacities in the four compartments. The concentration can easily be found as c = fZ. The Z values are shown in the diagram.

If the total emission in all phases is denoted M then:

$$M = \Sigma c_i V_i = f \Sigma Z_i V_i \qquad (8.20)$$

where c_i, V_i, and Z_i are concentration, volume, and fugacity capacity of sphere number i. Level 1 and 2 are usually sufficient to calculate the environmental risk of a chemical. For level 1 calculations, the fugacity capacities are found from Table 8.12, and Eq. (8.20) is applied to find f, because the total emission and the volumes of the spheres are known. The concentrations are then easily determined from $c_i = f \, Z_i$. The amounts in the spheres are found from the concentration x the volume of the spheres. Illustration 8.1 presents these calculations.

ILLUSTRATION 8.1

A chemical compound has a molecular weight of 200g/mole and a water solubility of 20 mg/L, which gives a vapor pressure of 1 Pa. The distribution coefficient octanol-water is 10,000 and the $K_{ac} = 4000$. How will an emission of 1000 moles be distributed in a region with an atmosphere of 6×10^8 m^3, a hydrosphere of 6×10^6 m^3, a lithosphere of 50.000 m^3 with a specific gravity of 1.5 kg/L, and an organic carbon content of 10%. Biota (fish) is estimated to be 10 m^3 (specific gravity 1.00 kg/L and a lipid content of 5%). The temperature is presumed to be 20°C.

Solution

Fugacity capacities:

$Z_a = 1/RT = 1/8.314*293 = 0.00041$ mol/m^3 Pa
$Z_w = (20/200)/1 = 0.1$ moles/m^3 Pa
$Z_s = 0.1 \times 0.1 \times 4000 = 40$ moles/m^3 Pa
$Z_{biota} = 0.1 \times 0.05 \times 10,000 = 50$ moles/m^3 Pa
$Z_i V_i = 0.00041 \times 6 \times 10^8 + 0.1 \times 6 \times 10^6 + 40 \times 50,000 + 10 \times 50 = 2846500$ moles /Pa
$f = M/\sum Z_i V_i = 1000/2846500 = 3.51 \times 10^{-4}$

Concentrations:

$c_a = f \, Z_a = 3.51 \times 10^{-4} \times 0.00041 = 1.44 \times 10^{-7}$ moles/m^3
$c_w = f \, Z_w = 3.51 \times 10^{-4} \times 0.1 = 3.51 \times 10^{-5}$ moles/m^3
$c_s = f \, Z_s = 3.51 \times 10^{-4} \times 40 = 1.404 \times 10^{-2}$ moles/m^3
$c_{biota} = f \, Z_{biota} = 3.51 \times 10^{-4} \times 50 = 1.755 \times 10^{-2}$ moles/m^3

Amounts:

$M_a = c_a V_a = 1.44 \times 10^{-7}$ mol/m^3 $\times$ 6 $\times$ 10^{8}m^3 = 86 moles
$M_w = c_w V_w = 3.51 \times 10^{-5}$ mol/m^3 x 6 $\times$ 10^6 m^3 = 211 moles
$M_s = c_s V_s = 1.404 \times 10^{-2}$ mol/m^3 $\times$ 50.000 m^3 = 702 moles
$M_{biota} = c_{biota} V_{biota} = 1.755 \times 10^{-2}$ mol/m^3 $\times$ 10 m^3 = 0.2 moles

The sum of the four amounts is 999.2, which is in good accordance with the total emission of 1000 moles.

Level 2 fugacity models presume a steady-state situation, but with a continuous advection to and from the phases and a continuous reaction (decomposition) of the considered chemical. Steady state implies that input = output + decomposition. The following equation is therefore valid:

$$E + \Sigma Gin_i x c_{i \text{ ind}} = \Sigma Gout_i x c_i + \Sigma V_i c_i k_i \qquad (8.21)$$

where E is the emission and Gin$_i$ i the advection into the phase i, $c_{i \text{ ind}}$ is the concentration in the inflow, Gout$_i$ is the outflow by advection, c_i + is the concentration in the phase, and $V_i c_i$ k$_i$ is the reaction of the considered component in phase i. As $c_i = f Z_i$, we get the following equation:

$$E + \Sigma Gin_i c_{i \text{ ind}} = f(\Sigma Gout_i Z_i + \Sigma V_i c_i k_i) \qquad (8.22)$$

f is the total amount of the component going into phase i divided by ($\Sigma Gout_i Z_i + \Sigma V_i Z_i k_i$). We can often presume that Gin$_i$ = Gout$_i$ denoted G$_i$. The concentration in the phase is usually f Z$_i$. The amount is correspondingly the concentrations in the phase multiplied by the volume. The turnover rate of the compound in phase i is f(G$_i$ Z$_i$ + V$_i$c$_i$ k$_i$). Illustration 8.2 presents these calculations.

ILLUSTRATION 8.2

In an area consisting of 10,000 m^3 atmosphere, 1000 m^3 of water, 100 m^3 of soil, and 10 m^3 of biota, the same chemical compound as mentioned in Illustration 8.2 is emitted. This means that the same fugacity capacities can be applied:

Fugacity capacities:

$Z_a = 1/RT = 1/8.314*293 = 0.00041$ moles/m^3 Pa
$Z_w = (20/200)/1 = 0.1$ moles/m^3 Pa
$Z_s = 0/1 \times 0.1 \times 4000 = 40$ moles/m^3 Pa
$Z_{biota} = 0.1 \times 0.05 \times 10,000 = 50$ moles/m^3 Pa

10,000 m³/24h of air with a contamination corresponding to a concentration of 0.01 moles/m³ and 10m³/24h of water with a concentration of the chemical on 1 mole/m³ is flowing into the area by advection. Within the area, an emission of 500 moles/24h takes place. Decomposition of the chemical takes place with a rate coefficient for air, water, soil, and biota of 0.001 1/24 h, 0.01 1/24h, and 0.1 1/24h for soil and biota. What will the concentration of the chemical be as a result of a steady-state situation in the various spheres?

Solution

The total amount of chemical entering the area is $500 + 100 + 10 = 610$ moles/24h.

The following table summarizes the calculations:

Phase Rate	Volume	Z_i	$G_i Z_i$	$V_i Z_i k_i$	c_i	M_i	conv.
Air	10,000	0.00041	4.1	0.0041	0.00055	5.5	5.48
Water	1000	0.1	1.0	1.0	0.134	134	2.67
Soil	100	40	0	400	53.5	5350	534.8
Biota	10	50	0	50	66.9	669	66.9
			5.1	451			609.9

f is the total in-flowing amount of the chemical divided by $(\Sigma G_i Z_i + \Sigma V_i Z_i k_i) = 610/456.1 = 1.337$. The concentrations are found as $c_i = f Z_i$. The total conversion/24h is 609.9 moles in good accordance with the total input of 610 moles.

Transfer rates between two phases by diffusion are expressed by the following equation (models per unit of area and time):

$$N = D * \Delta f, \tag{8.23}$$

where N is the rate of transfer, D is the diffusion coefficient, and Δf is the difference in fugacity. D is the total resistance for the transfer consisting of the resistances of the two phases in series. Notice that D may be found as $K*Z$, where K is the transfer coefficient and Z is the fugacity capacity defined earlier.

The so-called "unit world model" consists of six compartments: air, water, soil, sediment, suspended sediment, and biota. This simplified model attempts to identify the partition among these six compartments of toxic substances emitted to the environment. The volumes and densities of the unit world and the definition of fugacity capacities are given in Mackay (1991) and in Jørgnesen and Bendoricchio (2001). The average residence time, tr, due to reactions may be found by use of the following equation:

$$tr = M/E \tag{8.24}$$

and the overall rate constant, K, is E/M or 1/tr.

The third level is devoted to a steady-state, nonequilibrium situation, which implies that the fugacities are different in each phase. Equation (8.24) is used to account for the transfer. The D values may be calculated from quantities such as interface areas, mass transfer coefficients (as indicated above, D is the product of the transfer coefficient and the fugacity capacity: $D = K * Z$), release rate of chemicals into phases such as biota or sediment, and Z values, or by use of the estimation methods presented in Section 8.5.

Level 4 involves a dynamic version of level 3, where emissions and thus concentrations, vary with time. This implies that differential equations must be applied for each compartment to calculate the change in concentrations with time, for instance:

$$V_i * dc_i/dt = E_i - V_i * C_i * k_i - \Sigma D_{ij} * \Delta f_{ij} \tag{8.25}$$

Level 1 or 2 is usually sufficient, but if the environmental management problem requires the prediction of the (1) time taken for a substance to accumulate to a certain concentration in a phase after emission has started or (2) length of time for the system to recover after the emission has ceased, then the fourth level must be applied.

This approach has been widely used and a typical example is given by Mackay (1991). It concerns the distribution of PCB between air and water in the Great Lakes. Here k_H is 49.1 and the distribution coefficient for air/water ($= k_H/R*T$) was 0.02. The unit for C is mole/m^3. The fugacity capacity for water $= 1/k_H$ was 0.0204 and the fugacity capacity for air $= 1/ R*T = 0.000404$. The distribution coefficient between water and suspended matter in the water was estimated to be 100,000. As the concentration of suspended matter in the Great Lakes was $2*10^{-6}$ on a volume basis (approximately 4 mg/L, with a density of 2000 g/L), the fraction dissolved was $1/(1 + 0.2) = 0.833$.

9 Individual-Based Models

CHAPTER OUTLINE

A useful approach for modelling ecological systems of interacting organisms is through the use of individual-based models (IBMs). Individuals differ from each other in distinct ways and also from themselves during different stages of their life cycle. More important, they have self-directed motivation, can adapt to changing conditions, and can modify their environment through their actions. An IBM (also called agent-based models, ABM) allows the capture of this feedback within a modelling framework. Properties at higher levels — populations, communities, and ecosystems — emerge from these individual interactions and the interactions with their environment. Without self-direction and adaptation ecological systems would be much easier to model and understand. Such is the case with physical or chemical systems. Since individuals within ecological systems do have self-direction and the ability to adapt, IBMs are one way to capture this complexity.

9.1. History of Individual-Based Models

Early IBMs in ecology include a forest model (Botkin, Janak, & Wallis, 1972) and a fish cohort model (DeAngelis, Cox, & Coutant, 1980). The forest model, JABOWA, has successfully predicted species composition resulting from succession in a mixed-species forest, and has spawned

Fundamentals of Ecological Modelling. DOI: 10.1016/B978-0-444-53567-2.00009-0

a series of related models (Shugart, 1984; Liu & Ashton, 1995). The success of the fish cohort model was due to the inclusion of feedback processes such as cannibalism and competition, and it also precipitated a plethora of off-shoot models (Grimm & Railsback, 2005). Other early applications of ABMs originated in the artificial life literature, such as ECHO, Tierra, and Avida (Parrott, 2008).

The IBM approach was first formalized as a discipline in the article by Huston, DeAngelis, and Post (1988), and has developed considerably since then. What makes an IBM different from a population model? The first question to be addressed about this approach is what makes it different from the standard approaches already being employed in ecological modelling. For one thing, traditional population models were not able to answer specific questions central to ecology regarding mating, foraging, and dispersal because the models treated all individuals within the population as homogeneous; therefore, the entire population acted accordingly without any individual variation. Giving specific traits to each individual allowed for greater variation in the behavior of the population. Furthermore, in traditional models, the agents were unlikely to adapt their behavior throughout the length of the simulation. In IBMs, the heuristics that determine the individual behavior can be updated based on feedback from the success of previous interactions and encounters. Lastly, the environment can be altered by the actions of the individuals performing work on it to survive. In this manner, there is another level of feedback in which the organism influences the environment in which its future success is determined, exerting some degree of self-control on overall higher level system behavior. Such closed loop feedback is an important characteristic of systems ecology as expressed in network environ analysis (Patten 1978a, 1981) or niche construction (Odling-Smee, Laland, & Feldman, 2003).

Uchmanski and Grimm (1996) proposed four criteria that represent the core features of individuality, adaptability, and environmental feedback to consider what distinguishes an IBM from classical models:

1. The degree to which the complexity of the individual's life cycle is reflected in the model
2. Extent to which variability among individuals of the same age is considered

3. Whether real or integer numbers are used to represent the population size
4. Whether or not the dynamics of resources used by individuals are explicitly represented

The implementation of IBMs can affect the paradigm one has about ecology in general. This has led to a new approach called Individual-Based Ecology (IBE) in which the understanding of macroscopic organizational levels (populations, communities, ecosystems, and biosphere) arise from the interactions of microscopic components (agents and individuals). Characteristics of IBE have been proposed by Grimm and Railsback (2005):

- Systems are understood and modelled as collections of unique individuals.
- System properties and dynamics arise from the interactions of individuals with their environment and with each other.
- IBMs are a primary tool for IBE.
- IBE is based on theory.
- Observed patterns are a primary kind of information used to test theories and design models.
- Instead of being framed in the concepts of differential calculus, models are framed by complexity concepts such as emergence, adaptation, and fitness.
- Models are implemented and solved using computer simulations.
- Field and laboratory studies are crucial for developing IBE theory.

9.2. Designing Individual-Based Models

There are three primary aspects to consider when developing an IBM: (1) agent behavior, (2) agent-agent interactions, and (3) environment. The key to IBMs is developing them in a manner in which the adaptive traits can model behavior of real organisms. An adaptive trait is a rule or heuristic that allows the organism to make situation-specific decisions. The traits may be programmed or learned. They determine the choices that the organisms make during each encounter, and are often programmed using a series of IF-THEN statements and loops corresponding to the individual's specific conditions. Following the heuristics does not necessarily lead to an optimal behavior, since not all information is known to always make optimal decisions, but the behaviors are context-dependent and goal directed

(Grimm & Railsback, 2005). For example, rules describing foraging behavior describe how the agent responds to the local conditions (is food available or not) and the agent's internal goals (time since last feeding). The movement pattern may be programmed from a simple random walk function to a more complex environmental assessment and deliberate moves such as seeking a preferred food source, following subtle perceived differences in environmental gradients, or learning from previous encounters with the landscape. A conceptual model used in this instance is called beliefs-desires-intents (BDI), which models the hierarchical progression leading to certain actions. The beliefs contain the background information held by the agent (i.e., food is good, mating is necessary, run from predators, etc.), the desires are the goals, and the intents are the actions taken to achieve these ends (Parrott, 2008).

Agent-agent interactions may be direct such as mating, communication, predation, or resource competition, or indirect through modifications to the environment. An example of indirect interaction is the chemical or physical marking of an area as signals to ensuing agents upon that area. The end result is that group-level dynamics emerge from these agent-agent interactions.

The environment represents the local landscape on which the organisms move and interact. It is typical that the environment has variation but is regular enough for agent learning and adaptation. The environment is commonly modelled as a lattice or network. A lattice approach provides spatial variation such that each cell in the lattice may be heterogeneous and can include environmental variables as well as other agents. Network models forego some spatial capability to focus on the flows or interactions, such as trophic networks. An important, but not surprising, conclusion from IBM work is that the environment can have a substantial influence on the individual behavior and on the overall group dynamics (Parrott, 2008). This is consistent with the perspective of systems ecology, which also places high value on the role of environment, indirect interactions, and holism.

9.3. Emergent versus Imposed Behaviors

As stated previously, one of the important outcomes of IBMs is the unexpected macroscopic behavior that can be viewed from the results of the simulation. This occurs because the agent-agent interactions with adaptive traits and adaptive environmental variables allow for the emergence of novel system behavior. In addition to being unexpected, emergent behaviors can differ from the behavior of individuals and are holistic in the sense that the whole is more than the sum of the parts. Therefore, it is essential not to impose strict, unchanging attributes to the individual's choices. One way to view behavior is that if the attributes are derived from an understanding of process then there is more variability and freedom of option as the behavior unfolds. However, if the attributes are derived from strict empirical observations, that is, fixed parameters from field or laboratory experiments, then the outcome will be predictable since there can be no variation. For example, consider the case of egg production rate in fish in which temperature dependence has been documented (Secor & Houde, 1995). In one model, the rate is fixed based on empirically derived field studies and each individual has this trait. In the second model, the rate is a function of the temperature of the environment in which the individual inhabits at that time. In this manner, the results of the first model are imposed by the rigid constraint of the parameterization, whereas in the second model, variation and adaptability can lead to new patterns, such as clustering of high density populations around warmer pools. Another possibility in model development is to have intermediary outcomes so that the first stage might be imposed, such as egg production rate in each grid cell, but a second choice, based on movement across the environment, can allow for the same kind of clustering if there is a process preference for certain temperature ranges. Overall, the goal for IBMs is to develop rules that are process-based so that the organism can respond accordingly to different situations with flexibility. Therefore, it is important to know some factors that motivate, guide, and orient the behavior of the agents.

9.4. Orientors

A key question in formulating an IBM is determining the characteristics that comprise the individual's decision set. There are a primary set of survival and behavioral functions common to all agents (as modelled as complex adaptive systems). There have been proposals to holistically describe these tendencies in which these systems change over time. One approach worth mentioning identifies six fundamental orientors, which are meant to apply for all complex adaptive agents (Bossel, 1998, 1999). These include:

1. *Existence*: Attention to existential conditions is necessary to ensure the basic compatibility and immediate survival of the system in the normal environmental state.
2. *Effectiveness*: In its efforts to secure scarce resources (energy, matter, information) from, and to exert influence on, its environment, the system should on balance be effective.
3. *Freedom of action*: Ability to cope in various ways with the challenges posed by environmental variety.
4. *Security*: Ability to protect itself from the detrimental effects of variable, fluctuating, unpredictable, and unreliable environmental conditions.
5. *Adaptability*: Ability to change its parameters and/or structure in order to generate more appropriate responses to challenges posed by changing environmental conditions.
6. *Coexistence*: Ability to modify its behavior to account for behavior and interests (orientors) of other systems.

Orientors are defined as dimensions of concern, not specific goals, as they arise from the system interactions and are considered emergent system properties. They function as attractors of the system development and the six orientors are responsive to the six general properties of the environment.

1. *Normal environmental state*: The actual environmental state can vary around this state in a certain range.
2. *Scarce resources*: Resources (energy, matter, information) required for a system's survival are not immediately available when and where needed.
3. *Variety:* Many qualitatively different processes and patterns occur in the environment constantly or intermittently.
4. *Reliability*: Normal environmental state fluctuates in random ways, and the fluctuations may occasionally take it far from the normal state.
5. *Change*: In the course of time, the normal environmental state may gradually or abruptly change to a permanently different normal environmental state.

6. *Other systems*: Behavior of other systems changes the environment of a given system.

Bossel (1999) proposed a one-to-one relationship between the properties of the environment and the basic orientors of systems. Therefore, the system equipped to secure better overall orientor satisfaction will have better fitness, having a better chance for long-term survival and sustainability. The orientor approach provides some guidance for determining individual attributes that shape the choices according to a basic needs hierarchy.

9.5. Implementing Individual-Based Models

Many IBMs are created from scratch by the modelling team; however, it can be quite difficult and time-consuming to gather and analyze a large number of observations, equations, and parameters. Without a standard toolbox, such as from object-oriented programming, the developed software can be inefficient and not easily transparent. Alternatives for developing the model from scratch are using software libraries, such as Swarm and Repast, which are maintained by active user communities, or established modelling environments. These modelling environments, such as CORMAS and NetLogo, are more general programming platforms from which one can develop IBMs. They are also maintained by their developers and as teaching tools include tutorial support and examples, making them a good choice for beginners in the field. In any case, the field benefits from the extraordinary increase in computing power that every personal computer (PC) now has, which is sufficient to run most IBMs, although large models or sensitivity analyses may require PC clusters or other advanced computing power (Grimm, 2008).

One effort to add standardization to the IBM model development was the introduction of the ODD protocol by Grimm et al. (2006), which refers to three primary blocks: Overview, Design concepts, and Details. Within these three blocks there are seven elements. The overview block includes: (1) purpose, (2) state variables and scales, and (3) process overview and scheduling. This block lays out the model purpose and structure from which the model skeleton is apparent including the definition of the objects (state variables) and process scheduling. The second block, design concepts, with only one element – design concepts – links the study to the broader framework of complex adaptive systems. It should address issues of interaction types, adaptation, learning, emergence, and the role of stochasticity. The third block, details, includes three elements: (1) initialization, (2) input, and (3) submodels. This section includes all the model detail, such as initial conditions, equations, and parameters. The information should be sufficient for any reader to reconstruct the model and achieve the baseline simulations. In their paper, Grimm et al. (2006) referred to testing ODD on 19 different models (with specific examples therein), and, since then, the approach has been widely used in the IBM community.

An example following this protocol is given by Dur et al. (2009) to study the reproduction of egg-bearing copepods. The model was parameterized from laboratory and field experiments as well as data from the literature. It is a good application for IBM because the authors were able to model the detailed reproductive cycle of the organism (Figure 9.1). The IBM included attributes: location, number, age, longevity, embryonic development time,

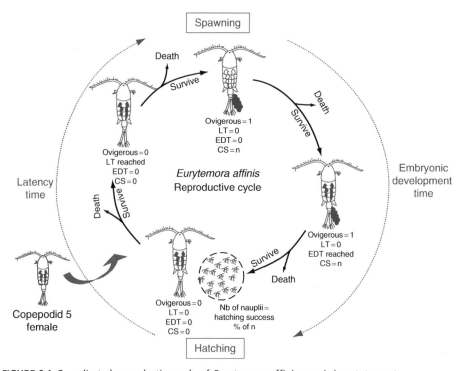

FIGURE 9.1 Complicated reproductive cycle of *Eurytemora affinis* permission statement.

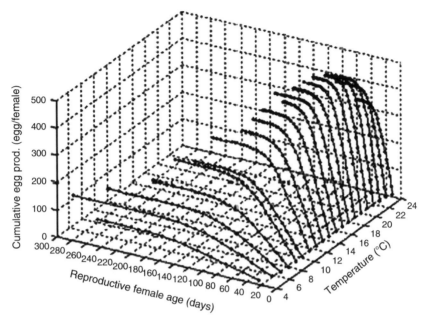

FIGURE 9.2 Results from an IBM developed using ODD shows that egg production is strongly affected by temperature with a maximum production at approximately 20°C.

latency time, spawning time, hatching time, ovigerous state of female, clutch size, and four intermediate parameters regarding the individual variability on longevity, latency, embryonic development time, and clutch size. The environment is represented by one attribute, temperature. Results showed that temperature effects are very important to daily egg production (Figure 9.2). For example, females at 4°C were able to produce only 16 clutches, whereas production reached a maximum of 30 clutches at 23°C. In this model, the emphasis of detail is on the life history of the reproducing individuals, not the environmental factors.

9.6. Pattern-Oriented Modelling

Due to the high complexity of IBMs, the results can be hard to understand. A new general strategy, pattern-oriented modelling (POM), has been developed to optimize model complexity and deal with uncertainty in model structure and parameters. A pattern is the macroscopic order that arises from the microscopic interactions from the system's internal organization and is an indicator that there is something more going on than simple random variation. Because of the emergence of higher order organization, it is necessary to develop approaches to recognize these patterns as different from the background. A pattern is a clearly identifiable structure in nature or data that is distinguishable from random variation indicating that underlying processes could be generating it (Grimm et al., 1996). In other words, the macroscopic pattern is generated by microscopic activity, such as the demographic interactions (dispersal, foraging, mating, etc.) and environmental constraints (topography, landscape, resources, climate, etc.). These patterns occur at a higher level than the processes that cause them. Comparing the observed processes with the model simulations that produced them, it is possible to restrict the parameter space available for uncertain or key features to detect the underlying processes. For example, Swannack et al. (2009) used POM to estimate life history characteristics of amphibian populations. Specifically, they compared simulation results to observations from four population-level patterns: population size, adult sex ratio, proportion of toads returning to their natal pond, and mean maximum distance moved. The models (11 of 16) that did not fit the observed patterns were rejected (Figure 9.3).

Table 3-summary of results from 650, 10-year, monte carlo simulations based on each of 16 versions of the model						
Model version	Juvenile survival	Male survival	Population size	Sex ratio	Percentage at natal pond	Maximum distance moved
Field	?	0.15–0.27[a]	225[a]	5.5[b]	0.73[c]	900–1900[d]
1	0.005	0.15	0.02	0.00	0.73	1133
2	0.005	0.27	0.10	41.01	0.72	1184
3[e]	0.0075	0.15	3.32	5.76	0.66	1289
4	0.0075	0.27	7.69	15.81	0.65	1308
5[e]	0.009	0.15	45.35	6.67	0.62	1355
6	0.009	0.27	63.22	15.49	0.61	1399
7[e]	0.0095	0.15	86.52	7.33	0.60	1382
8	0.0095	0.27	123.05	15.81	0.60	1429
9[e]	0.01	0.15	164.88	7.65	0.59	1427
10	0.01	0.27	209.69	15.83	0.59	1460
11[e]	0.0105	0.15	290.00	7.85	0.58	1461
12	0.0105	0.27	365.29	16.31	0.57	1511
13	0.015	0.15	46814.61	4.48	0.70	1751
14	0.015	0.27	48967.53	5.98	0.71	1761
15	0.02	0.15	42581.38	1.92	0.84	1665
16	0.02	0.27	43335.67	2.72	0.84	1676

Results include mean (1) final population size, (2) final adult sex ratio, (3) percentage of toads at their natal pond at the time of their death or at the end of a simulation, and (4) maximum distance moved (m) by an individual toad during a simulation. Different versions of the model represent different combinations of annual survival estimates (probabilities) for juveniles and adult males. ? represents no field data available.

[a] Swannack (2007).
[b] Swannack, Grant, and Forstner (2007).
[c] Breden (1987).
[d] Price (2003).
[e] Versions of the model that generate reasonable patterns in all 4 system attributes.

FIGURE 9.3 Shows table from Swannack et al. (2009) in which the 16 model runs compare observations with simulation results.

The remaining models had a similar feature that population depends heavily on juvenile survival and provided a narrow range for the juvenile survival parameter. Values of juvenile survival below 0.01 had populations too small and those with values 0.015 and higher were too high. The model was very sensitive to this parameter. This is a very good application of using POM to identify key parameters and to provide a range for acceptable values.

9.7. Individual-Based Models for Parameterizing Models

Whereas the previous example used POM to test the uncertainty of certain parameters, a growing tendency is to supplement the paucity of certain field data with simulated data to parameterize and evaluate population models. One such approach is the use of a data set generated by IBMs. Two such examples are presented next.

Hilker, Hinsch, and Poethke (2006) used an IBM to parameterize a patch-matrix model (PMM) and a grid-based model (GBM). They first constructed an IBM (in this specific case, agents represent three different grasshopper species in varying landscapes with demographic and environmental stochasticity). From this model, they generated a long-term set of simulated data and extracted from this short-term "snapshot" data, which are used as estimators within the PMM and GBM (Figure 9.4). Specifically, they wanted parameter estimates for grasshopper movement regarding nest and mate radius as well as patch and matrix distance over a range of three mobility types. They used snapshot data from two or five years to correspond to typical field studies (amount of years ended up not having a big impact on the model performance). The best result was obtained with the inclusion independent migration data (such as from mark-recapture experiments). Overall, the authors were able to demonstrate the IBM as a general model that can be used to relate IBM-simulated parameters to emergent behavior at the metapopulation level.

In another example, Gilioli and Pasquali (2007) also used an IBM for estimating population parameters. In this case, the IBM is applied to egg production of a fruit fly. Specifically, an IBM simulates the number of eggs produced by the adults and a compartmental model simulates stage-structured population dynamics. The IBM allows for a precise description of the physiological age-structure (eggs, larvae, and pupae)

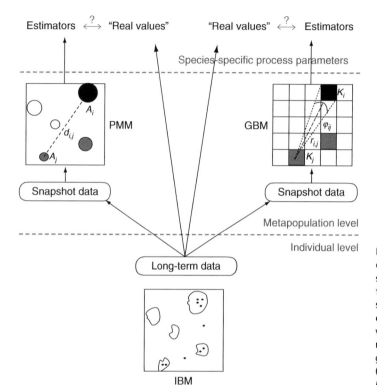

FIGURE 9.4 Conceptual diagram of using snapshot data taken from a long-term IBM simulation for estimating parameter values in a patch-matrix model (PMM) and the grid-based model (GBM). *(Reprinted from Hilker et al., 2006.)*

and time distribution such as recruitment and emergence profiles. The IBM also contributes to the estimation of age-structured mortality and fecundity parameter values. The combination of a microscopic (IBM) and macroscopic (compartmental) models provides a more detailed prediction of population dynamics and good agreement with the observed data. Overall, the use of IBMs for parameterizing models is becoming a more common approach.

9.8. Individual-Based Models and Spatial Models

While there are many different examples of IBM applications to address ecological questions, let us end this chapter with one further example that combines the IBM approach with a spatially explicit model (Chapter 11). Overall, we see there is a lot of synergy between the ability to model individual agents and the spatially explicit landscape on which they interact. Wallentin, Tappeiner, Strobla, and Tasserd (2008) constructed an IBM to understand alpine tree line dynamics. This model is used to test the effects of climate change on a forest community in the Austrian Central Alps. Due to a warming climate, the leading edge of the tree line from spontaneous forest regeneration is climbing to higher elevations. Forest regeneration is influenced by seed dispersal characteristics and land use changes (i.e., availability of migration into abandoned alpine pastures). The model construction involves six steps: (1) deriving landscape features from remote sensing data, (2) building the model, (3) parameterization based on ecological processes, (4) scenario runs, (5) validation, and (6) sensitivity analysis (Figure 9.5). The model includes as main processes recruitment, growth, and mortality. Recruitment is a function of distance to seed trees, land cover type, and elevation. Growth follows a standard sigmoid curve and mortality is impacted by age and density. The model iterates each year through the processes of recruitment, growth, and mortality. Establishment of new seedlings depends on distance to the nearest seed tree, ground vegetation, and elevation. A tree dies if the survival probability based

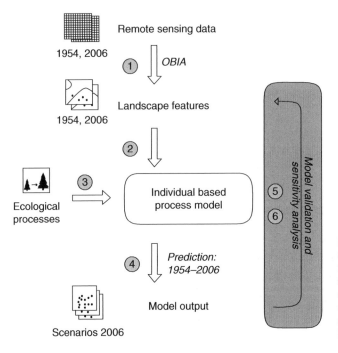

FIGURE 9.5 Model flow diagram of incorporating remote sensing data for studying the impact of climate on alpine tree line migration using an object based image analysis (OBIA) and individual-based model (IBM).

on tree age and tree density is smaller than random mortality values. Results from the model show the upward movement in elevation of the tree line, which is a good prediction of the observed forest regeneration trend during the study period from 1954 to 2006. Overall, the maximum elevation rose almost 150 m and the mean elevation about 90 m. This is a good example of how a spatially explicit IBM can be used to model population dynamics in response to changing environmental conditions, such as climate change.

9.9. Example

To give the reader a clearer idea of how to construct an IBM, in this section we present an IBM recently developed by Chon, Jørgensen, and Cho, (2010) for studying how individual survival is dependent on the dynamic relation between the gene-individual-population. At the lowest scale the genes are under different constraints regarding the metabolic efficiency and toxin susceptibility (Chon et al., 2010). In this model, the individuals move around in 2D space and compete for food, such that the entire population acquires the most adaptable genes (concerning combination of metabolic efficiency and toxin resistance) over the long run. The individual attributes are controlled by the gene information, which in turn determines the gene levels of the entire population.

Individual attributes include age, health score, and location (x and y coordinates). Food and toxins were present in the grid as environmental factors. Individuals on the same location as food or toxins would consume them and their health would be affected accordingly (positive for food and negative for toxin). Food and toxins were both resupplied regularly to the matrix. Variables in the model include total population densities and densities in different types of gene information in the population (Chon et al., 2010).

The model, programmed in Visual Basic, uses a lattice grid size of 800×800 units and was run for 7000 time steps. Each interior site (i, j) (where $i = 2, \ldots, n - 1$ and $j = 2, \ldots, n - 1$) has 8 immediate neighbor cells $(i - 1, j - 1)$, $(i - 1, j)$, $(i - 1, j + 1)$, $(i, j - 1)$, $(i, j + 1)$, $(i + 1, j - 1)$, $(i + 1, j)$, and $(i + 1, j + 1)$. Individuals move across this landscape according to a random walk (one unit per time step). If a nutrient is located at one of the neighbor lattices, then the individual moves to that lattice. In the case of multiple food items in the individual's nearest neighbors, the movement selection is made randomly. If there are no food items in the nearest neighbors, then the individuals move at random. Toxin exposure occurs randomly (Chon et al., 2010).

Two different genes carry information regarding the metabolic efficiency and toxin susceptibility, and both were determined at fixed rates with low, medium, or high levels (e.g., 0.1, 0.25, and 0.5 for metabolic efficiency). This information was converted to phenotypic properties through health scores. The maximum score of metabolic efficiency and toxin susceptibility was assumed to be 20 points. The health scores accumulate according to the food uptake and toxin exposure. If the health score drops below zero, then the individual dies. When the health score is greater than a fixed threshold and the age is older than three time steps, reproduction occurs. Reproduction can occur by asexual fission or conjugation if a neighboring cell is occupied. In each iteration, only gene type is randomly selected for exchange (see Chon et al., 2010).

The model was initialized with food occupying 20% of the total lattice and replenished at regular intervals in 10% of the empty spaces in each 100 time step. Toxin was also present initially in 20% of the total lattice but was resupplied at a rate of 1% of the remaining empty space (after resupply of nutrients) in each 100 time step. A range for the model parameters, metabolic efficiency (ME) and toxin susceptibility (TS), was determined to obtain balanced population densities, which occurred expectedly in the range of higher ME (i.e., efficiency in metabolism) and lower TS (i.e., higher resistance to toxins). Two similar sets of conditions (C1 and C2) for different genetic values were provided to ME and TS as follows.

The first condition (C1):

Type A: ME; 0.5; TS; 0.4
Type B: ME; 0.4; TS; 0.3
Type C: ME; 0.3; TS; 0.2

The second condition (C2):

Type A: ME; 0.5; TS; 0.4
Type B: ME; 0.3; TS; 0.25
Type C: ME; 0.1; TS; 0.1

Chon et al. (2010) found that the overall changes in population size showed common patterns through simulation. Population densities increased rapidly with consumption of initial nutrients, peaking at around 100 iterations. The population size decreased as nutrients were depleted, and reached the minimal size at around 200 iterations. Afterwards, population size periodically changed in the range of 400 to 800 individuals along with resupply of the nutrients at 100 iteration intervals (Figure 9.6). The Determination of dominant types of gene information appeared to be critical when the population size was minimized due to nutrient depletion.

The overall change in fitness due to reproduction was also modelled. The case without conjugation did not allow for gene recombination and is not discussed here. It is noted that the case with the best initial parameter values (type A-A) had the highest population and the one

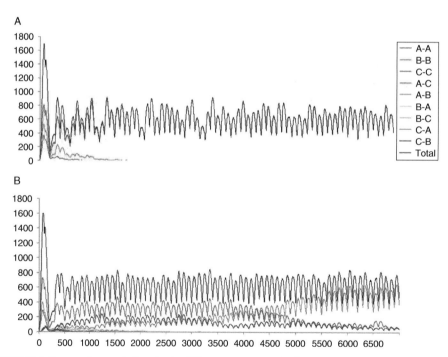

FIGURE 9.6 Changes in population size in different gene types in the gene-individual-population relationships: (a) without conjugation and (b) with conjugation = 25%.

with the worst values (type C-C) had the lowest or went extinct. Greater diversity was found in the species composition when conjugation was allowed for gene exchange between individuals. The amount of mixing depended on level of conjugation (which ranged from 0 to 100%). The dominant types changed depending on the different simulation conditions, C1 and C2. For condition C1, Type A-C, which is most suitable for both ME and TS, appeared as the first dominant type, followed by A-B and A-A at conjugation = 25%. For condition C2, however, type A-B was most dominant, followed by A-A and A-C. The overall diversity changed with increasing conjugation (Figure 9.7). In conclusion, the authors found by using an IBM that overall biomass and eco-exergy (see Chapter 10) increased with conjugation.

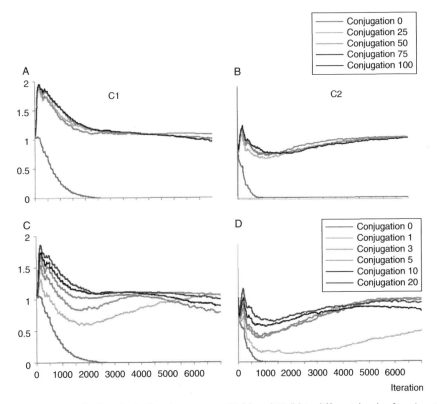

FIGURE 9.7 Changes in diversity indices in averages. C1 (a) and C2 (b) in different levels of conjugation = 0, 25, 75, and 100 %, and C1 (c) and C2 (d) in different levels of conjugation = 1, 3, 5, 10, and 20%.

9.10. Conclusions

Individual-based models have filled a natural gap in the ecological modelling toolbox. They allow more detail and flexibility for individual action than the traditional compartment modelling approach. The key factors in an IBM are: (1) the inclusion of individual variation including detail about the life history and age classes, (2) the possibility for agents to adapt and learn (i.e., update in real time the interaction rules) from experiences, and (3) the modification of the environment by the behavior of the individual. Libraries of data now exist for use in IBMs, development of a standard protocol for developing IBMs, and software platforms that are available for IBMs. Many applications of this new approach have been implemented and their use will continue to grow in the future.

Problems

1. What is an individual-based ecology? How does the interplay of microscopic and macroscopic levels influence ecological characteristics?
2. The three main features of an IBM are: (1) agent behavior, (2) agent-agent interactions, and (3) the environment. Explain how each of these could be modelled.
3. Explain the ODD protocol introduced to standardized IBM studies.
4. Develop a conceptual model for an IBM representing a forest ecosystem. Include a description of the spatial variation in the species distribution and how the different sized structures could be modelled. What are some important traits that should be considered in the model?
5. Adaptability is more likely to lead to emergent system properties. Explain why and how this could be modelled.
6. Results from IBMs are often most useful when analyzed at a higher scale of observation. Explain how POM is used to identify these structures. Give an example of how it could lead to estimation of model parameters.
7. What role do Geographical Information Systems (GISs) play in the development and implementation of IBMs?
8. Explain the difference between a metapopulation model and an IBM. Which circumstances would each one be best used?

10
Structurally Dynamic Models

CHAPTER OUTLINE

10.1. Introduction

Ecological models attempt to capture the characteristics of ecosystems. However, ecosystems differ from most systems because they are extremely adaptive, self-organized, and have a large number of feedback mechanisms. The real challenge of modelling is to answer this question: How can we construct models that are able to reflect these dynamic characteristics? This chapter attempts to answer this question by the use of structurally dynamic models. Section 10.2 focuses on the characteristics of ecosystems and Section 10.3 is devoted to the development of structurally dynamic models (SDMs) or variable parameter models, which are sometimes called the fifth generation of models. The thermodynamic variable eco-exergy (work capacity defined for ecosystems; the definition and presentation are given in Section 10.3) has been applied to develop SDMs in 21 cases (Zhang et al., 2010). The 21 case studies are:

1–8. Eight eutrophication models of six different lakes
   9. A model to explain the success and failure of biomanipulation based on removal of planktivorous fish
 10. A model to explain under which circumstances submerged vegetation and phytoplankton are dominant in shallow lakes

Fundamentals of Ecological Modelling. DOI: 10.1016/B978-0-444-53567-2.00010-7

11. A model of Lake Balaton, which was used to support the intermediate disturbance hypothesis
12–15. Small population dynamic models
16. A eutrophication model of the Lagoon of Venice
17. A eutrophication model of the Mondego Estuary
18. An ecotoxicological model focusing on the influence of copper on zooplankton growth rates
19. A model of Darwin's finches
20. A model of the interaction between parasites and birds
21. A model of Lake Fure in Denmark

Sections 10.4–10.6 present three illustrative examples of SDMs; namely 19, 9 and 18 from the previous list of case studies using eco-exergy as a goal function to develop SDMs. The use of this model type will most likely increase in the future in our endeavor to make more adaptive models because reliable predictions can only be made by models with a correct description of ecosystem properties. If our models do not properly describe adaptation and possible shifts in species composition, then the prognoses will inevitably be more incorrect.

10.2. Ecosystem Characteristics

Ecology deals with irreducible systems (Wolfram 1984a,b, Jørgensen, 1990, 1992a,b, 2002, Jørgensen & Fath, 2004b). We cannot design simple experiments to reveal a relationship that can, in all detail, be transferred from one ecological situation and one ecosystem to another situation in another ecosystem. This may be possible with Newton's laws of gravity, because the relationship between forces and acceleration is reducible. The relationship between force and acceleration is also linear, but growth of living organisms is dependent on many interacting factors, which again are functions of time. Feedback mechanisms simultaneously regulate all the factors and rates, interact, and are also functions of time (Straskraba, 1979, 1980).

Table 10.1 shows the hierarchy of regulation mechanisms that are operating at the same time. From this example the complexity alone clearly prohibits the reduction to simple relationships that can be used repeatedly. An ecosystem has so many interacting components that it is impossible to examine all of these relationships. Even if we could,

Table 10.1 The Hierarchy of Regulating Feedback Mechanisms

Level	Explanation of Regulation Process	Exemplified by Phytoplankton Growth
1.	Rate by concentration in medium	Uptake of phosphorus in accordance with phosphorus concentration
2.	Rate by needs	Uptake of phosphorus in accordance with intracellular concentration
3.	Rate by other external factors	Chlorophyll concentration in accordance with previous solar radiation
4.	Adaptation of properties	Change of optimal temperature for growth
5.	Selection of other species	Shift to better fitted species
6.	Selection of other food web	Shift to better fitted food web
7.	Mutations, new sexual recombinations, and other shifts of genes	Emergence of new species or species properties

From Jørgensen, 1988.

it would be impossible to separate one relationship and examine it carefully to reveal its details because the relationship is different when it works in nature with interactions from many other processes and from when it is examined in a laboratory with the relationship separated from the other ecosystem components. The observation that it is impossible to separate and examine processes in real ecosystems corresponds to the examinations of organs that are separated from the organisms in which they are working. Their functions are completely different when separated from their organisms and examined in a laboratory from when they are placed in their right context and in "working" condition.

These observations are indeed expressed in ecosystem ecology — "everything is linked to everything" or "the whole is greater than the sum of the parts" (Pascal and repeated by Allen & Starr, 1982). These expressions imply that it may be possible to examine the parts by reduction to simple relationships, but when the parts are put together they form a whole that behaves differently from the sum of the parts. This statement requires a more detailed discussion of how an ecosystem works. The latter statement is correct because of the evolutionary potential that emerges from living systems. The ecosystem contains the possibility of becoming something different, that is, adapting and evolving. The evolutionary potential is linked to the existence of microscopic freedom, represented by stochasticity and nonaverage behavior, resulting from the diversity, complexity, and variability of its elements.

Underlying the taxonomic classification is the microscopic diversity, which only adds to the complexity to such an extent that it will be completely impossible to cover all the possibilities and details of the observed phenomena. We attempt to capture at least a part of the reality using models. It is not possible to use one or a few simple relationships, but a model seems the only useful tool when we are dealing with irreducible systems. However, using one model is far from realistic. Using many models simultaneously to capture a more complete image of reality seems the only possible way to deal with complex living systems.

This has been acknowledged by holistic ecology or systems ecology, whereas the more reductionistic style of ecology attempts to understand ecological behavior by analysis of one or a few processes, which are related to one or two components. The results of analyses are expanded to be used in the more reductionistic approaches as a basic explanation of observations in real ecosystems, but such an extrapolation is often invalid and leads to false conclusions. Both analyses and syntheses are needed in ecology, and the analysis is a necessary foundation for the synthesis, but it may lead to wrong scientific conclusions to stop at the analysis. Analysis of several interacting processes may give a correct result of the processes under the analyzed conditions, but the conditions in ecosystems are constantly changing and even if the processes were unchanged (which they very rarely are), it is not possible to oversee the analytical results of so many simultaneously working processes. Our brain simply cannot calculate what will happen in a system where, for example, six or more interacting processes are working simultaneously.

So, reductionism does not consider that the:

1. Basic conditions determined by the external factors for our analysis are constantly changing (one factor is typically varied by an analysis, while all the others are assumed constant) in the real world and the analytical results are not valid in the system context.
2. Interaction from all of the other processes and components may change the processes and the properties of all biological components

significantly in the real ecosystem so the analytical results are invalid.

3. Direct overview of the many simultaneously working processes is not possible and wrong conclusions may result if an overview is attempted.

Therefore, a tool is needed to oversee and synthesize the many interacting processes in an ecosystem. The synthesis may just be "putting together" the various analytical results, but afterward we need to make changes to account for the fact that the processes are working together and become more than the sum of the parts. In other words, there is a synergistic effect or a symbiosis. In Chapter 6, Section 6.4, it was mentioned how important the indirect effects are compared to the direct effects in an ecological network and the emergence of network mutualism.

Modelling can be used as a synthesizing tool. It is our hope that a further synthesis of knowledge will enable us to attain a system-wide understanding of ecosystems and help us cope with the environmental problems that are threatening human survival.

A massive scientific effort is needed to teach scientists how to cope with ecological complexity or even with complex systems in general. Which tools should we use to attack these problems? How do we use the tools most efficiently? Which general laws are valid for complex systems with many feedbacks and particularly for living systems? Do all hierarchically organized systems with many hierarchically organized feedbacks and regulations have the same basic laws? What do we need to add to these laws for living systems?

Many researchers have advocated a holistic approach to ecosystem science (e.g., E. P. Odum, 1953; Ulanowicz, 1980, 1986, 1995). Holism is the description of the system level properties of an ensemble, rather than simply an exhaustive description of all the components. It is thought that by adopting a holistic viewpoint, certain properties become apparent and other behaviors that otherwise would be undetected become visible.

It is, however, clear from this discussion that the complexity of ecosystems has set the limitations for our understanding and for the possibilities of proper management. We cannot capture the complexity and all its details, but we can understand why ecosystems are complex and set up a realistic strategy for gaining sufficient knowledge about the system — not knowing all the details, but still understanding and knowing the mean behavior and the important reactions of the system, particularly to specified impacts. It means that we can only try to reveal the basic properties behind the complexity.

We have no other choice than to go holistic. The results from the more reductionistic ecological tests are essential in our effort "to get to the root" of the system properties of ecosystems, but we need systems ecology, which consists of many new ideas, approaches, and concepts, to follow the path to the root of the basic system properties of ecosystems. In other words, we cannot find the properties of ecosystems by analyzing all the details because there are simply too many, but we can try to reveal the system properties of ecosystems by examining the entire system.

The number of feedbacks and regulations is extremely high, which makes it possible for the living organisms and populations to survive and reproduce in spite of changes in external conditions.

These regulations correspond to levels 3 and 4 in Table 10.1. Numerous examples can be found in the literature. If the actual properties of the species are changed, then the regulation is called adaptation. Phytoplankton, for instance, is able to regulate its chlorophyll concentration according to available solar radiation. If more chlorophyll is needed because the radiation is insufficient to guarantee growth, then more chlorophyll is produced by the phytoplankton. The digestion efficiency of the food for many animals depends on the abundance of food. The same species may be a different size in different environments, depending on what is most beneficial for survival and growth. If nutrients are scarce, then phytoplankton becomes smaller and vice versa. In this latter case, the change in size is a result of a selection process, which is made possible because of the distribution in size.

The feedbacks are constantly changing, that is, the adaptation itself is adaptable because if a regulation is insufficient, another regulation process higher in the hierarchy of feedbacks (see Table 10.1) will take over. The change in size within the same species is limited. When this limitation has been reached, other species will take over. This implies that the processes and the components, as well as the feedbacks, can be replaced, if needed, to achieve better utilization of the available resources.

Three different concepts have been used to explain the functioning of ecosystems:

1. The individualistic or Gleasonian concept assumes populations respond independently to an external environment.

2. The superorganism or Clementsian concept views ecosystems as organisms of a higher order and defines succession as ontogenesis of this superorganism (Margalef, 1968, 1991). Ecosystems and organisms are different in one important aspect. Ecosystems can be dismantled without destroying them; they are just replaced by others, such as agroecosystems, human settlements, or other succession states. Patten (1981) pointed out that the indirect effects in ecosystems are significant compared to the direct ones, while in organisms, the direct linkages are most dominant. An ecosystem has more linkages than an organism, but most of them are weaker. This makes the ecosystem less sensitive to the presence of all the existing linkages. It does not imply that the linkages in ecosystems are insignificant and do not play a role in ecosystem behavior. The ecological network is of great importance in an ecosystem, but the many and indirect effects give the ecosystem buffer capacities to deal with minor changes.
The description of ecosystems as superorganisms therefore seems insufficient.

3. The hierarchy theory (Allen & Star, 1982) insists that the higher level systems have emergent properties that are independent of the properties of their lower level components. This compromise between the two other concepts seems consistent with our observations in nature.

The hierarchical theory is a very useful tool to understand and describe complex "medium number" systems, such as ecosystems (O' Neill et al., 1975).

During the last decades, there has been a debate over whether "bottom-up" (limitation by resources) or "top-down" (control by predators) effects primarily control system dynamics. The conclusion of this debate seems that *both* effects control the dynamics of the system. Sometimes the effect of the resources may be most dominant, sometimes the higher levels control the dynamics of the system, and sometimes both effects determine the dynamics of the system. This conclusion is nicely presented in *Plankton Ecology* by Sommer (1989).

The ecosystem and its properties emerge as a result of many simultaneous and parallel focal-level processes influenced by even more remote environmental features. It means that the ecosystem will be seen by an observer to be factorable into levels. Features of the immediate environment are enclosed in entities of yet a larger scale and so on. This implies that the environment of a system includes historical factors, as well as immediately cogent ones (Patten, 1981; Jørgensen & Fath, 2004b). The history of the ecosystem and its components is important for the behavior and further development of the ecosystem. This is one of the main ideas behind Patten's indirect effects; the indirect effects account for the "history," while the direct effects only reflect the immediate interactions. The importance of the history of the ecosystem and its components emphasizes the need for a dynamic approach and supports the idea that we will never observe the same situation in an ecosystem twice. The history will always be "between" two similar situations. Therefore, as previously mentioned, the equilibrium models may fail in their conclusions, particularly when we want to look into reactions on the system level.

10.2.1. Ecosystems Show a High Degree of Heterogeneity in Space and Time

An ecosystem is a very dynamic system. All of its components, particularly the biological ones, are steadily changing and their properties are steadily modified, which is why an ecosystem never returns to the same situation. Every point is different from any other point, offering different conditions for the various life forms. This enormous heterogeneity explains why biodiversity is so plentiful on Earth. There is an ecological niche for "everyone" and "everyone" may be able to find a niche where he best fits to utilize the resources.

Ecotones, the transition zones between two ecosystems, offer a particular variability in life conditions, which often results in a particular richness of species diversity. Studies of ecotones have recently drawn much attention from ecologists because they have pronounced gradients in the external and internal variables. This gives a clearer picture of the relation between external and internal variables.

Margalef (1991) claimed that ecosystems are anisotropic; they exhibit properties with different values when measured along axes in different directions. This means that the ecosystem is not homogeneous in relation to properties concerning matter, energy, and information, and that the entire dynamics of the ecosystem work toward increasing these differences.

These variations in time and space make it particularly difficult to model ecosystems and to capture their essential features. However hierarchy theory applies these variations to develop a natural hierarchy as a framework for ecosystem descriptions and theory. The strength of hierarchy theory is that it facilitates the studies and modelling of ecosystems.

10.2.2. Ecosystems and Their Biological Components Evolve Steadily and Over the Long Term Toward Higher Complexity

Darwin's theory describes the competition among species and states that those species best fitted to the prevailing conditions in the ecosystem will survive. Darwin's theory can, in other words, describe the changes in ecological structure and the species composition, but cannot directly be applied quantitatively in ecological modelling (see the next section).

All species in an ecosystem are confronted with the question: How is it possible to survive or even grow under the prevailing conditions? The prevailing conditions are considered as all factors that influence the species, that is, all external and internal factors including those originating from other species. This explains coevolution, as any change in the properties of one species will influence the evolution of the other species. The environmental stage on which the selection plays out is comprised of all the interacting species, each influencing another.

All natural external and internal factors of ecosystems are dynamic; the conditions are steadily changing, and there are always many species waiting in the wings ready to take over if they are better fitted to the emerging conditions than the species dominating under the present conditions. There is a wide spectrum of species representing different combinations of properties available for the ecosystem. The question remains: Which of these species are best able to survive and grow under the present conditions and which species are best able to survive and grow under the conditions one time step further, two time steps further, and so on? The necessity in Monod's (1971) sense is given by the prevailing conditions — species must have genes or phenotypes (properties) that match these conditions to be able to survive. But the natural external factors and the genetic pool available for the test may change randomly or by "chance."

Steadily, new mutations (misprints are produced accidentally) and sexual recombinations (genes are mixed and shuffled) emerge and steadily produce new material to be tested by the question: Which species are best fitted under the prevailing conditions?

These ideas are illustrated in Figure 10.1. The external factors are steadily changed and some even relatively fast and partly at random, such as the meteorological or climatic factors. The species within the system are selected among the species available and represented by the genetic pool, which again is slowly, but surely, changed randomly, or by chance. The selection in Figure 10.1 includes level 4 of Table 10.1. It is a selection of the organisms that possess the properties best fitted to the prevailing organisms according to the frequency distribution.

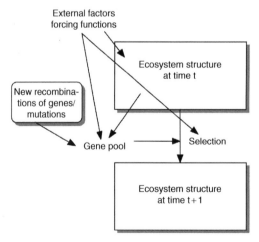

FIGURE 10.1 Conceptualization of how the external factors steadily change the species composition. The possible shifts in species composition are determined by the gene pool, which is steadily changed due to mutations and new sexual recombinations of genes. The development is, however, more complex. This is indicated by arrows from " structure" to "external factors" and "selection" to account for the possibility that the species can modify their own environment and their own selection pressure along with an arrow from "structure" to "gene pool" to account for the possibilities that species can, to a certain extent, change their own gene pool.

Ecological development includes the changes over time in nature caused by the dynamics of the external factors, which gives the system sufficient time to modify its structure and behavior.

Evolution, on the other hand, is related to the genetic pool. It is the result of the relation between the dynamics of the external factors and the dynamics of the genetic pool. The external factors steadily change the conditions for survival, and the genetic pool steadily comes up with new solutions to the problem of survival.

Species are continuously tested against the prevailing conditions (external as well as internal factors) and the better they fit, the better they are able to maintain and even increase their biomass. The specific rate of population growth may even be used as a measure for fitness (Stenseth, 1986). But the property of fitness must be heritable to have any effect on the species composition and the ecological structure of the ecosystem in the long run.

Natural selection has been criticized for being a tautology: Fitness is measured by survival, therefore survival of the fittest means survival of the survivors. However, the entire Darwinian theory including the previously listed three assumptions, cannot be conceived as a tautology, but may be interpreted as follows: Species offer different solutions to survive under prevailing conditions, and the species that have the best combinations of properties to match the conditions also have the highest probability of survival and growth.

Human changes in external factors, that is, anthropogenic pollution, have created new problems because new genes, and hence organisms, fitted to these changes do not develop overnight, while most natural changes have occurred many times previously and the genetic pool is therefore prepared and fitted to meet the natural changes. Life is able to meet most natural changes, but not all of the human changes, because they are new and untested in the ecosystem.

Evolution moves the system toward increasing complexity in the long run. Fossil records have shown a steady increase of species diversity. There may be destructive forces, such as pollution or natural catastrophes, for a short time, but the probability that (1) new and better genes are developed and (2) new ecological niches are utilized will increase with time. The probability will even (again excluding the short time perspective) increase faster and faster, as the probability is roughly proportional to the amount of genetic material on which the mutations and new sexual recombinations can be developed.

It is equally important to note that a biological structure is more than an active nonlinear system. In the course of its evolution, the biological structure is continuously changed so that its structural map is modified. The overall structure thus becomes a representation of all the information received. Through its complexity, biological structure represents a synthesis of the information with which it has been in communication (Schoffeniels, 1976).

Evolution is maybe the most discussed topic in biology and ecology and millions of pages have been written about evolution and its ecological implications. Today, the facts of evolution are taken for granted and the interest has shifted to more subtle classes of fitness/selection; that is, toward understanding the complexity of the evolutionary processes. One of these classes concerns traits that influence not only the fitness of the individuals possessing them, but also the entire population. These traits overtly include social behaviors, such as aggression or cooperation, and activities that, through some modification of the biotic and abiotic environment feedback, affect the population at large, such as pollution and resource depletion.

It can be shown that many observations support the various selection models used to describe selection in nature. For example, kin selection has been observed in bees, wasps, and ants (Wilson, 1978). Prairie dogs endanger themselves (altruism) by conspicuously barking to warn fellow dogs of an approaching enemy (Wilson, 1978), and a parallel behavior is observed for a number of species.

Coevolution explains the interactive processes among species. It is difficult to observe coevolution, but it is easy to understand that it plays a major role in the entire evolution process. For example, coevolution of herbivorous animals and plants is an illustrative example. The plants develop toward better seed dispersal and a better defense toward herbivorous animals. In the latter case, selected herbivorous animals are able to cope with the defense. Therefore, the plants and the herbivorous animals will coevolve. Coevolution means that the evolution process cannot be described as reductionistic, but that the entire system is evolving. A holistic description of the system evolution is needed.

Having presented some main features of ecosystem development over time, the next crucial question should be: How can we account for these properties in modelling? Some preliminary results on how to consider levels 4–6 of dynamics (see Table 10.1) will be presented in the next section.

10.3. How to Construct Structurally Dynamic Models and Definitions of Exergy and Eco-exergy

If we follow the modelling procedure proposed in Figure 2.2, then a model that describes the processes in the focal ecosystem will be attained, but the parameters will represent the properties of the state variables as they exist in the ecosystem during the examination period. They are not necessarily valid for another period because we know that an ecosystem can regulate, modify, and change them if needed as a response to changes in the existing conditions determined by the forcing functions and the interrelations between the state variables. Our present models have rigid structures and a fixed set of parameters, so no changes or replacements of the components are possible. We need to introduce parameters (properties) that can change according to changing forcing functions and general conditions for the state variables (components) to optimize continuously the ability of the system to move away from thermodynamic equilibrium. So, we may hypothesize levels 5 and 6 in the regulation hierarchy shown in Table 10.1 that can be accounted for in our model by a current change of parameters, according to an ecological goal function. The idea is to test if a change of the most crucial parameters produces a higher goal function of the system and, if that is the case, to use that set of parameters.

The structurally dynamic model can account for the change in species composition as well as the ability of the species (i.e., the biological components of our models) to change their properties (i.e., to adapt to the existing conditions imposed on the species). The SDM is able to capture structural changes. They are called the next, or fifth, generation of ecological models to underline that they are radically different from previous modelling approaches and can do more; namely, describe changes in species composition.

It could be argued that the ability of ecosystems to replace present species with other (level 6 in Table 10.1), better fitted species, can be considered by constructing models that encompass all actual species for the entire period that the model attempts to cover. This approach has two essential disadvantages. First, the model becomes very complex, because it will contain many state variables for each trophic level. Therefore, the model will contain many more parameters that have to be calibrated and validated and, as presented in Sections 2.5 and 2.6, this will introduce a high uncertainty to the model and render the application of the model very case specific (Nielsen 1992a,b). In addition, the model will still be rigid and not allow continuously changing parameters, even without changing the species composition (Fontaine, 1981).

Bossel (1992) used his six basic orientors, or requirements, to develop a system model, which can describe the system performance properly. The six orientors are:

1. Existence. The system environment must not exhibit any conditions that may move the state variables out of its safe range.
2. Efficiency. The exergy gained from the environment should exceed the exergy expenditure over time.
3. Freedom of action. The system reacts to the inputs (forcing functions) with a certain variability.
4. Security. The system has to cope with the different threats to its security requirement with appropriate but different measures. These measures either aim at internal changes in the system or at particular changes in the forcing functions (external environment).
5. Adaptability. If a system cannot escape the threatening influences of its environment, then the one remaining possibility consists of changing the system to cope better with the environmental impacts.
6. Consideration of other systems. A system must respond to the behavior of other systems. The fact that these other systems may be of importance to a particular system should be considered with this requirement.

Bossel (1992) applied maximization of a benefit or satisfaction index based upon balancing weighted surplus orientor satisfactions on a common satisfaction scale. The approach is used to select the model structure of continuous dynamic systems and is able to account for the ecological structural properties as presented in Table 10.1. This approach seems very promising, but has unfortunately not been widely applied to ecological systems.

Straskraba (1979) used biomass maximization as a governing principle. His model computes the biomass and adjusts one or more selected parameters to achieve the maximum biomass at every instance. It has a routine that computes the biomass for all possible combinations of parameters within a given realistic range. The combination that gives the maximum biomass is selected for the next time step and so on. This is an example of an early structurally dynamic model.

Exergy has been used widely as a goal function in ecological models, and a few of the available case studies will be presented and discussed in this section. Exergy has two pronounced advantages as a goal function. (1) Exergy is defined far from thermodynamic equilibrium, and (2) it relates to the state variables, which are easily determined or measured, as opposed to being derived from the flows. As exergy is not a generally used thermodynamic function, we need to explain this concept before we can go any further.

Exergy expresses energy with a built-in measure of quality like energy. Exergy accounts for natural resources and can be considered as fuel for any system that converts energy and matter in a metabolic process (Schrödinger, 1944). Ecosystems consume energy, and an exergy flow through the system is necessary to keep the system functioning — living systems operate far-from-equilibrium. Exergy measures the distance from a reference condition in energy terms, as will be further explained in this section.

Exergy, Ex, is defined by the following equation:

$$Ex = T_o * NE = T_o * I = T_o * (S_{eq} - S) \qquad (10.1)$$

where T_o is the temperature of the environment; I is the thermodynamic information, defined as NE; and NE is the negentropy of the system, that is, $= (S_{eq} - S) =$ the difference between the entropy for the system at thermodynamic equilibrium and the entropy at the present state.

Exergy differences can be reduced to differences of other, better known, thermodynamic potentials, which may facilitate the computations of exergy in some relevant cases.

As noted, the exergy of the system measures the contrast — it is the difference in free energy if there is no difference in pressure, as may be assumed for an ecosystem — against the surrounding environment. If the system is in equilibrium with the surrounding environment, then the exergy is zero.

Since the only way to move systems away from equilibrium is to perform work on them, and since the available work in a system is a measure of the ability, we have to distinguish between the system and its environment or thermodynamic equilibrium. For ecosystems, the prebiotic "inorganic soup" has been used as the reference. Therefore it is reasonable to use the available work, that is, the exergy, as a measure of the distance from thermodynamic equilibrium.

Let us translate Darwin's theory into thermodynamics (Section 10.2), applying exergy as the basic concept. Survival implies biomass maintenance, and growth means biomass increase. It costs exergy to construct biomass and biomass therefore possesses exergy, which is transferable to support other exergetic (energetic) processes. Survival and growth can therefore be measured using the thermodynamic concept exergy, which may be understood as the free energy relative to the environment (Eq. 10.1).

Darwin's theory may therefore be reformulated in thermodynamic terms as follows: *The prevailing conditions of an ecosystem steadily change and the system will continuously select the species and thereby the processes that can contribute most to the maintenance or even growth of the exergy of the system.*

Ecosystems are open systems and receive an inflow of solar energy. The solar energy carries low entropy, while the radiation away from the ecosystem carries high entropy.

If the power of the solar radiation is W and the average temperature of the system is T_1, then the exergy gain per unit of time, ΔEx is:

$$\Delta Ex = T_1 * W \left(\frac{1}{T_0} - \frac{1}{T_2} \right), \tag{10.2}$$

where T_0 is the temperature of the environment and T_2 is the temperature of the sun. This exergy flow can be used to construct and maintain structure far away from equilibrium.

Notice that the thermodynamic translation of Darwin's theory requires that populations have the properties of reproduction, inheritance, and variation. The selection of the species that contributes most to the exergy of the system under the existing conditions requires that there are enough individuals with different properties that a selection can take place; it means that the reproduction and the variation must be high and that once a change has taken place due to better fitness, it can be conveyed to the next generation.

Notice also that the change in exergy is not necessarily ≥ 0, it depends on the changes of the resources of the ecosystem. The proposition claims, however, that the ecosystem tendency is to move toward the highest possible exergy level under the given circumstances and with the available genetic and species pool (Jørgensen & Mejer, 1977, 1979). Compare Figure 10.2, where the nutrient concentrations of a lake ecosystem decrease and the exergy increases. It is not possible to measure exergy directly, but it is possible to compute it if the composition of the ecosystem is known. Jørgensen and Mejer (1979) showed, by the use of thermodynamics, that the following equation is valid for the components of an ecosystem:

$$Ex = RT \sum_{i=1}^{i=n} \left(C_i \ln \frac{C_i}{C_{eq,i}} - (C_i - C_{eq,i}) \right) \tag{10.3}$$

where R is the gas constant; T is the temperature of the environment (Kelvin); and C_i represents the ith component expressed in a suitable unit, (for phytoplankton in a lake, C_i could be milligrams of a focal nutrient in the phytoplankton per liter of lake water); $C_{eq,i}$ is the concentration of the ith component at thermodynamic equilibrium, which can be found in Morowitz (1968); and n is the number of components. $C_{eq,i}$ is a very small concentration of organic components corresponding to the probability of forming a complex organic compound in an inorganic soup (at thermodynamic equilibrium). Morowitz (1968) calculated this probability and found that for proteins, carbohydrates, and fats, the

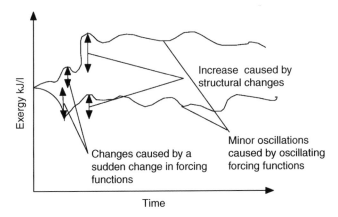

FIGURE 10.2 Exergy response to increased and decreased nutrient concentration.

concentration is about 10^{-86} µg/L, which may be used as the concentration at thermodynamic equilibrium.

The idea of the new generation of models presented here is to continuously locate a new set of parameters (limited for practical reasons to the most crucial, i.e., sensitive parameters) better fitted for the prevailing conditions of the ecosystem. "Fitted" is defined in the Darwinian sense by the ability of the species to survive and grow, which may be measured by the use of exergy (Jørgensen, 1982, 1986, 1988, 1990; Jørgensen & Mejer, 1977, 1979). Figure 10.3 shows the proposed

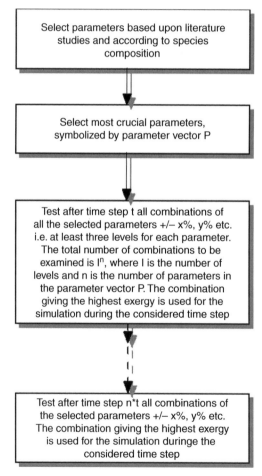

FIGURE 10.3 The procedure used for the development of SDMs.

modelling procedure, which has been applied in the cases presented in Sections 10.4–10.7.

Exergy has previously been tested as a "goal function" for ecosystem development (i.e., Jørgensen, 1986; Jørgensen & Mejer, 1979). However, in all these cases, the model applied did not include the "elasticity" of the system obtained by use of variable parameters; therefore the models did not reflect real ecosystem properties. A realistic test of the exergy principle would require the application of variable parameters.

Exergy is defined as the work the system can perform when it is brought into equilibrium with the environment or another well-defined reference state. If we presume a reference environment for a system at thermodynamic equilibrium — meaning that all the components are: (1) inorganic, (2) at the highest possible oxidation state signifying that all free energy has been utilized to do work, and (3) homogeneously distributed in the system (meaning no gradients) — then the situation illustrated in Figure 10.4 is valid. It is possible to distinguish between chemical exergy and physical exergy. The chemical energy embodied in organic compounds and biological structure contributes most to the exergy content of ecological systems.

Temperature and pressure differences between systems and their reference environments are small in contribution to overall exergy and, for present purposes, can be ignored. We will compute the exergy based entirely on chemical energy: $\Sigma_i(\mu_c - \mu_{c,o})N_i$, where i is the number of exergy-contributing compounds, and c and μ_c are the chemical potential relative to that at a reference inorganic state, $\mu_{c,o}$. Our (chemical) exergy index for a system will be taken with reference to the same system at the same temperature and pressure, but in the form of a prebiotic environment without life, biological structure, information, or organic molecules — the so called inorganic soup.

As $(\mu_c - \mu_{co})$ can be found from the definition of the chemical potential, replacing activities by concentrations we obtain the following expression for chemical exergy:

$$Ex = RT \sum_{i=1}^{i=n} C_i \ln \frac{C_i}{C_{eq,i}} \cdot [ML^2T^{-2}] \tag{10.4}$$

R is the gas constant, T is the temperature of the environment and system (Figure 10.4), c_i is the concentration of the ith component expressed in

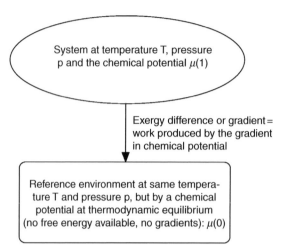

FIGURE 10.4 Illustration of the exergy concept used to compute the exergy index for an ecological model. Temperature and pressure are the same for the both the system and the reference state, which implies that only the difference in chemical potential can contribute to the exergy.

suitable units, $c_{i,eq}$ is the concentration of the ith component at thermodynamic equilibrium, and n is the number of components. The quantity $c_{i,eq}$ represents a very small, but nonzero, concentration (except for $i = 0$, which is considered to cover the inorganic compounds), corresponding to the very low probability of forming complex organic compounds spontaneously in an inorganic soup at thermodynamic equilibrium. The chemical exergy contributed by components in an open system is given by (Jørgensen & Meyer, 1979; Jørgensen, 1982, 2002):

$$Ex = RT \sum_{i=0}^{n} \left[c_i \ln \left(\frac{c_i}{c_{i,eq}} \right) - (c_i - c_{i,eq}) \right] \cdot \left[ML^2 T^{-2} \right] \qquad (10.5)$$

The problem in applying these equations is related to the magnitude of $c_{i,eq}$. Contributions from inorganic components are usually very low and can in most cases be neglected. Exergy can be calculated from the elementary composition of the organisms. For our purposes, this is, however, unsatisfactory because compositionally similar higher and lower organisms would have the same exergy, which would not account for the exergy embodied in information. The problem of assessing $c_{i,eq}$ has been discussed and a possible solution proposed by Jørgensen (1997, 2002) and Jørgensen et al. (2000). The essential arguments are repeated here. The chemical potential of dead organic matter, indexed $i = 1$, can be expressed from classical thermodynamics as:

$$\mu_1 = \mu_{1,eq} + RT \ln \frac{c_1}{c_{1,eq}}, [ML^2T^{-2}\ moles^{-1}] \tag{10.6}$$

where μ_1 is the chemical potential. The difference $\mu_1 - \mu_{1,eq}$ is known for detritus organic matter, which is a mixture of carbohydrates, fats, and proteins.

By using this particular exergy based on the same system at thermodynamic equilibrium as a reference, the eco-exergy becomes dependent only on the chemical potential of the numerous biochemical components.

It is possible to distinguish in Eq. (1) between the contribution to the eco-exergy from the information and from the biomass. We define p_i as c_i/A, where:

$$A = \sum_{i=1}^{n} c_i \tag{10.7}$$

is the total amount of matter density in the system. With introduction of this new variable, we get:

$$Ex = A\,RT \sum_{i=1}^{n} p_i \ln \frac{p_i}{p_{io}} + A \ln \frac{A}{A_o} \tag{10.8}$$

As $A \approx A_o$, eco-exergy becomes a product of the total biomass A (multiplied by RT) and Kullback measure:

$$K = \sum_{i=1}^{n} p_i \ln \left(\frac{p_i}{p_{io}} \right) \tag{10.9}$$

where p_i and p_{io} are probability distributions, *a posteriori* and *a priori* to an observation of the molecular detail of the system. It means that K expresses the amount of information that is gained as a result of the observations. For different organisms that contribute to the eco-exergy of the ecosystem, the eco-exergy density becomes c RT ln (p_i/p_{io}), where c is the concentration of the considered organism. RT ln (p_i/p_{io}), denoted β, is found by calculating the probability to form the considered organism at thermodynamic equilibrium, which would require that organic matter is formed and that the proteins (enzymes) controlling the life processes in the considered organism have the right amino acid sequence. These calculations can be seen in Jørgensen and Svirezhev (2005). In the latter reference, the latest information about the β values for various organisms is presented (see Table 10.2). For humans, the β value is 2173, when the eco-exergy is expressed in detritus equivalent or 18.7 times as much, or 40635 kJ/g if the eco-exergy should be expressed as kJ and the concentration unit g/unit of volume or area. One hypothesis, apparently confirmed by observation, is that the β values increase as a result of evolution. To mention a few β values from Table 10.2: bacteria 8.5, protozoa 39, flatworms 120, ants 167, crustaceans 232, mollusks 232, fish 499, reptiles 833, birds 980, and mammals 2127. Evolution has resulted in an increasingly more effective transfer of what we could call the classical work capacity to the work capacity of the information. A β value of 2.0 means that the eco-exergy embodied in the organic matter and the information are equal. As the β values become much bigger than 2.0 the information eco-exergy becomes the most significant part of the eco-exergy of organisms.

In accordance with the previously presented interpretation of Eqs. (10.8) and (10.9), it is now possible to find the eco-exergy density for a model as:

$$\text{Eco-exergy density} = \sum_{i=1}^{i=n} \beta_i c_i \qquad (10.10)$$

The eco-exergy due to the "fuel" value of organic matter (chemical energy) is about 18.7 kJ/g (compared with coal: about 30 kJ/g and crude oil: 42 kJ/g). It can be transferred to other energy forms, such as mechanical work directly, and be measured by bomb calorimetry, which requires destruction of the sample (organism). The information eco-exergy = $(\beta - 1) \times$ biomass or density of information eco-exergy = $(\beta - 1) \times$ concentration. The information eco-exergy controls the function of the many biochemical processes. The ability of a living system to do work is contingent upon its functioning as a living dissipative system. Without the information eco-exergy, the organic matter could only be used as fuel similar to fossil fuel. Because of the information eco-exergy, organisms are able to make a network of the sophisticated biochemical processes that characterize life. The eco-exergy (of which the major part is embodied in the information) is a measure of the organization (Jørgensen & Svirezhev,

Table 10.2 ß values = Exergy Content Relatively to the Exergy of Detritus

Early Organisms	Plants		Animals
Detritus		1.00	
Virus		1.01	
Minimal cell		5.8	
Bacteria		8.5	
Archaea		13.8	
Protists	Algae	20	
Yeast		17.8	
		33	Mesozoa, Placozoa
		39	Protozoa, amoebae
		43	Phasmida (stick insects)
Fungi, molds		61	
		76	Nemertina
		91	Cnidaria (corals, sea anemones, jelly fish)
	Rhodophyta	92	
		97	Gatroticha
Porifera, sponges		98	
		109	Brachiopoda
		120	Platyhelminthes (flatworms)
		133	Nematoda (round worms)
		133	Annelida (leeches)
		143	Gnathostomulida
	Mustard weed	143	
		165	Kinorhyncha
	Seedless vascular plants	158	
		163	Rotifera (wheel animals)
		164	Entoprocta
	Moss	174	
		167	Insecta (beetles, fruit flies, bees, wasps, bugs, ants)
		191	Coleodiea (Sea squirt)
		221	Lepidoptera (buffer flies)
		232	Crustaceans, Mollusca, bivalvia, gastropodea
		246	Chordata
	Rice	275	
	Gynosperms (incl. pinus)	314	
		322	Mosquito
	Flowering plants	393	
		499	Fish
		688	Amphibia
		833	Reptilia
		980	Aves (birds)
		2127	Mammalia
		2138	Monkeys
		2145	Anthropoid apes
		2173	*Homosapiens*

Jørgensen, Ladegaard, Debeljak, and Marques, 2005.

2005). This is the intimate relationship between energy and organization that Schrödinger (1944) was struggling to find.

Eco-exergy is a result of copying again and again in a long chain of copies where only minor changes are introduced for each new copy. The energy required for the copying process is very small, but it requires a lot of energy to come to the "mother" copy through evolution from prokaryotes to human cells. To cite Margalef (1969, 1991, 1995) in this context:

> evolution provides for cheap — unfortunately often in exact — copies of messages or pieces of information.

The information concerns the degree of uniqueness of entities that exhibit one characteristic complexion that may be described.

Eco-exergy has successfully been used to develop structurally dynamic models in 21 case studies so far. The eco-exergy goal function is found using Eq. (10.10), while the β values are found using Table 10.2.

The application is based on what may be considered thermodynamic translation of survival of the fittest. Biological systems have many possibilities for moving away from thermodynamic equilibrium, and it is important to know along which pathways among the possibilities a system will develop. This leads to the following hypothesis, which is sometimes denoted the ecological law of thermodynamics (Jørgensen & Fath, 2004b). If a system receives an input of exergy, then it will utilize this exergy to perform work. The work performed is first applied to maintain the system (far) away from thermodynamic equilibrium where exergy is lost by transformation into heat at the temperature of the environment. If more exergy is available, then the system is moved further away from thermodynamic equilibrium, which is reflected in growth of gradients. If more than one pathway to depart from equilibrium is offered, then the one yielding the highest eco-exergy storage (denoted Ex) will tend to be selected. In other words, among the many ways for ecosystems to move away from thermodynamic equilibrium, the one maximizing dEx/dt under the prevailing conditions will have a propensity to be selected.

This hypothesis is supported by several ecological observations and case studies (Jørgensen & Svirezhev, 2005; Jørgensen & Fath, 2004; Jørgensen, 2008b). Survival implies maintenance of the biomass, and growth means increase of biomass and information. It costs exergy to construct biomass and gain information and biomass and information possess exergy. Survival and growth can therefore be measured using the thermodynamic concept eco-exergy, which may be understood as the work capacity the ecosystem possesses.

10.4. Development of Structurally Dynamic Model for Darwin's Finches

The development of an SDM for Darwin's finches illustrates the advantages of SDMs very clearly (see details in Jørgensen & Fath, 2004). The model reflects the available knowledge, which in this case is comprehensive and sufficient to validate even the ability of the model to describe the changes in the beak size as a result of climatic changes, causing changes in the amount, availability, and quality of the seeds that make up the main food item for the finches. The medium ground finches, *Geospiza fortis*, on the island Daphne Major, were selected for this modelling case due to very detailed case-specific information found in Grant (1986). The model has three state variables: seed, Darwin's finches adult, and Darwin's finches juvenile. The juvenile finches are promoted to adult finches 120 days after birth. The mortality of the adult finches is expressed as a normal mortality rate plus an additional mortality rate due to food shortage and an additional mortality rate caused by a disagreement between bill depth and the size and hardness of seeds. Due to a particular low precipitation from 1977 to 1979, the population of the medium ground finches declined significantly and the beak size increased about 6% at the same time. An SDM was developed to describe this adaptation of the beak size due to bigger and harder seeds as a result of the low precipitation.

The beak depth can vary between 3.5 and 10.3 cm according to Grant (1986). The beak size is furthermore equal to the square root of D*H, where D is the diameter and H is the hardness of the seeds. Both D and H are dependent on the precipitation, particularly from January to April. The coordination or fitness of the beak size with D and H is a survival factor for the finches. The fitness function is based on the seed handling time and it influences the mortality as stated above, but it also impacts the number of eggs laid and the mortality of the juveniles. The growth rate and mortality rate of the seeds is dependent on the precipitation and the temperature, which are forcing functions known as f(time). The food shortage is calculated from the food required by the finches (which is known according to Grant, 1986) and the actual available food according to the state function seed. How the food shortage influences the mortality of the adults and juveniles can be found in Grant (1986). The seed biomass and the number of finches are known as a function of time for the period 1975–1982 (Grant, 1986). The observations of the state variables from 1975 to 1977 were applied for calibration of the model, focusing on the following parameters:

1. The influence of the fitness function on: (a) the mortality of adult finches, (b) the mortality of juvenile finches, and (c) the number of eggs laid.
2. The influence of food shortage on the mortality of adult and juvenile finches is known (Grant, 1986). The influence is therefore calibrated within a narrow range of values.
3. The influence of precipitation on the seed biomass (growth and mortality).

All other parameters are known from the literature (Grant, 1986).

The eco-exergy density is calculated (estimated) as $275 \times$ the concentration of seed $+ 980 \times$ the concentration of finches (see Table 10.2). Every 15 days, it is decided if a feasible change in the beak size, taking the generation time and the variations in the beak size into consideration, will give a higher exergy. If it is feasible, then the beak size is changed accordingly. The modelled changes in the beak size were confirmed by the observations. The model results of the number of Darwin's finches are compared with the observations in Figure 10.5. The standard deviation between modelled and observed values was 11.6 %. The validation and the correlation coefficient, r^2, for modelled versus observed values, is 0.977. The results of a nonstructural dynamic model would not be able to predict the changes in the beak size, therefore giving values that are too low for the number of Darwin's finches because their beak would not adapt to the lower precipitation yielding harder and bigger seeds. The calibrated model not using the eco-exergy optimization for the SDMs in the validation period 1977–1982 resulted in complete extinction of the finches. A nonstructurally dynamic model — a normal biogeochemical model — could not describe the impact of the low precipitation, while the SDM gave an approximately correct number of finches and could describe the increase of the beak at the same time.

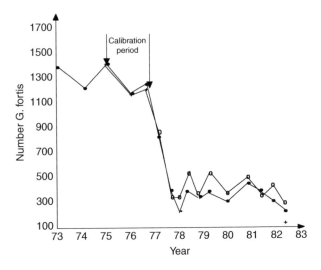

FIGURE 10.5 The observed number of finches (shown with solid dots) from 1973 to 1983, compared with the simulated result (shown with open circles). 1975 and 1976 were used for calibration and 1977 and 1978 for the validation. (The x-axis should indicate 1973 – 1983).

10.5. Biomanipulation

The eutrophication and remediation of a lacustrine environment do not proceed according to a linear relationship between nutrient load and vegetative biomass, instead they display a sigmoid trend with delay (as shown in Figure 10.6). The hysteresis reaction is completely in

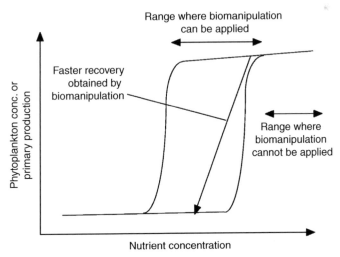

FIGURE 10.6 The hysteresis relation between nutrient level and eutrophication measured by the phytoplankton concentration is shown. The possible effect of biomanipulation is also shown. An effect of biomanipulation can hardly be expected above a certain concentration of nutrients, as indicated on the diagram. The biomanipulation can only give the expected results in the range where two different structures are possible.

accordance with observations (Hosper, 1989; Van Donk et al., 1989) and it can be explained by structural changes (De Bernardi, 1989; Hosper, 1989; Sas, 1989; De Bernardi & Giussani, 1995). A lake ecosystem shows a marked buffering capacity to increasing nutrient level that can be explained by an increasing removal rate of phytoplankton by grazing and settling. Zooplankton and fish abundance are maintained at relatively high levels under these circumstances. At a certain level of eutrophication, it is not possible for zooplankton to increase the grazing rate further, and the phytoplankton concentration will increase very rapidly by slightly increasing concentrations of nutrients. When the nutrient input is decreased under these conditions, a similar buffering capacity to variation is observed. The structure has now changed to a high concentration of phytoplankton and planktivorous fish, which causes a resistance and delay to a change where the second and fourth trophic levels become dominant again.

Willemsen (1980) distinguished two possible conditions:

1. A bream state characterized by turbid water, high eutrophication, low zooplankton concentration, absent of submerged vegetation, large amount of breams, while pike is hardly found at all.
2. A pike state, characterized by clear water and low eutrophication. Pike and zooplankton are abundant and there are significantly fewer bream.

The presence of two possible states in a certain range of nutrient concentrations may explain why biomanipulation has not always been used successfully. According to the observations referred to in the literature, success is associated with a total phosphorus concentration below 50 µg/L (Lammens, 1988) or at least below 100–200 µg/L (Jeppesen et al., 1990), while disappointing results are often associated with phosphorus concentration above this level of more than approximately 120 µg/L (Benndorf, 1987, 1990) with a difficult control of the standing stocks of planktivorous fish (Shapiro, 1990; Koschel et al., 1993).

Scheffer (1990) used a mathematical model based on catastrophe theory to describe these shifts in structure. However, this model does not consider the shifts in species composition, which is of particular importance for biomanipulation. The zooplankton population undergoes a structural change when we increase the concentration of nutrients; for example, from a dominance of calanoid copepods to small caldocera and rotifers (according to De Bernardi & Giussani, 1995 and Giussani & Galanti, 1995). Hence, a test of SDMs could be used to give a better understanding of the relationship between concentrations of nutrients and the vegetative biomass and to explain possible results of biomanipulation. This section refers to the results achieved by an SDM that aims to understand the previously described changes in structure and species compositions (Jørgensen & De Bernardi, 1998). The applied model has 6 state variables: (1) dissolved inorganic phosphorus; (2) phytoplankton, phyt.; (3) zooplankton, zoopl.; (4) planktivorous fish, fish 1; (5) predatory fish, fish 2; and (6) detritus. The forcing functions are the input of phosphorus, in P, and the throughflow of water determining the retention time. The latter forcing function also determines the outflow of detritus and phytoplankton. The conceptual diagram is similar to Figure 2.1, except that only phosphorus is considered as nutrient, as it is presumed that phosphorus is the limiting nutrient.

Simulations have been carried out for phosphorus concentrations in the inflowing water of 0.02, 0.04, 0.08, 0.12, 0.16, 0.20, 0.30, 0.40, 0.60, and 0.80 mg/L. For each of these cases, the model was run for any combination of a phosphorus uptake rate of 0.06, 0.05, 0.04, 0.03, 0.02, and 0.01 1/24h and a grazing rate of 0.125, 0.15, 0.2, 0.3, 0.4, 0.5, 0.6, 0.8, and 1.0 1/24h. When these two parameters were changed, simultaneous changes of phytoplankton and zooplankton mortalities were made according to allometric principles (Peters, 1983). The parameters for phytoplankton growth rate (uptake rate of phosphorus) and mortality and for zooplankton growth rate and mortality are made variable to account for the dynamics in structure.

The settling rate of phytoplankton was made proportional to the (length)^2. Half of the additional sedimentation when the size of phytoplankton increases corresponding to a decrease in the uptake rate was allocated to detritus to account for resuspension or faster release from the sediment. A sensitivity analysis revealed that exergy is most sensitive to changes in these six selected parameters, which also represent the parameters that change significantly by size. The 6 levels selected from the previous list represent an approximate range in size for phytoplankton and zooplankton respectively.

For each phosphorus concentration, 54 simulations were carried out to account for all combinations of the two key parameters. Simulations over 3 years (1100 days) were applied to ensure that steady state, limit cycles, or chaotic behavior would be attained. This SDM approach presumed that the combination with the highest exergy should be selected to represent the process rates in the ecosystem. If exergy oscillates during the last 200 days of the simulation, then the average value for the last 200 days was used to decide on which parameter combination would give the highest exergy. The combinations of the two parameters, the uptake rate of phosphorus for phytoplankton and the grazing rate of zooplankton giving the highest exergy at different levels of phosphorus inputs, are plotted in Figures 10.7 and 10.8. The uptake rate of phosphorus for phytoplankton is gradually decreasing when the phosphorus concentration increases. As seen, the zooplankton grazing rate changes at the phosphorus concentration 0.12 mg/l from 0.4 1/24h to 1.0 1/24h, i.e. from larger species to smaller species, which is according to the expectations.

Figure 10.9 shows the eco-exergy, named on the diagram information, with an uptake rate according to the results in Figure 10.7 and a grazing rate of 1.0 1/24h (called information 1) and 0.4 1/24h (called information 2). Below a phosphorus concentration of 0.12 mg/L, information 2 is slightly higher, while information 1 is significantly higher above this concentration. The phytoplankton concentration increases

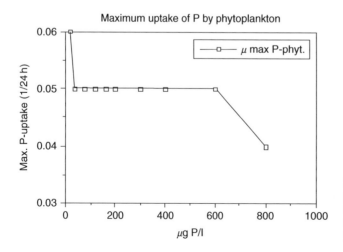

FIGURE 10.7 The maximum growth rate of phytoplankton obtained by the SDM approach is plotted versus the phosphorus concentration.

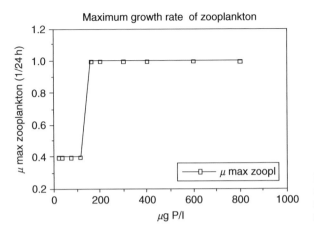

FIGURE 10.8 The maximum growth rate of zooplankton obtained by the SDM approach is plotted versus the zooplankton concentration.

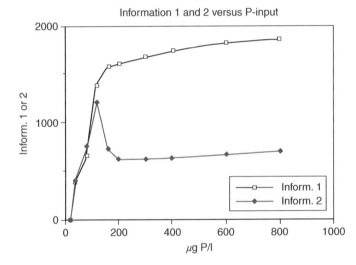

FIGURE 10.9 The exergy is plotted versus the phosphorus concentration. Information 1 corresponds to a maximum growth rate of 1 1/24h and information 2 corresponds to a maximum zooplankton growth rate of 0.4 1/24h. The other parameters are the same for the two plots, including the maximum phytoplankton growth rate (uptake of P) is taken from Figure 10.7 as a function of the phosphorus concentration.

for both parameter sets with increasing phosphorus input, as shown Figure 10.10, while the planktivorous fish shows a significantly higher level by a grazing rate of 1.0 1/24h when the phosphorus concentration is ≥ 0.12 mg/L (= valid for the high exergy level). Below this concentration, the difference is minor. The concentration of fish 2 is higher for case 2 corresponding to a grazing rate of 0.4 1/24h for phosphorus concentrations below 0.12 mg/L. Above this value, the differences are minor, but at a phosphorus concentration of 0.12 mg/L the level is

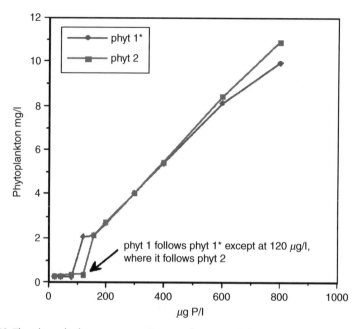

FIGURE 10.10 The phytoplankton concentration as a function of the phosphorus concentration for parameters corresponding to "information 1" and "information 2"; see Figure 10.9. The plot named "phyt 1*" coincides with "phyt 2," except for a phosphorus concentration of 0.12 mg/L, where the model shows limit cycles. At this concentration, information 1* represents the higher phytoplankton concentration, while information 2 represents the lower phytoplankton concentration. Notice that the structural dynamic approach can explain the hysteresis reactions.

significantly higher for a grazing rate of 1.0 1/24h, particularly for the lower exergy level, where the zooplankton level is also highest.

If it is presumed that eco-exergy can be used as a goal function in ecological modelling, then the results seem to explain why we observe a shift in grazing rate of zooplankton at a phosphorus concentration in the range of 0.1–0.15 mg/L. The ecosystem selects the smaller species of zooplankton above this level of phosphorus because it means a higher level of the eco-exergy, which can be translated to a higher rate of survival and growth. It is interesting that this shift in grazing rate produces only a small rise in the level of zooplankton, while the exergy index level rises significantly higher by this shift, which may be translated as survival and growth for the entire ecosystem. Simultaneously, a shift from a zooplankton, predatory fish dominated system to a system dominated by phytoplankton and particularly by planktivorous fish takes place.

It is interesting that the levels of eco-exergy and the four biological components of the model for phosphorus concentrations at or below 0.12 mg/L parameter combinations are only slightly different for the two parameter combinations. This explains why biomanipulation is more successful in this concentration range. Above 0.12 mg/L the differences are much more pronounced and the exergy index level is clearly higher for a grazing rate of 1.0 1/24h. It should therefore be expected that the ecosystem, after the use of biomanipulation, easily falls back to the dominance of planktivorous fish and phytoplankton. These observations are consistent with the general experience of success and failure of biomanipulation.

An interpretation of the results points toward a shift at 0.12 mg/L, where a grazing rate of 1.0 1/24h yields limit cycles. It indicates an instability and probably an easy shift to a grazing rate of 0.4 1/24, although the exergy level is on average highest for the higher grazing rate. A preference for a grazing rate of 1.0 1/24h at this phosphorus concentration should therefore be expected, but a lower or higher level of zooplankton is dependent on the initial conditions.

If the concentrations of zooplankton and fish 2 are low and high for fish 1 and phytoplankton, that is, the system is coming from higher phosphorus concentrations, then the simulation produces with high probability a low concentration of zooplankton and fish 2. When the system is coming from high concentrations of zooplankton and of fish 2, the simulation illustrates with high probability a high concentration of zooplankton and fish 2, which corresponds to an eco-exergy index level slightly lower than obtained by a grazing rate of 0.4 1/24h. This grazing rate will therefore still persist. As it also takes time to recover the population of zooplankton and particularly of fish 2; and in the other direction of fish 1, these observations explain the presence of hysteresis reactions.

This model is considered to have general applicability and has been used to discuss the general relationship between nutrient level and vegetative biomass and the general experiences by application of biomanipulation. When the model is used in specific cases, it may be necessary to include more details and change some of the process descriptions to account for the site specific properties, which is according to general modelling strategy. It could be considered to include two state variables to cover zooplankton, one for the bigger and one for the smaller species. Both zooplankton state variables should have a current change of the grazing rate according to the maximum value of the goal function.

This model could probably also be improved by introducing size preference for the grazing and the two predation processes, which is in accordance with numerous observations. In spite of these shortcomings of the applied model, it has been possible to give a qualitative description of the response to changed nutrient level and biomanipulation, and even to indicate an approximately correct phosphorus concentration where the structural changes may occur. This may be due to an increased robustness by the SDM approach.

Ecosystems are very different from physical systems mainly due to their enormous adaptability. It is therefore crucial to develop models that are able to account for this property, if we want reliable model results. The use of goal functions such as eco-exergy to simulate fitness offers a good way to develop a new generation of models, which are able to consider the adaptability of ecosystems and to describe shifts in species composition. The latter advantage is probably the most important because a description of the dominant species in an ecosystem is often more essential than assessing the level of the focal state variables.

It is possible to model competition between a few species with very different properties, but the SDM approach makes it feasible to include more species even with only slightly different properties, which is impossible by the usual modelling approach (see also the unsuccessful attempt by Nielsen, 1992a,b). The rigid parameters of the various species make it difficult for the species to survive under changing circumstances. After some time, only a few species will still be present in the model, which is different in reality, where more species survive because they are able to adapt to the changing circumstances. It is important to capture this feature in our models. The SDMs seem promising when applied in lake management, as this type of model could explain our experiences with biomanipulation. It has the advantage compared with catastrophe models, which can also be used to explain success and failure of biomanipulation that it is able also to describe the shifts in species composition expressed by the size.

10.6. An Ecotoxicological Structurally Dynamic Models Example

The conceptual diagram of the ecotoxicological model used to illustrate an SDM is shown in Figure 10.11. This model is presented by Jørgensen (2009) in Devillers (2009). The model software STELLA was used for the model simulation results. Copper is an algaecide causing an increase in the mortality of phytoplankton (Kallqvist & Meadows, 1978) and a decrease in the phosphorus uptake and photosynthesis. Copper also reduces the carbon assimilation of bacteria. The literature changes these three model parameters: growth rate of phytoplankton,

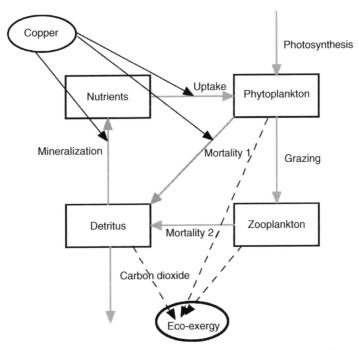

FIGURE 10.11 Conceptual diagram of an ecotoxicological model focusing on the influence of copper on the photosynthetic rate, phytoplankton mortality rate, and the mineralization rate. The boxes are the state variables, the thick gray arrows symbolize processes, and the thin black arrows indicate the influence of copper on the processes and the calculation of eco-exergy from the state variables. Due to the change in these three rates, it is advantageous for the zooplankton and the entire ecosystem to decrease its size. The model is therefore made structurally dynamic by allowing zooplankton to change their size and the specific grazing rate and the specific mortality rate according to allometric principles. The size yielding the highest eco-exergy is currently found.

mortality of phytoplankton, and mineralization rate of detritus with increased copper concentration (Havens, 1999). As a result, the zooplankton is reduced in size, which, according to allometric principles, means an increased specific grazing rate and specific mortality rate. It has been observed that the size of zooplankton in a closed system (e.g., a pond) is reduced to less than half the size at a copper concentration of 140 mg/m^3 compared with a copper concentration less than 10 mg/m^3 (Havens, 1999). In accordance with allometric principles (Peters, 1983), it would result in a more than doubled grazing and mortality rate.

The model shown in Figure 10.11 was made structurally dynamic by varying the zooplankton size and using an allometric equation to determine the corresponding specific grazing and mortality rates. This equation expresses that the two specific rates are inversely proportional to the linear size (Peters, 1983). Different copper concentrations from 10 mg to 140 mg/m^3 are found by the model in which zooplankton size yields the highest eco-exergy. In accordance to the presented SDM approach, it is expected that the size yielding the highest eco-exergy would be selected. The results of the model runs are shown in Figures 10.12, 10.13, and 10.14. The specific grazing rate, the size yielding the highest eco-exergy, and the eco-exergy are plotted versus the copper concentration in these three figures.

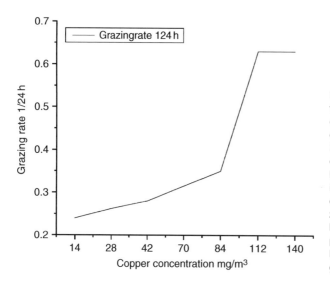

FIGURE 10.12 The grazing rate that yields the highest eco-exergy is shown at different copper concentrations. It increases more rapidly as the copper concentration increases. But at a certain level, it is impossible to increase the eco-exergy further by changing the zooplankton parameters because the amount of phytoplankton becomes the limiting factor for zooplankton growth.

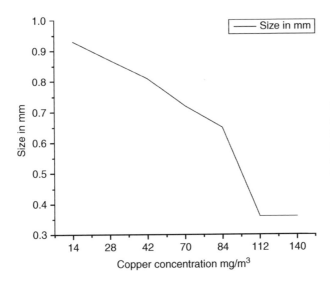

FIGURE 10.13 The zooplankton size that yields the highest eco-exergy is plotted versus the copper concentration. The size decreases more rapidly as the copper concentration increases. But at a certain level, it is impossible to increase the eco-exergy further by changing the zooplankton size because the amount of phytoplankton becomes the limiting factor for zooplankton growth.

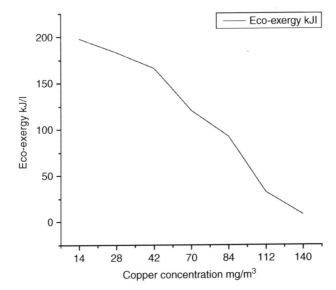

FIGURE 10.14 The highest eco-exergy obtained when varying the zooplankton size is plotted versus the copper concentration. The eco-exergy decreases almost linearly with increasing copper concentration. The discrepancy from a nearly linear plot may be due to model uncertainty, discontinuous change of the copper concentration, and the zooplankton size.

As expected, the eco-exergy, even at the zooplankton size, yields the highest eco-exergy that decreases with increased copper concentration due to the toxic effect on phytoplankton and bacteria.

From the literature, we see the selected size at 140 mg/m^3 is less than 0.4 mm, which is less than one half the size (0.93 mm) at 10 mg/m^3 (see

Figure 10.13). The eco-exergy decreases from 198 kJ/L at 10 mg/m^3 to 8 kJ/L at 140 mg/m^3. The toxic effect of the copper, in other words, results in an eco-exergy reduction to about 4% of the original eco-exergy level, which is a very significant toxic effect. If the zooplankton was not adaptable to the toxic effect by changing its size and the parameters, then the reduction in eco-exergy would have been even more pronounced already at a lower copper concentration. It is therefore important for the model results that the model is made structurally dynamic and accounts for the change of parameters when the copper concentration is changed.

Zooplankton is changing in size in the presented ecotoxicological case. It is an advantage that SDMs can approximately predict the changes in species' properties, but it is an even more important advantage that the state variables are predicted closer to the observations by the SDMs than by biogeochemical models because the organisms are able to adapt to the existing conditions. The toxic effect of copper would have been more pronounced if a nonstructurally dynamic model was applied, which would inevitably have illustrated concentrations of zooplankton that were too small.

Problems

1. Discuss why it would be beneficial to apply an SDM to describe the consequences of (a) global warming, (b) invading species, and (c) an oil spill.
2. Explain why it would not be beneficial to apply an SDM for construction of a subsurface wetland (e.g., the model of a subsurface wetland presented in Chapter 7)
3. Explain why it would advantageous to develop new model types that would be a hybrid of (a) IBM and SDMs, (b) ANN and SDMs, and (c) spatial models and SDMs.
4. Which factors determine the interval between two optimizations of eco-exergy? How could we quantify these factors?
5. Under what circumstances would it be sufficient to use optimization of biomass for the description of the structural changes instead of eco-exergy?

11[grid icon]

Spatial Modelling

CHAPTER OUTLINE

11.1. Introduction

Ecology is a spatial science (Tilman & Kareiva, 1997). Therefore, it is important to be able to model the distribution, movement, and dispersal of species and individuals across a varied and variable landscape. The methods and techniques introduced in Chapter 9 for individual-based models (IBMs) are similar to those used in spatial modelling. This chapter gives an overview, with examples from early methods, for spatial modelling in ecology leading to the state-of-the-art models with current applications.

Compartment models are zero-dimensional models, because all of the processes occur in one place without distinguishing any spatial relations between the compartments. This simplifies the system for ease of finding mathematical solutions, but also obscures the complex reality of ecological systems. The assumption is more reasonable in lake models where one can assume the system represents a continuously stirred tank reactor (CSTR), but even that has severe limitations, as lakes can have vertical (stratification) and horizontal (spatial) variation. Approaches have been developed to overcome this constraint by specifically adding spatial dimensions to the interactions of the ecological components. The rise of Geographical Information Systems (GIS) and remote sensing has

Fundamentals of Ecological Modelling. DOI: 10.1016/B978-0-444-53567-2.00011-9

contributed greatly to the tools available for explicitly representing ecological space. Landscape ecology and spatial modelling are two of the most common tools to benefit from this contribution (Turner and Gardner, 1991). As a general rule, these models are more complex because they add the additional spatial dimensions and require additional knowledge about how movement occurs on the landscape.

As with all models, their application to ecology helps formalize our understanding and develop theory about how ecological processes interact across spatial patterns. According to Turner, Gardner, and O'Neill (2001), there are three general conditions for which spatial models are important:

1. When spatial pattern may be one of the independent variables in the analysis
2. When predicting spatial variation of an attribute through time
3. When the question involves biotic interactions that generate patterns

The first condition refers to questions such as: How do species forage differentially across a landscape of variable resources? How do nutrient inputs respond to vegetation variation across a watershed? The second condition deals specifically with the change in time of the landscape, such as questions of succession or following disturbance. The third often deals with homogeneous space and varying organism traits resulting in the emergence of heterogeneous distributions across the landscape and are generally modelled using cellular automata. Landscape models cover a broad diversity of types and applications; the most frequent subject includes single species metapopulation dynamics influenced by factors such as fragmentation, corridors, dispersal, and invasion. Models that represent disturbance and vegetation dynamics are also common. An area that has seen recent attention and is still in need of more is the integrated models of ecological and socioeconomic processes. The rapidly increasing availability of GIS tools and software has greatly aided the development of spatial models, but one must not let the technological advances outpace the ecological understanding or what is left will be a technically advanced, but unreliable model. One key for successful spatial models is that the model equations should include ecological processes, rather than just correlations, so that the individuals can change over time given different environmental conditions. In this manner, with a dynamic spatial model, the interactions of the ecological species can lead to unexpected or emergent behavior as is often observed in nature. In the next section, we review the early contributions to spatial models in ecology.

11.1.1. Concepts and Terms

A key concept in the area of spatial modelling is *scale*, which refers to the spatial extent of the ecological processes. It is important to choose an appropriate scale related to the specific question at hand, because the processes that affect the different organisms may influence them differently depending on the scale. In fact, many processes operate at multiple scales.

The presence of spatial *patterns* is a key feature of organisms distributed on a landscape. The patterns arise as a result of the ecological processes and the behavioral response of the organisms. Patterns can be classified into three broad distribution categories: (1) *gradients*, which show a smooth directional change over space; (2) *patches*, which show clusters of homogeneous features separated by gaps; and (3) *noise*, which are the random fluctuations not explained by the model. Identification of the pattern can be accomplished with two methods, point pattern and surface pattern analyses. The first category describes the type of distribution and what processes may have caused the pattern. Nearest neighbor method is a common approach to implement this. The second category deals with spatially continuous data and statistical techniques such as correlograms or variograms, which can be used to quantify the magnitude and intensity of the spatial correlation in the data. *Spatial autocorrelation* is an important concept because it identifies the likelihood that samples taken close to each other are more similar than would have occurred by random chance. Positive spatial autocorrelation occurs when the values of samples are more similar than expected by chance and they are negatively spatially correlated otherwise. Most ecological data show some spatial autocorrelation. This tends to decrease with distance. Closer objects tend to have more positive autocorrelation than those further apart since the phenomena that shape species behavior — environmental factors, communication, or interactions — are more similar with proximity.

Acquisition and handling of spatial data are necessary when dealing with spatial models. The availability of spatial data has exploded, thanks to the development of advanced satellite *remote sensing*. The first Earth observation satellite, the Television and Infrared Observation Satellite (TIROS), was launched in 1960 and used mainly for television signal transmission and weather monitoring. Now, such satellites are used for identifying land cover, crop management, forest management, water management, ice cover analysis, national security, and so forth. Some of the most common satellites are from the United States (Landsat), France (SPOT), and India (IKONOS). Prior to the advent of this technology, aerial photographs served as the source for spatial data, and is still used today for gathering information about a specific time and place when a specific resolution is needed, such as for ground truthing satellite data.

In 1972, the U.S. government launched the first in a series of Landsat satellites. The program began at National Aeronautics Space Administration (NASA), but was transferred to the National Oceanic and Atmospheric Association (NOAA) and is now managed by the United States Geological Society (USGS). Since the first launch, there have been six additional satellites (Table 11.1), although the latest was over a decade ago and has had some technical problems. Only two remain active — Landsat 5 and 7. Landsat 5, intended for a 3-year mission, has been sending data for over 25 years at a maximum transmission bandwidth of 85 Mbit/s. It was developed as a backup to Landsat 4 and carries a Thematic Mapper (TM) and Multi-Spectral Scanner (MSS). It orbits at

Table 11.1 Satellite Chronology of the U.S. Landsat Program

Satellite	Launch date	Status
Landsat 1	July 23, 1972	Terminated January 6, 1978
Landsat 2	January 22, 1975	Terminated January 22, 1981
Landsat 3	March 5, 1978	Terminated March 31, 1983
Landsat 4	July 16, 1982	Terminated 1993
Landsat 5	March 1, 1984	Still functioning
Landsat 6	October 5, 1993	Failed to reach orbit
Landsat 7	April 15, 1999	Still functioning, but with faulty scan line corrector

Table 11.2   Spectral Bands of the MSS Sensor

Band	Wavelength (μm)
1	0.45–0.52
2	0.52–0.60
3	0.63–0.69
4	0.76–0.90

an altitude of 705 km and takes 16 days to scan the entire Earth. The MSS has four bands (Table 11.2) and scans at a resolution of about 76 m. This scanner was placed on the first four Landsat satellites, but has been phased out due to the improved TM sensor. TM sensors have seven bands of image data (Table 11.3) with resolution of about 30 m. It is a useful tool for identifying ground cover types as well as albedo and its relation to global climate change. In Landsat 7, the TM was upgraded to what was called an Enhanced Thematic Mapper Plus (ETM+). Landsat 7 also orbits at 705 km and takes 16 days to scan the entire Earth's surface. Resolution of the ETM+ is 15 m in the panchromatic band and 60 m in one thermal infrared channel. Images from the Landsat satellites are in false color (e.g., Figures 11.1 and 11.2) and must be managed and classified according to the user's interest. Data from these satellites are available from the USGS at http://landsat.gsfc.nasa.gov/data/where.html. Much of the archived ETM+, TM, and MSS data are available for free. These data are used in applications such as Google Earth and NASA World Wind.

Table 11.3   Thematic Mapper Bands

Band	Wavelength (μm)	Resolution (m)
1	0.45–0.52	30
2	0.52–0.60	30
3	0.63–0.69	30
4	0.76–0.90	30
5	1.55–1.75	30
6	10.4–12.5	120
7	2.08–2.35	30

FIGURE 11.1 False infrared color image of Washington DC (15 m resolution) taken from Landsat 7. *(As a work of the United States Government, the image is in the public domain.)*

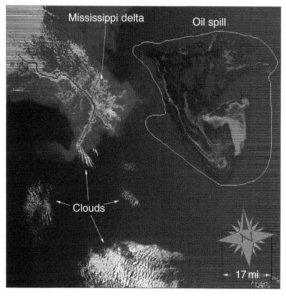

FIGURE 11.2 False infrared color image from Landsat 7 of the Mississippi Delta showing the oil spill following the explosion of the Deepwater Horizon Offshore drilling rigs taken on May 1, 2010. *(As a work of the United States Government, the image is in the public domain.)*

With such abundant data, it is necessary to have platforms in which to import, store, manipulate, and present them. As stated previously, the rise of GIS has met this challenge. Essentially, it is the computational merging of cartography and database technology. A GIS is any tool that allows the user to integrate, store, edit, analyze, share, and display spatial data. The boundaries and specifics depend on the application, but in all cases it provides for the analysis of the spatial information. There are many types of image processing software, such as IDRISI, EASI/PACE, ENVI, LCCS, ER Mapper, ERDAS Imagine, and GRASS GIS. One purpose of this software is to convert the "raw" data from satellite images into specific land use classifications. There has been a number of different land cover classification systems used, such as the "Global Land Cover Classification Collection 1988." This classification system distinguishes 14 land cover classes: 0) water, 1) evergreen needleleaf forest, 2) evergreen broadleaf forest, 3) deciduous needleleaf forest, 4) deciduous broadleaf forest, 5) mixed forest, 6) woodland, 7) wooded grassland, 8) closed shrubland, 9) open shrubland, 10) grassland, 11) cropland, 12) bare ground, and 13) urban and built. An unsupervised or supervised process can be used to classify data. The unsupervised process group's structure is based on similar signals and is useful when previous knowledge of the area is not available and it minimizes the opportunity for human error. The disadvantage is the lack of control over the classification process, which may result in groups that do not correspond to physical real-world data. A supervised process uses known samples to "train" the identification process of unknown pixels. This approach is more time-consuming and assumes a good working knowledge of the area, but gives greater control over the classification process.

The combination of remote sense data and GIS has greatly contributed to the development and implementation of spatial models in ecology. Next we discuss some of the early pioneer applications, as well as the current state-of-the art technology.

11.2. Spatial Ecological Models: The Early Days

One of the main questions addressed through spatial models is the distribution and movement of material or energy across the landscape. To model this movement, there must be a spatial grid over which the movement occurs, as well as a set of rules for the movement to follow. In the absence of such rules, it is assumed that the movement is random across the landscape. This approach is used to generate neutral models for comparison (Caswell, 1976). The neutral model generates random patterns, assuming that species do not interact or react differently to differences in abiotic factors. In reality, the movement is constrained by physical and biological processes as described in the following section. Therefore, the utility of a neutral model is to compare how far from this unrealistic situation is the actual system at hand. The further from this baseline, the more articulated or organized the system, which can then be used as a measure of distance from "equilibrium."

One alternative to the zero-dimensional model was to apply reaction diffusion theory to the distribution of movement along a spatial gradient. This approach uses mathematical models to explain how species disperse across a landscape based on the concentration gradient, which causes the organisms to spread out across the landscape from areas of high population to areas of low population. This provided a mechanism for spatial distribution, but did not capture relevant ecological processes in which organisms more actively make choices about moving across the landscape. Therefore, it was unable to explain observed patchiness and adequately represent the behavior of discrete individuals.

One of the first attempts to combine process-based compartmental modelling with spatial considerations was by Sklar, Costanza, and Day (1985), when they studied the habitat succession of the Atchafalaya delta/Terrebonne marsh area in Louisiana. They divided the area in fixed, equal-sized, square cells (today hexagonal grids are common). The choice to use a finite element method with a fixed grid is appropriate for systems with fixed hydrologic structure. Variable sized mesh grid is used in some hydrodynamic modelling, whereas the grid approach is used in global atmospheric circulation models. Within each cell was a two-compartment, dynamic, nonlinear simulation model representing suspended sediment and bottom sediments with exchanges between them. Furthermore, to make it spatially dynamic, each cell was connected to each adjacent cell by exchange of water and materials. In this first version of the model, there was allowable exchange of salt, sediment, and water across the grids, but not movement of organisms. The spatial extent includes 1,162,641 grid cells representing 50 m^2 each. Initial conditions and parameter values were taken from data and high

Table 11.4 Range of Values for Variable to Classify Habitat Types

Habitat Type	Variable	Range
Upland	Water	0 –8000 m^3
	Salt	0– 30‰
	Bottom sediments	<510 cm^3
Fresh marsh	Water	20–10000 m^3
	Salt	0–5‰
	Bottom sediments	<480– 510 cm^3
Brackish marsh	Water	20–10000 m^3
	Salt	5–15‰
	Bottom sediments	<480–510 cm^3
Salt marsh	Water	20–10000 m^3
	Salt	10–30‰
	Bottom sediments	<480–510 cm^3
Open water	Water	<1000 m^3
	Salt	0–30‰
	Bottom sediments	<480 cm^3

altitude photographs from 1956. They classified the habitat types into five categories: upland, fresh marsh, brackish marsh, salt marsh, or open water depending on the values of the three main variables (Table 11.4). The model simulation was run until the model condition was stable. Constant inputs and a series of IF-THEN statements were used to determine if the cell had switched to a new habitat type based on the water level, salinity, and depth of bottom sediments. The model was validated using photographic data from 1978 (later 1983).

Building on this simple water exchange model, the authors added ecological processes of primary production and decomposition (Costanza et al., 1990). The improved model, called CELSS (short for Coastal Ecological Landscape Spatial Simulation), consisted of seven state variables: water volume, salt, susseds, nitrogen, biomass, detritus, and elevation. The cell size was adjusted to 1 km^2 and the model consisted of 2479 interconnected cells. Simulation results showed a strong similarity to photographic data. The model was then used to consider five different climate scenarios modelled out to the year 2033. All runs showed a marked decrease in all classification types except "open water," which dominated the future landscape due to the sea level rise as a result of increasing climate. This example provided a powerful new approach to combining process-based ecological models linked together across a landscape.

11.3. Spatial Ecological Models: State-of-the-Art

The generalized grid approach introduced by Sklar and Costanza has been adopted and modified for many other studies. The journal *Ecological Modelling* recently published a special issue titled: *Spatially Explicit Landscape Modelling: Current Practices and Challenges* (Volume 220, Issue 24, 2009). It contained a review article and 17 research papers describing the current state-of-the art methodologies, models, and applications. Three examples from that issue, plus one additional one, are provided in the following sections — forest succession, savanna succession, agricultural succession, and fish habitat suitability along a river corridor.

11.3.1. Example 1: Forest Succession After Blowdown

Rammig, Fahse, Bugmann, and Bebi (2006) developed a spatially explicit model to simulate forest succession of Norway spruce in the Swiss Alps following a windstorm blowdown that occurred in 1990. Significant damage was done to the spruce, flattening an area of approximately 128 ha. Following the blowdown, a monitoring program went into place to track the changes in vegetation within the affected area. The spatially explicit model was developed according to the ODD protocol (see Chapter 9). The area within the model was divided into 100×100 grid cells each with a cell size of 1 m^2. Within each cell was an individual based tree-regeneration model. Cells were classified according to 1 of 12 micro-site types, depending on the site characteristics at time t = 0. The sites were defined by factors such as disturbed soil, fallen logs, decaying wood, and different herbaceous vegetation layers. The condition of the micro-site influences the ability for spruce establishment and growth. The model state variables are the number and height of Norway spruce in each cell. The model process overview and scheduling is given in Table 11.5. The first step is to assess the change in the micro-site condition. This is followed by the dispersal of seeds to new sites, and then the germination and establishment of spruce on the new sites. Lastly, the growth and mortality of the spruce is modelled using a vegetation growth model for individual trees. The model parameters were taken from the literature and from a 10-year observation record at the blowdown site. The model time step was one year. The model was run for 50 years and repeated 100

Table 11.5   Process Overview and Scheduling of Spatial Model Following the ODD Protocol

Processes (in Order)	
1) Changes in micro-sites	New micro-sites assigned according to transition probabilities and neighborhood rules
2) Seed dispersal	Random seed distribution, number depends on occurrence of mast years
3) Germination	Norway spruce may establish in each cell depending on micro-site specific germination probabilities
4) Growth	Modified Bertalanffy growth equation
5) Mortality	Intraspecific competition within spruce and interspecific competition between spruce and herb layer

times to gain estimates for mean outputs values. This model was able to adequately simulate the regeneration dynamics of the spruce forest. Of particular interest were the management implications regarding actions such as the clearing of fallen logs and the expectation for the length of duration for recovery of the forest stands.

In a subsequent study, Rammig and Fahse (2009) modified the original model to construct a nonspatial point model from the original one in order to test and compare the predictions made by the spatial and nonspatial models. The goal was to determine the added value of using a spatial model compared to a nonspatial model. Is the extra complexity worth the effort, or can a nonspatial model give comparable results? Therefore, this second model can be considered an extended sensitivity analysis to the first model. The model was derived from the original model, allowing the use of the same parameter values, thus providing a more direct comparison. The main modification to the model involved using a random draw for determining the nearest neighbors. In the original model, the influence of each cell was felt from the eight nearest neighbors. In the revised model, those eight neighbors are drawn randomly form the landscape, eliminating a direct role for spatial proximity (Figure 11.3).

Figure 11.3 illustrates that the selection of cells influencing the local micro-site conditions are spatially selected in the original model and randomly selected in the follow-up model. This allows the investigation of the role of space without changing the overall model structure (Rammig & Fahse, 2009).

The new model produced results that generally overestimated the number of trees for all of the different height classes (Figure 11.4), in

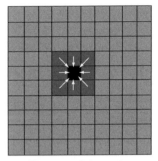

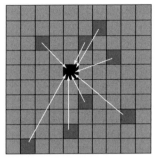

The state of the center cell is determined by the 8 next neighbours

The state of the center cell is determined by 8 randomly drawn cells

FIGURE 11.3 Selection of cells that influence the local micro-site conditions are spatially selected in the original model and randomly selected in the follow-up model. This allows for the investigation of the role of space without changing the overall model structure. *(From Rammig & Fahse, 2009.)*

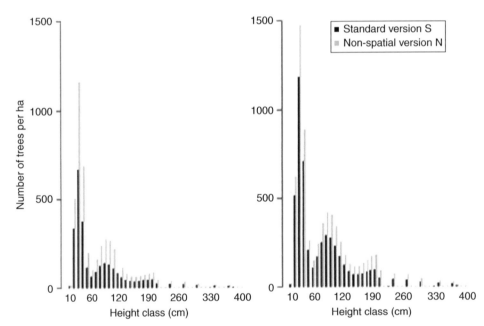

FIGURE 11.4 Results from the nonspatial model overestimate the number of trees in each height class. *(From Rammig & Fahse, 2009.)*

which the total number of trees changed from just over 2600 stems in the spatial model to almost 5000 stems of height 1 to 400 cm in the non-spatial model. This difference is attributed to the role that favorable regeneration sites have on the model; spruce recruitment takes place on favorable sites, such as those dominated by rowan, that tended to be clustered together on the landscape and, hence, in the spatial model. Thus, the spruce regeneration was clustered in these regions in the spatial model, but not constrained by this distribution in the nonspatial model since the rowan sites could be randomly influencing any cells in the nonspatial model. The overall chance of selecting a rowan cell as a neighbor increased and the Norway spruce establishment and growth benefited from this unrealistic circumstance.

This approach shows the sensitivity of the model to the spatial considerations and has important management implications, because the non-forested areas in the high mountain regions pose a risk for increased rockfalls and avalanches. Therefore, it is important to understand the impact of clearing fallen vegetation from the blowdown zone to maximize the regeneration and the ability of the forest to protect the surrounding villages from natural disasters.

11.3.2. Example 2: Long-Term Savanna Succession

Savanna ecosystems are heterogeneous environments characterized by the presence of trees, bushes, and grasses. Nutrient and soil moisture availability are usually the limiting factors affecting the biomass growth in savannas, and overall biomass is impacted by competition, fire, grazing, and harvesting. There is a hypothesis that savannas are naturally patchy environments due to these constraints. Moustakas et al. (2009) developed a spatially explicit savanna succession model to better understand long-term savanna dynamics as well as test the patchy landscape hypothesis. This model is described in detail in the following section.

Similar to the previous example, the grid lattice is 100×100 cells, but here each cell represents approximately $3\,km^2$ for an overall coverage of about $90,000\,km^2$. The state variables of the model are the number of individual trees, bushes, and grass biomass. The vegetation model on each grid cell includes biotic and abiotic factors. The latter factors include temperature and soil moisture characteristics. The former covers the wide range of ecological processes controlling the vegetation biomass dynamics, such as growth and germination factors, and competition terms as well as mortality and grazing/harvesting values (Figure 11.5). Overall, the model has more than 50 parameters initialized from the literature and from field observations. The authors selected a model with

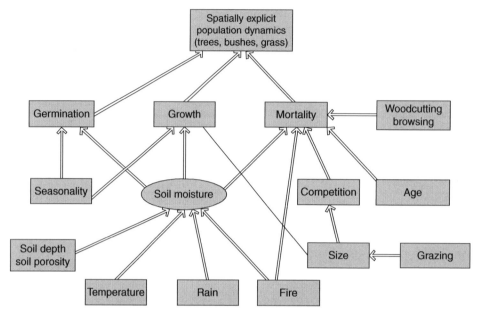

FIGURE 11.5 Conceptual diagram of the vegetation dynamics model. Arrows show the influence of the starting process to the one connected. *(From Moustakas et al., 2009.)*

high complexity, but the well-known processes of the plant dynamics make it possible to have high confidence in the model results. The initialization of state variable values, such as size and age of each tree, was randomly chosen from within reasonable ecological ranges. Because of this random initialization, the model took some time to pass through a transient phase and reach stable conditions. Therefore, while the model is run for 2100 years, the first 100 years of the simulation were excluded from the model results as this time period was used to bring the model to stable conditions. The cells are updated on a daily time step and vegetation growth depends on the soil moisture and season.

The authors considered application of the model to two different regions: an arid savanna (122 mm/year average precipitation and thin soils) in western Namibia and a mesic savanna (780 mm/year average precipitation and brown calcerous soils) located on the Serengeti Plains of Tanzania. They were interested in improving savanna succession to help with management of the ecosystem.

After the model is initialized, during each daily time step the cell site characteristics are updated based on temperature and precipitation

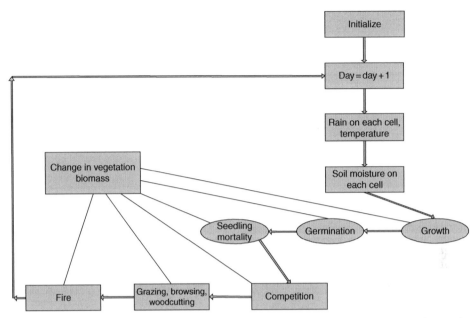

FIGURE 11.6 Flow chart showing the process for the spatially explicit model. *(From Moustakas et al., 2009.)*

which in turn affects the soil moisture. The vegetation dynamics (growth, germination, and mortality) then respond to these environmental conditions. Competition, grazing, and fire further influence the overall change in vegetation biomass (Figure 11.6).

Model results show that total biomass followed a cyclic behavior with grass for about 3 years, bushes for about 50 years, and trees for about 200 years. Tree biomass was similar under both precipitation conditions, but bush biomass doubled in the mesic environment and grass biomass increased sevenfold due to the greater precipitation and soil moisture. Over time, the open savannas were encroached by woody vegetation, which then eventually gave way to a transition back to open areas again. The long-term period for this dynamic was ~230 years for the mesic environment and ~300 years for the arid environment (Figure 11.7). The patchiness of the landscape is affected by fire, grazing, and harvesting, yet these actions did not prevent the observed vegetation cycles. Typical management practice is to remove the woody material through controlled burns or grazing that encroaches on the open savannas, but these results show that this transition to a

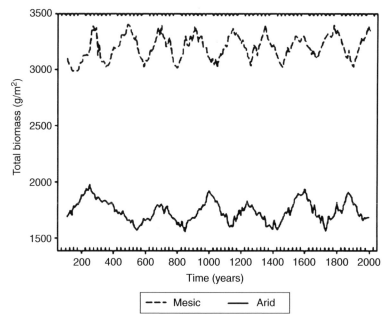

FIGURE 11.7 Model results covering 2000 years showing the long-term cyclical trend in woody biomass structure. *(From Moustakas et al., 2009.)*

woody dominated ecosystem is a naturally cyclic process and that total eradication of the woody material should be avoided unless the aim is to convert the savanna ecosystem solely into grasslands for pastoral reasons.

A model of this type, which incorporates general ecological knowledge with site-specific parameterization, is useful in projecting the landscape changes over time. The model could be applied to other savannas, but would need to be reparameterized to local conditions. More interestingly, this model could be applied to the current locations, but under the conditions of climate change. In other words, human changes to the concentration of greenhouse gases result in new temperature and precipitation patterns. While this model was only looking at the long-term dynamic under current conditions, it would be easy enough to extend the analysis over a long time period for a changing climatic regime. Better understanding of these important ecosystems under the changed climate conditions is important information as the global community grapples with the issue of reducing greenhouse gas emissions. In fact, these models will be called upon to simulate the new climatic conditions and assess the impacts of these changes on the global biosphere.

11.3.3. Example 3: Ecosystem Indicators to Assess Agricultural Landscape Succession

Müller, Schrautzer, Reiche, and Rinker (2006) introduced an ecosystem-oriented indicator set and applied it to a landscape level to assess retrogressional succession in a wetland ecosystem. The model combines field-based measurements linked with GIS for the Bornhöved Lakes District of Northern Germany. It runs 30 years to derive the indicator levels and results of carbon and nitrogen compounds, which demonstrate a shift of the landscape from a sink function to a source. Overall, the indicator set is derived to assess ecosystem structure, such as biodiversity and number of specialized species, as well as ecosystem function measured in terms of energy balance, water balance, and matter balance. For the classification, ecosystem types were characterized according to their soil and vegetation structures. Model results were validated by measurements in the main research area. The classification focused on ecosystems of Histosols and mineral soils. The most common ecosystem type was cultivated fields on mineral soils (64.4%) followed by grasslands and beech forests on mineral soils (13.6 and 4.1%). Wet and drained alder carrs and other alder carrs on Histosol accounted for 1.7% and 2.9% of the study area, respectively, and weakly drained wet grasslands on Histosol made up the remaining 1.2% of cover (Figure 11.8).

The functional variables were calculated using the Water and Substance Simulation Model (WAMOD). It describes processes of water nitrogen and carbon at each location and for lateral transfers of water and nutrients. Results indicate a decrease in species richness both in the wetland ecosystems and on the mineral soils. The overall changes represent a retrogressive succession in loss of specialization. The functional characteristics also show signs of retrogressive succession by a shift toward greater net primary production but less carbon storage. Therefore, the system has switched from a carbon sink to a carbon source as drainage and land use intensity increase (Müller et al., 2006). Nitrogen leaking is also observed to be higher in the mineral soils

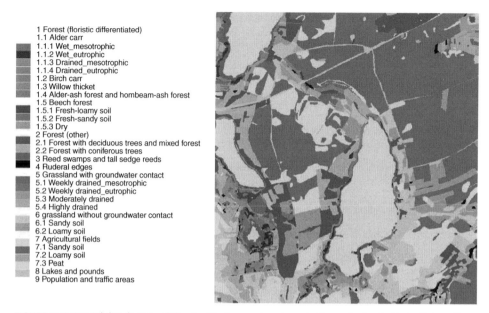

1 Forest (floristic differentiated)
1.1 Alder carr
1.1.1 Wet_mesotrophic
1.1.2 Wet_eutrophic
1.1.3 Drained_mesotrophic
1.1.4 Drained_eutrophic
1.2 Birch carr
1.3 Willow thicket
1.4 Alder-ash forest and hombeam-ash forest
1.5 Beech forest
1.5.1 Fresh-loamy soil
1.5.2 Fresh-sandy soil
1.5.3 Dry
2 Forest (other)
2.1 Forest with deciduous trees and mixed forest
2.2 Forest with coniferous trees
3 Reed swamps and tall sedge reeds
4 Ruderal edges
5 Grassland with groundwater contact
5.1 Weekly drained_mesotrophic
5.2 Weekly drained_eutrophic
5.3 Moderately drained
5.4 Highly drained
6 grassland without groundwater contact
6.1 Sandy soil
6.2 Loamy soil
7 Agricultural fields
7.1 Sandy soil
7.2 Loamy soil
7.3 Peat
8 Lakes and pounds
9 Population and traffic areas

FIGURE 11.8 Spatial distribution of the classified ecosystem types in the watershed of Lake Belau. *(From Müller et al., 2006.)*

than in the Histosol, which contributes to eutrophication of the aquatic systems. Along the retrogressive successional gradient, the water budget tendency of decreasing biotic water use with increasing land use intensity is due to lower proportion of transpiration in evapotranspiration. Results from the study are presented in the amoeba-shaped diagram showing the differences in indicator values for four wetland types of ecosystems (Figure 11.9). In the alder carrs, the consequences of eutrophication are greater than that of draining. In the wet grasslands, the differences are higher, because the extensively drained area for agricultural purposes shows the greater deviation.

11.3.4. Example 4: Fish Habitat Along a River Corridor

A final example in this chapter uses an aquatic model of fish habitat suitability. The spatial scale of a river could be collapsed to one linear dimension following the flow along the river, but here the river is wide compared to cell size in order to better track the fish movement and habitat preference. Therefore, it is quite common for such aquatic spatial models to be three-dimensional to account for the water depth as

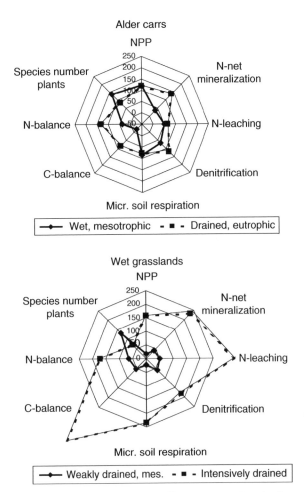

FIGURE 11.9 An amoeba diagram to compare four different stages of wetland retrogression. On the top, alder carrs are depicted; the bottom shows two wet grassland ecosystems. The selected ecosystem types represent the starting points of the retrogressions as well their end points. 100 (%) refers to the average values of the whole data set. *(From Müller et al., 2006).*

well. Hatten and Parsley (2009) developed a spatial model for white sturgeon habitat in the Columbia River. The river ecosystem is impacted by heavy volumes of commerce shipping and the subsequent dredging necessary to maintain open channels. In particular, there is concern that white sturgeon — an important ecological, sport, and commercial fish — mortality is impacted from the material deposition on the sturgeon habitat. The authors developed a spatial model to test the impact of dredging material on sturgeon rearing habitat. They (1) used GIS and survey data to compile a river bathymetry database; (2) developed a habitat model for sturgeon using fish location data from 2002; (3) constructed a habitat suitability map, verified with fish location data from 2003; and (4) simulated the effects of in-water dredge deposition on sturgeon habitat by reducing water depths.

Knowledge about sturgeon ecology was taken from earlier studies showing that the fish are constantly moving with diel migrations preferring deeper zones in the daytime and shallower during night. They are benthic foragers and occupy a broad range of conditions as habitat generalists. However, they are impacted by water depth, bottom slope, and roughness; all of which are affected by the presence of the dredge fill material (Figure 11.10).

A 10 m resolution digital elevation map of the riverbed was created based on a bathymetric survey of the region in 2003. Fish population numbers were obtained from acoustic telemetry data gathered in 2002 and 2003. They tracked between 19 and 33 sturgeon, recording 74,000 locations in 2002 and 88,000 in 2003. From this, 5000 random values from 2002 were used to develop the model and 5000 random locations from 2003 were used to validate the model. The information was compiled in a GIS framework to determine the overall habitat suitability for sturgeon populations under existing conditions and then scenarios in which the fill increased by levels of 3 m increments (Figure 11.11).

The changes due to the increased fill measurements can be beneficial or harmful to the habitat rating resulting in four change classes: (1) low suitability, no change; (2) high suitability becomes low; (3) low suitability

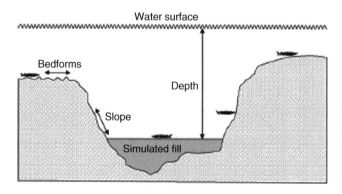

FIGURE 11.10 Fill from dredging affects the water depth, bedform roughness, and bottom slope, all of which impact sturgeon habitat. *(From Hatten & Parsley, 2009.)*

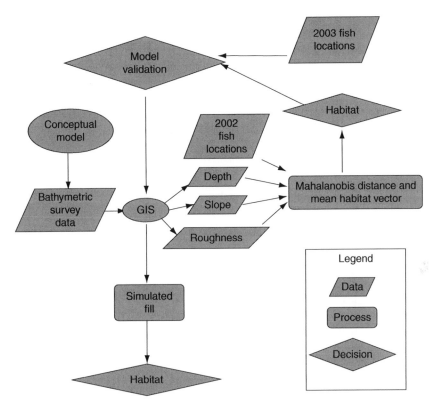

FIGURE 11.11 Conceptual model of steps and processes for the white sturgeon spatial model and the impact of fill on sturgeon habitat. *(From Hatten & Parsley, 2009.)*

becomes high; and (4) high suitability, no change. Figure 11.12 shows that many regions experienced no change, but the central channel transitioned mostly from high suitability to low suitability. The model results showed that there was little change in the area ($<1\%$) of suitable sturgeon habitat for fill levels up to 9 m. When this was increased to 12 m, there was a 12% decrease in suitable area and for 15 m fill, sturgeon habitat decreased by 44%.

   This example provides another use for ecological models. Using this model, researchers were able to get a good estimation of the impact of fill without having to conduct a full-scale field trial to assess the changes. This has clear management implications, because the model indicates that the river has some absorptive buffer capacity to mitigate against low amounts of fill material (<9 m), but as this value increases,

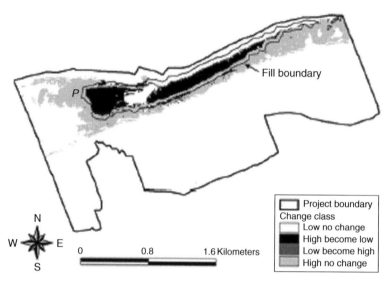

FIGURE 11.12 Estimated changes in sturgeon habitat under different dredge fill scenarios. The largest change is seen in the habitat that transitions high suitability to low suitability. *(From Hatten & Parsley, 2009.)*

the impact on sturgeon habitat increases noticeably, so a good policy recommendation is to keep fill levels below the critical threshold.

Problems

1. Explain some ecological processes that make it necessary to use spatially explicit models.
2. When would it be appropriate to use (a) 0-D model, (b) 1-D model (c) 2-D model, and (d) 3-D model?
3. Management questions often drive the construction of the ecological models. What are some pros and cons of this approach?
4. Explain how a nearest neighbor grid cell approach represents spatial distribution on the landscape. What grid cell shape is most appropriate and why?
5. Explain the difference between gradients, patches, and noise.
6. Does spatial autocorrelation typically increase or decrease with distance? Why?
7. How are image data from Thematic Mapper Plus converted to ecological classifications?

References

Ahlgren, I. (1973). Limnologiska studier av Sjøn Norrvikan. 111. Avlastningens effekter. *Scripta Limnologica Upsaliencia*, No. 333.

Alcamo, et al. (1990). *The Rains Model of Acidification* (p. 366). Dordrecht, The Netherlands: Kluwer and Laxenburg, Austria: IIASA.

Allen, P. M. (1976). Evolution, population dynamics, and stability. *Proceedings of the National Academy of Sciences, 73*, 665–668.

Allen, T. F. H., & Starr, T. B. (1982). *Hierarchy: Perspectives for Ecological Complexity* (p. 310). Chicago, USA: University of Chicago Press.

Allesina, S., & Bondavalli, C. (2004). Wand: An ecological network analysis user-friendly tool. *Environmental Modelling and Software, 19*(4), 337–340.

Andreasen, I. (1985). *A General Ecotoxicological Model for the Transport of Lead Through the System: Air-Soil(Water)-Grass-Cow-Milk. Thesis at DIA-K* (p. 57). Kongens Lyngby: Technical University of Denmark.

Aoyama, I., Inoue, Y., & Inoue, Y. (1978). Simulation analysis of the concentration process of trace heavy metals by aquatic organisms from the viewpoint of nutrition ecology. *Water Research, 12*, 837–842.

ApSimon, H., Goddard, A. J. H., & Wrigley, J. (1976). Estimating the possible transfrontier Consequences of accidental releases: the MESOS model for long range atmospheric dispersal. Seminar on radioactive releases and their dispersion in the atmosphere following a hypothetical reactor accident, Risø, Denmark, April 1980, CEC, Luxembourg pp. 819–842.

Bartell, S. M., Gardner, R. H., & O'Neill, R. V. (1984). The fates of aromatics model. *Ecological Modelling, 22*, 109–123.

Bartell, S. M., Gardner, R. H., & O'Neill, R. V. (1992). *Ecological Risk Estimation*. Boca Raton, Florida: Lewis Publishers.

Beck, M. B. (1987). Water quality modeling: A review of the analysis of uncertainty. *Water Resources Research, 23*(8), 1393–1442.

Benndorf, J. (1987). Food-web manipulation without nutrient control: A useful strategy in lake restoration? *Schweizerische Zeitschrift fur Hydrologie, 49*, 237–248.

Benndorf, J. (1990). Conditions for effective biomanipulation. Conclusions derived from whole-lake experiments in Europe. *Hydrobiologia, 200/201*, 187–203.

Beyer, J. E. (1981). *Aquatic Ecosystem: An Operational Research Approach* (p. 315). Seattle and London: University of Washington Press.

Bierman, V. J., Jr. (1976). Mathematical model of the selective enhancement of blue-green algae by nutrient enrichment. In R. P. Canale (Ed.), *Modelling Biochemical Processes in Aquatic Ecosystems* (pp. 1–31). Michigan: Ann Arbor Sciences Publishers.

Bonner, J. T. (1965). *Size and Cycle. An Essay on the Structure of Biology* (p. 219). New Jersey: Princeton University Press.

Bossel, H. (1992). Real structure process description as the basis of understanding ecosystems. In *Workshop Ecosystem Theory*, October, 14–17, 1991 (pp. 261–276). Kiel. *Special issue of Ecological Modelling* 63.

Bossel, H. (1998). Ecological orientors: Emergence of basic orientors in evolutionary self-organization. In F. Müller & M. Leupelt (Eds.), *Eco Targets, Goal Functions, and Orientors* (pp. 19–33). New York: Springer.

Bossel, H. (1999). *Indicators for sustainable development: Theory, method, applications.* Winnipeg: International Institute for Sustainable Development.

Bosserman, R. W. (1980). Complexity measures for assessment of environmental impact in ecosystem networks. In *Proceedings of the Pittsburgh Conference on Modelling and Simulation.* Pittsburgh, Pennsylvania, April 20–23, 1980.

Bosserman, R. W. (1982). Structural comparison for four lake ecosystem models. In L. Troncale (Ed.), *A General Survey of Systems Methodology, Proceedings of the Twenty- Sixth Annual Meeting of the Society for General Systems Research* (pp. 559–568). Washington, D.C., January 5–9, 1982.

Botkin, D. B., Janak, J. F., & Wallis, J. R. (1972). Some ecological consequences of a computer model of forest growth. *Journal of Ecology, 60,* 849–873.

Brandes, M., Chowdry, N. A., & Cheng, W. W. (1974). *Experimental Study on Removal of Pollutants from Domestic Sewage by Underdrained Soil Filters.* Chicago: National Home Sewage Disposal Symposium on Agricultural Engineering.

Brandes, R. J. (1976). *An Aquatic Ecological Model for Texas Bays and Estuaries.* Austin, Texas: Water Resources Engineers, Inc. For the Texas Water Development Board.

Breck, J. E., DeAngelis, D. L., Van Winkle, W., & Christensen, S. W. (1988). Potential importance of spatial and temporal heterogeneity in pH, Al, and Ca in allowing survival of a fish population: A model demonstration. *Ecological Modelling, 41,* 1–16.

Caswell, H. (1976). Community structure: A neutral model analysis. *Ecological Monographs, 46,* 327–354.

Chon, T. S., Jørgensen, S. E., & Cho, W. S. (2010). The role of conjugation in the gene-individual-population relationships in ecotoxicological systems under the constraints of toxin susceptibility and metabolic efficiency. in press.

Christensen, T. H. (1981). *The Application of Sludge as Soil Conditioner* (vol. 3, pp. 19–47). Copenhagen: Polyteknisk Forlag.

Christensen, T. H. (1984). Cadmium soil sorption at low concentrations, 1) Effect of time, cadmium load, pH, and calcium and 2) Reversibility, effect of changes in solute composition, and effect of soil ageing. *Water, Air and Soil Pollution, 21,* 105–125.

Christensen, V. (1991). On ECOPATH, Fishbyte, and fisheries management. *Fishbyte, 9*(2), 62–66.

Christensen, V., & Pauly, D. (1992). The ECOPATH II – a software for balancing steady-state ecosystem models and calculating network characteristics. *Ecological Modelling, 61,* 169–185.

Christensen, V., & Pauly, D. (Eds.), (1993). *Trophic models of aquatic ecosystems.* ICLARM Conference Proceedings 26, Manila, Philippines.

Chubin, R. G., & Street, J. J. (1981). Adsorption of cadmium on soil constituents in the presence of complexing agents. *Journal of Environmental Quality, 10,* 225–228.

Costanza, R., & Sklar, F. H. (1985). Articulation, accuracy and effectiveness of mathematical models: A review of freshwater wetland applications. *Ecological Modelling, 27,* 45–69.

Costanza, R., Sklar, F. H., & White, M. L. (1990). Modelling coastal landscape dynamics. *Bio-Science, 40,* 91–107.

Cowan, C. E., Mackay, D., Feijtel, T. C. J., can de Meent, D., di Guardo, A., Davies, J., & Mackay, N. (Eds.), (1995). *The Multi-Media Fate Model: A vital tool for predicting the fate of chemicals.* Pensacola FL: SETAC Press.

Cox, J. L. (1970). Accumulation of DDT residues in Triphoturus mexicanus from the Gulf of California. *Nature, 227,* 192–193.

Dahl-Madsen, K. I., & Strange-Nielsen, K. (1974). Eutrophication models for ponds. *Vand, 5,* 24–31.

Dame, R. F., & Patten, B. C. (1981). Analysis of energy flows in an intertidal *oyster reef. Mar, Ecol. Prog. Ser. 5,* 115–124.

DeAngelis, D. L., Cox, D. K., & Coutant, C. C. (1980). Cannibalism and size dispersal in young-of-the-year largemouth bass: Experiment and model. *Ecological Modelling, 8,* 133–148.

De Bernardi, R. (1989). Biomanipulation of aquatic food chains to improve water quality in eutrophic lakes 195-215. In O. Ravera (Eds.), *Ecological Assessment of Environmental Degradation, Pollution, and Recovery* (p. 356). Amsterdam: Elsevier Science Publications.

De Bernardi, R., & Giussani, G. (1995). Biomanipulation: Bases for a Top-down Control 1-14. In R. De Bernardi, & G. Giussani (Eds.), *Guidelines of Lake Management, Volume 7, Biomanipulation in Lakes and Reservoirs* (p. 211). ILEC and UNEP.

De Luna, J. T., & Hallam, T. G. (1987). Effect of toxicants on populations: A qualitative approach IV. Resource-consumer-toxicant models. *Ecological Modelling, 35,* 249–273.

Devillers, J. (2009). *Ecotoxicological Modelling* (p. 399). New York: Springer-Verlag.

Dillon, P. J., & Kirchner, W. B. (1975). The effects of geology and land use on the export of phosphorus from watersheds. *Water Research, 9,* 135–148.

Dillon, P. J., & Rigler, F. H. (1974). A test of a simple nutrient budget model predicting the phosphorus concentration in lake water. *Journal of the Fisheries Research Board of Canada, 31,* 1771–1778.

Dur, G., Souissia, S., Devrekera, D., Ginot, V., Schmitta, F. G., & Hwang, J. S. (2009). An individual-based model to study the reproduction of egg bearing copepods: Application to *Eurytemora affinis* (*Copepoda Calanoida*) from the Seine estuary, France. *Ecological Modelling, 220,* 1073–1089.

EPA Denmark. (1979). *The Lead Contamination in Denmark* (p. 145). Copenhagen.

Fagerstrøm, T., & Aasell, B. (1973). Methyl mercury accumulation in an aquatic food chain. A model and implications for research planning. *Ambio, 2,* 164–171.

Fath, B. D. (2004). Network analysis applied to large-scale cyber-ecosystems. *Ecological Modelling, 171,* 329–337.

Fath, B. D., & Borrett, S. R. (2006). A Matlab® function for network environ analysis. *Environmental Modelling and Software, 21,* 375–405.

Fath, B. D., & Halnes, G. (2007). Cyclic energy pathways in ecological food webs. *Ecological Modelling, 208,* 17–24.

Fath, B. D., & Patten, B. C. (1999). Review of the foundations of network environ analysis. *Ecosystems, 2*, 167–179.

Fath, B. D., Patten, B. C., & Choi, J. S. (2001). Complementarity of ecological goal functions. *Journal of Theoretical Biology, 208*, 493–506.

Felmy, A. R., Brown, S. M., Onoshi, Y., Yabusaki, S. B., Argo, R. S., Girvin, D. C., et al. (1984). *Modelling the Transport, Speciation, and Fate of Heavy Metals, AquaticSystems* (p. 4). Project Summary, EPA-600/S3-84-033, April 1984. Athens, Georgia: US EPA, Environmental Research Laboratory. (EPA Project Officer: R.B. Ambrose).

Fenchel, T. (1974). Intrinsic rate of natural increase: The relationship with body size. *Oecologia, 14*, 317–326.

Findeisen, W., Lastebrov, A., Lande, R., Lindsay, J., Pearson, M., & Quade, E. S. (1978). *A sample glossary of systems analysis*. Working Paper WP-78-12 Laxenburg, Austria: International Institute for Applied Systems Analysis.

Flather, C. H. (1992). *Pattern of Avian Species-Accumulation Rates Among Eastern Forested Landscapes*. Ph.D dissertation. Fort Collins, Colorado: Colorado State University.

Flather, C. H. (1996). Fitting species-accumulation functions and assessing regional land use impacts on avian diversity. *Journal of Biogeography, 23*, 155–168.

Fomsgaard, I. (1997). Modelling the mineralisation kinetics for low concentrations of pesticides in surface and subsurface soil. *Ecological Modelling, 102*, 175–208.

Fomsgaard, I., & Kristensen, K. (1999). Influence of microbial activity, organic carbon content, soil texture, and soil depth on mineralisation rates of low concentrations of 14-C mecoprop — Development of a predictive model. *Ecological Modelling, 122*, 45–68.

Fontaine, T. D. (1981). A self-designing model for testing hypotheses of ecosystem development. In D. Dubois (Ed.), *Progress in Ecological Engineering and Management by Mathematical Modelling* (pp. 281–291). Proc. 2nd Int. Conf. State-of-the-Art Ecological Modelling, 18–24 April 18–24, 1980. Liege, Belgium.

Forrester, J. W. (1961). *Industrial Dynamics*. Cambridge, MA: MIT Press.

France, J., & Thornley, J. H. M. (1984). *Mathematical Models in Agriculture* (p. 333). Burlington, MA: Butterworths.

Gard, T. C. (1990). A stochastic model for the effect of toxicants on populations. *Ecological Modelling, 51*, 273–280.

Gause, G. F. (1934). *The Struggle of Existence* (p. 133). New York: Hafner.

Gilioli, G., & Pasquali, S. (2007). Use of individual-based models for population parameters estimation. *Ecological Modelling, 200*, 109–118.

Gillett, J. W., et al. (1974). *A conceptual model for the movement of pesticides through the environment* (p. 79). Corvallis, OR: National Environmental Research Center, U.S. Environmental Protection Agency, OR Report EPA 600/3-74-024.

Giussani, G., & Galanti, G. (1995). Case Study: Lake Candia (Northern Italy) 135-146. In R. De Bernardi & G. Giussani (Eds.), *Guidelines of Lake Management, Volume 7. Biomanipulation in Lakes and Reservoirs* (p. 211). Shiga, Japan: ILEC and UNEP.

Grant, P. R. (1986). *Ecology and Evolution of Darwin's Finches* (p. 492). Reprinted in 1999. New Jersey: Princeton University Press.

Grimm, V. (2008). Individual Based Models. In S. E. Jorgensen & B. Fath (Eds.), *Encyclopedia of Ecology* (Vol. 1, pp. 1959–1968). Amsterdam: Elsevier.

Grimm, V., et al. (2006). A standard protocol for describing individual-based and agent-based models. *Ecological Modelling, 198,* 115–126.

Grimm, V., Frank, K., Jeltsch, F., Brandl, R., Uchmanski, J., & Wissel, C. (1996). Pattern-oriented modelling in population ecology. *The Science of the Total Environment, 183,* 151–166.

Grimm, V., & Railsback, S. F. (2005). *Individual Based Modelling and Ecology* (p. 482). Princeton, New Jersey: Princeton University Press.

Gromiec, M. J., & Gloyna, E. F. (1973). *Radioactivity transport in water.* Final Report No. 22 to U.S. Atomic Energy Commission, Contract AT (11-1)-490.

Gryning, S. E., & Batchvarova, E. (2000). *Air Pollution Modelling and its Applications XIII. Kluwer Proceedings from a conference held in Varna, Bulgaria, 1998* (p. 810). Dordrecht, the Netherlands: Academic.

Halfon, E. (1983). Is there a best model structure? II. Comparing the model structures or different fate models. *Ecological Modelling, 20,* 153–163.

Halfon, E. (1984). Error analysis and simulation of Mirex behavior in Lake Ontario. *Ecological Modelling, 22,* 213–253.

Halfon, E. (1986). Modelling the fate of Mirex and Lindane in Lake Ontario, off the Niagara River Mouth. *Ecological Modelling, 33,* 13.

Halfon, E., Unbehauen, H., & Schmid, C. (1979). Model order estimation and system identification Theory to the modelling of 32P kinetics within the trophogenic zone of a small lake. *Ecological Modelling, 6,* 1–22.

Hannon, B. (1973). The structure of ecosystems. *Journal of Theoretical Biology, 41,* 535–546.

Hansen, J. A., & Tjell, J. C. (1981). *The Application of Sludge as Soil Conditioner* (Vol. 2, pp. 137–181). Copenhagen: Polyteknisk Forlag.

Hanski, I. (1994). Metapopulation dynamics in fragmented landscapes. *Trends in Ecology and Evolution, 9,* 131–195.

Hanski, I. (1999). *Metapopulation in Ecology* (p. 422). Cambridge, UK: Oxford University Press.

Harris, J. R. W., Bale, A. J., Bayne, B. L., Mantoura, R. C. F., Morris, A. W., Nelson, L. A., et al. (1984). A preliminary model of the dispersal and biological effect of toxins in the Tamar estuary, England. *Ecological Modelling, 22,* 253–285.

Hatten, J. R., & Parsley, M. J. (2009). A spatial model of white sturgeon rearing habitat in the lower Columbia River, USA. *Ecological Modelling, 220,* 3638–3646.

Havens, K. E. (1999). Structural and functional responses of a freshwater community to acute copper stress. *Environmental Pollution, 86,* 259–266.

Herendeen, R. A. (1981). Energy intensity in ecological and economic systems. *Journal of Theoretical Biology, 91,* 607–620.

Higashi, M., Patten, B. C., & Burns, T. P. (1993). Network trophic dynamics: The modes of energy utilization in ecosystems. *Ecological Modelling, 66,* 1–42.

Hilker, F. M., Hinsch, M., & Poethke, H. J. (2006). Parameterizing, evaluating, and comparing metapopulation models with data from individual-based simulations. *Ecological Modelling, 199,* 476–485.

Hippe, P. W. (1983). Environ analysis of linear compartmental systems: The dynamic, time-invariant case. *Ecological Modelling, 19*, 1–26.

Holling, C. S. (1959). Some characteristics of simple types of predation and parasitism. *Canadian Entomologist, 91*, 385–398.

Holling, C. S. (1966). The functional response of invertebrate predators to prey density. *Memoirs of the Entomological Society of Canada, 48*, 1–87.

Hosper, S. H. (1989). Biomanipulation, new perspective for restoring shallow, eutrophic lakes in The Netherlands. *Hydrobiological Bulletin, 73*, 11–18.

Huston, M., DeAngelis, D. L., & Post, W. M. (1988). New computer models unify ecological theory. *BioScience, 38*, 682–691.

Hutchinson, G. E. (1970). The biosphere. *Scientific American, 223*(3), 44–53.

Hutchinson, G. E. (1978). *An Introduction to Population Ecology.* New Haven: Yale University Press.

Jeffers, N. R. J. (1978). *An Introduction to Systems Analysis with Ecological Applications.* London, England: E. Arnold Publishers.

Jensen, K., & Tjell, J. C. (1981). *The Application of Sludge as Soil Conditioner* (Vol. 3, pp. 121–147). Copenhagen: Polyteknisk Forlag.

Jeppesen, E. J., et al. (1990). Fish manipulation as a lake restoration tool in shallow, eutrophic temperate lakes. Cross-analysis of three Danish Case Studies. *Hydrobiologia, 200/201*, 205–218.

Jørgensen, S. E. (1976a). A eutrophication model for a lake. *Journal of Ecological Modelling, 2*, 147–165.

Jørgensen, S. E. (1976b). An ecological model for heavy metal contamination of crops and ground water. *Ecological Modelling, 2*, 59–67.

Jørgensen, S. E. (1979). Modelling the distribution and effect of heavy metals in an aquatic ecosystem. *Ecological Modelling, 6*, 199–222.

Jørgensen, S. E. (1981). Application of exergy in ecological models. In D. Dubois (Ed.), *Progress in Ecological Modelling* (pp. 39–47). Liege: Cebedoc.

Jørgensen, S. E. (1982). A holistic approach to ecological modelling by application of thermodynamics. In W. Mitsch, et al. (Eds.), *Systems and Energy.* Ann Arbor, Michigan: Ann Arbor Press.

Jørgensen, S. E. (1986). Structural dynamic model. *Ecological Modelling, 31*, 1–9.

Jørgensen, S. E. (1988). Use of models as experimental tools to show that structural changes are accompanied by increased exergy. *Ecological Modelling, 41*, 117–126.

Jørgensen, S. E. (1990). Ecosystem theory, ecological buffer capacity, uncertainty, and complexity. *Ecological Modelling, 52*, 125–133.

Jørgensen, S. E. (Ed.), (1991). *A model for the distribution of chromium in Abukir Bay. Modelling in Environmental Chemistry* (Vol. 17). Amsterdam: Elsevier.

Jørgensen, S. E., Nors Nielsen, S., & Jørgensen, L. A. (1991). Handbook of Ecological Parameters and Ecotoxicology. Amsterdam: Elsevier, Published as CD under the name ECOTOX, with L.A. Jørgensen as first editor in 2000.

Jørgensen, S. E. (1992a). Parameters, ecological constraints and exergy. *Ecological Modelling, 62*, 163–170.

Jørgensen, S. E. (1992b). Development of models able to account for changes in species composition. *Ecological Modelling, 62*, 195–208.

Jørgensen, S. E. (1997). *Integration of Ecosystem Theories: A Pattern*. (2nd revised ed., p. 388). Dordrecht, The Netherlands: Kluwer Academic Publishers.

Jørgensen, S. E. (2000). *Pollution Abatement* (p. 488). Amsterdam: Elsevier.

Jørgensen, L. A., Jørgensen, S. E., & Nielsen, S. N. (2000). Ecotox. CD.

Jørgensen, S. E. (2002). *Integration of Ecosystem Theories: A Pattern*. (3rd ed., p. 428). The Netherlands, Dordrecht: Kluwer Academic Publishers.

Jørgensen, S. E. (2008a). *Evolutionary Essays* (p. 230). Amsterdam: Elsevier.

Jørgensen, S. E. (2008b). An overview of the model types available for development of ecological models. *Ecological Modelling, 215*, 3–9.

Jørgensen, S. E. (2009). The application of structurally dynamic models in ecology and ecotoxicology. In J. Devillers (Ed.), *Ecotoxicological Modelling* (pp. 377–394). New York: Springer-Verlag.

Jørgensen, S. E., & Bendoricchio, G. (2001). *Fundamentals of Ecological Modelling*. (3rd ed., p. 628). Amsterdam: Elsevier.

Jørgensen, S. E., & Fath, B. D. (2004). Application of thermodynamic principles in ecology. *Ecological Complexity, 1*(4), 267–280.

Jørgensen, S. E., Chon, T., & Recknagel, F. (2009). *Handbook of Ecological Modelling and Informatics* (p. 432). Southampton: WIT Press.

Jørgensen, S. E., Costanza, R., & Xu, F. (2005, 2010). *Handbook of Ecological Indicators for Assessment of Ecosystem Health*. (2nd ed., p. 439). Boca Raton, Florida: CRC Press.

Jørgensen, S. E., & De Bernardi, R. (1998). The use of structural dynamic models to explain the success and failure of biomanipulation. *Hydrobiologia, 379*, 147–158.

Jørgensen, S. E., & Fath, B. D. (2004). Modelling the selective adaptation of Darwin's finches. *Ecological Modelling, 176*, 409–418.

Jørgensen, S. E., Fath, B., Bastiononi, S., Marques, M., Müller, F., Nielsen, S. N., et al. (2007). *A New Ecology. Systems Perspectives* (p. 288). Amsterdam: Elsevier.

Jørgensen, S. E., Halling-Sørensen, B., & Mahler, H. (1997). *Handbook of Estimation Methods in Ecotoxicology and Environmental Chemistry* (p. 230). Boca Raton, Florida: Lewis Publishers.

Jørgensen, S. E., Halling-Sørensen, B., & Nielsen, S. N. (1995). *Handbook of Environmental and Ecological Modelling* (p. 672). Boca Raton, Florida: CRC Lewis Publishers.

Jørgensen, S. E., Jacobsen, O. S., & Hoi, I. (1973). A prognosis for a lake. *Vatten, 29*, 382–404.

Jørgensen, S. E., Jørgensen, L. A., Kamp Nielsen, L., & Mejer, H. F. (1981). Parameter estimation in eutrophication modelling. *Ecological Modelling, 13*, 111–129.

Jørgensen, S. E., Kamp-Nielsen, L., & Jacobsen, O. S. (1975). A submodel for anaerobic mud-water exchange of phosphate. *Ecological Modelling, 1*, 133–146.

Jørgensen, S. E., Kamp-Nielsen, L., Jørgensen, L. A., & Mejer, H. (1982). An environmental management model of the Upper Nile lake system. *ISEM Journal, 4*, 5–72.

Jørgensen, S. E., Ladegaard, N., Debeljak, M., & Marques, J. C. (2005). Calculations of exergy for organisms. *Ecological Modelling, 185*, 165–175.

Jørgensen, S. E., Löffler, H., Rast, W., & Straskraba, M. (2004). *Lake and Reservoir Management* (p. 503). Amsterdam: Elsevier.

Jørgensen, S. E., Lützhøft, H. C., & Halling-Sørensen, B. (1998). Development of a model for environmental risk assessment of growth promoters. *Ecological Modelling, 107*, 63–72.

Jørgensen, S. E., & Mejer, H. F. (1977). Ecological buffer capacity. *Ecological Modelling, 3*, 39–61.

Jørgensen, S. E., & Mejer, H. F. (1979). A holistic approach to ecological modelling. *Ecological Modelling, 7*, 169–189.

Jørgensen, S. E., Mejer, H. F., & Friis, M. (1978). Examination of a lake. *Journal of Ecological Modelling, 4*, 253–279.

Jørgensen, S. E., Nielsen, S. N., & Mejer, H. F. (1995). Emergy, environ, exergy, and ecological modelling. *Ecological Modelling, 77*, 99–109.

Jørgensen, S. E., Nors Nielsen, S., & Jørgensen, L. A. (1991). *Handbook of Ecological Parameters and Ecotoxicology*. Amsterdam: Elsevier, Published as CD under the name ECOTOX, with L.A. Jørgensen as first editor in 2000.

Jørgensen, S. E., Patten, B. C., & Straskraba, M. (2000). Ecosystem emerging: 4. Growth. *Ecological Modelling, 126*, 249–284.

Jørgensen, S. E., & Svirezhev, Y. (2004). *Toward a Thermodynamic Theory for Ecological Systems* (p. 366). Amsterdam: Elsevier.

Kallqvist, T., & Meadows, B. S. (1978). The toxic effects of copper on algae and rotifers from a soda lake. *Water Research, 12*, 771–775.

Kamp-Nielsen, L. (1975). A kinetic approach to the aerobic sediment-water exchange of phosphorous in Lake Esrom. *Ecological Modelling, 1*, 153–160.

Kauppi, P., Kämäri, J., Posch, M., Kauppi, L., & Matzner, E. (1986). Acidification of forest soils: model development and application for analysing impacts of acidic deposition in Europe. *Ecological Modelling, 33*, 231–253.

Kazancı, C. (2007). EcoNet: A new software for ecological modelling, simulation, and network analysis. *Ecological Modelling, 208*, 1, 3–8.

Kirchner, T. B., & Whicker, F. W. (1984). Validation of PATHWAY, a simulation model of the Transport of radionuclides through agroecosystems. *Ecological Modelling, 22*, 21–45.

Knudsen, G., & Kristensen, L. (1987). *Development of a Model for Cadmium Uptake by Plants*. Master's thesis at University of Copenhagen.

Koestler, A. (1967). *The Ghost in the Machine*. New York: Macmillan.

Kohlmaier, G. H., Sire, E. O., Brohl, H., Kilian, W., Fishbach, U., Plochl, M., et al. (1984). Dramatic development in the dying of German spruce-fir forests: In search of possible cause effect relationships. *Ecological Modelling, 22*, 45–65.

Kooijman, S. A. L. M. (2000). *Dynamic Energy and Mass Budget in Biological Systems*. (2nd ed., p. 424). Cambridge: Cambridge University Press, UK.

Koschel, R., Kasprzak, P., Krienitz, L., & Ronneberger, D. (1993). Long term effects of reduced nutrient loading and food-web manipulation on plankton in a stratified Baltic hard water lake. *Verhandlungen der Internationale Vereinigung für Theoretische und Angewandte Limnologie, 25*, 647–651.

Kristensen, P., Jensen, P., & Jeppesen, E. (1990). *Eutrophication Models for Lakes* (p. 120). Research Report C9. Copenhagen: DEPA.

Lam, D. C. L., & Simons, T. J. (1976). Computer model for toxicant spills in Lake Ontario. In J. O. Nriago (Ed.), *Metals Transfer and Ecological Mass Balances, Environmental Biochemistry* (vol.2, pp. 537–549). Ann Arbor, Michigan: Ann Arbor Science.

Lammens, E. H. R. R. (1988). Trophic interactions in the hypertrophic Lake Tjeukemeer: Top-down and bottom-up effects in relation to hydrology, predation, and bioturbation, during the period 1974-1988. *Limnologica (Berlin), 19*, 81–85.

Lassiter, R. R. (1978). *Principles and constraints for predicting exposure to environmental pollutants.* Corvallis: U.S. Environmental Protection Agency, OR Report EPA 118-127519.

Legovic, T. (1997). Toxicity may affect predictability of eutrophication models in coastal sea. *Ecological Modelling, 99,* 1–6.

Lehman, J. T., Botkin, D. B., & Likens, K. E. (1975). The assumptions and rationales of a computer model of phytoplankton population dynamics. *Limnology and Oceanography, 20,* 343–364.

Leontief, W. W. (1951). *The Structure of American Economy, 1919–1939: An Empirical Application of Equilibrium Analysis.* New York: Oxford University Press.

Leslie, P. H. (1945). On the use of matrices in certain population mathematics. *Biometrika, 33,* 183–212.

Leung, D. K. (1978). *Modelling the bioaccumulation of pesticides in fish.* Troy, NY: Center for Ecological Modelling, Polytechnic Institute, Report 5.

Levine, S. (1980). Several measures of trophic structure applicable to complex food webs. *Journal of Theoretical Biology, 83,* 195–207.

Levins, R. (1969). Some demographic and genetic consequences of environmental heterogeneity for biological control. *Bulletin of the Entomological Society of America, 15,* 237–240.

Lewis, E. G. (1942). On the generation and growth of a population. *Sankhya, 6,* 93–96.

Liebig, J. (1840). *Chemistry in its Application to Agriculture and Physiology.* London: Taylor and Walton.

Likens, G. E. (Ed.), (1985). *An Ecosystem Approach to Aquatic Ecology: Mirror Lake and Its Environment* (p. 516). New York: Springer-Verlag.

Lindeman, R. L. (1942). The trophic dynamic aspect of ecology. *Ecology, 23,* 399–418.

Liu, J., & Ashton, P. S. (1995). Individual-based simulation models for forest succession and management. *Forest Ecology and Management, 73,* 157–175.

Lombardo, P. S. (1972). *Mathematical Model of Water Quality in Rivers and Impoundments.* Palo Alto, California: Hydrocomp. Inc.

Longstaff, B. C. (1988). Temperature manipulation and the management of insecticide resistance in stored grain pests. A simulation study for the rice weevil, Sitophilus oryzae. *Ecological Modelling, 43,* 303–313.

Lotka, A. J. (1956). *Elements of Mathematical Biology* (p. 465). New York: Dover.

Mackay, D. (1991). *Multimedia Environmental Models* (p. 257). *The Fugacity Approach.* Boca Raton, Florida: Lewis Publishers.

Mackay, D., Shiu, W. Y., & Ma, K. C. (1991, 1992). *Illustrated Handbook of Physical-Chemical Properties and Environmental Fate for Organic Chemicals* (vol. I). Mono-aromatic Hydrocarbons. Chloro-benzenes and PCBs. 1991, Volume II. Polynuclear Aromatic Hydrocarbons, Polychlorinated Dioxines, and Dibenzofurans, 1992, and Volume III. Volatile Organic Chemicals. 1992. New York: Lewis Publishers.

Margalef, R. (1968). *Perspectives in Ecological Theory* (p. 122). Chicago, IL: University of Chicago Press.

Margalef, R. (1969). Diversity and stability: A practical proposal and a model for interdependence. In G. M. Woodwell & H. H. Smith (Eds.), *Diversity and Stability in Ecological Systems. Brookhaven Symposia in Biology, No. 22* (pp 25–38). Upton, NY: Brookhaven National Laboratory.

Margalef, R. (1991). Networks in ecology. In M. Higashi & T. P. Burns (Eds.), *Theoretical Studies of Ecosystems: The Network Perspective* (pp. 41–57). Cambridge: Cambridge University Press, UK.

Margalef, R. A. (1995). Information theory and complex ecology. In B. C. Patten & S. E. Jørgensen (Eds.), *Complex Ecology* (pp. 40–50). New Jersey: Prentice Hall PTR.

Matthies, M., Behrendt, H., & Münzer, B. (1987). *EXSOL Modell für den Transport und Verbleib von Stoffen im Boden*. GSF-Bericht 23/87 Neuherberg.

Mauersberger, P. (1983). General principles in deterministic water quality modelling. In G. T. Orlob (Ed.), *Mathematical Modelling of Water Quality: Streams, Lakes and Reservoirs* (pp. 42–115). *International Series on Applied System Analysis, 12*. New York: Wiley.

Mauersberger, P. (1985). Optimal control of biological processes in aquatic ecosystem. *Gerlands Beitr. Geiophysik., 94*, 141–147.

May, R. M. (1974). Ecosystem patterns in randomly fluctuating environments. *Progress in Theoretical Biology, 3*, 1–50.

May, R. M. (1975). Patterns of species abundance and diversity. In M. L. Cody & J. M. Diamond (Eds.), *Ecology and Evolution of Communities* (pp. 81–120). Cambridge, MA: Harvard University. Press.

May, R. M. (1977). *Stability and Complexity in Model Ecosystems* (3rd ed.). New Jersey: Princeton University Press.

McMahon, T. A., Denison, D. J., & Fleming, R. (1976). A long distance transportation model incorporating washout and dry deposition components. *Atmospheric Environment, 10*, 751–760.

Miller, D. R. (1979). Models for total transpon. In G. C. Butler (Ed.), *Principles of Ecotoxicology SCOPE* (Vol. 12, pp. 71–90). New York: Wiley.

Miller, J. G. (1978). *Living Systems* (p. 102). New York: McGraw-Hill.

Milne, G. W. A. (1994). *CRC Handbook of Pesticides*. Boca Raton, FL: CRC Press.

Mogensen, B. (1978). *Chromium pollution in a Danish fjord*. Licentiate Thesis. Copenhagen: Royal Danish School of Pharmacy.

Mogensen, B., & Jørgensen, S. E. (1979). Modelling the distribution of chromium in a Danish firth. In S. E. Jørgensen (Ed.), *Proceedings of 1st International Conference on State of the An in Ecological Modelling* (pp. 367–377). Copenhagen, 1978. Copenhagen: International Society for Ecological Modelling.

Monod, J. (1971). *Chance and Necessity: An Essay on the Natural Philosophy of Modern Biology*. New York: Alfred A. Knopf.

Monte, L. (1998). Prediction the migration of dissolved toxic substances from catchments by a collective model. *Ecological Modelling, 110*, 269–280.

Morgan, M. G. (1984). Uncertainty and quantitative assessment in risk management. In J. V. Rodricks & R. G. Tardiff (Eds.), *Assessment and Management of Chemical Risks*. Chapter 8. ACS Symposium Series 239. Washington, D.C.: American Chemical Society.

Morioka, T., & Chikami, S. (1986). Basin-wide ecological fate model for management of chemical hazard. *Ecological Modelling, 31*, 267.

Morowitz, H. J. (1968). *Energy flow in biology. Biological Organisation as a Problem in Thermal Physics* (p. 179). New York: Academic Press. See review by H. T. Odum, Science, *164*, 683–684, 1969.

Moustakas, A., Sakkos, K., Wiegand, K., Ward, D., Meyer, K. M., & Eisingerd, D. (2009). Are savannas patch-dynamic systems? A landscape model. *Ecological Modelling, 220*, 3576–3588.

Müller, F., Schrautzer, J., Reiche, E. W., & Rinker, A. (2006). Ecosystem based indicators in retrogressive successions of an agricultural landscape. *Ecological Indicators, 6*, 63–82.

Nagra synpunkter angaende limnoplanktons okologi med sarskild hansyn till fytoplankton. Svensk Botanisk Tidskrift 13, 129–163.

Naumann, E. (1919). Nagra synpunkter angaende limnoplanktons okologi med sarskild hansyn till fytoplankton. *Svensk Botanisk Tidskrift, 13*, 129–163.

Nielsen, S. N. (1992a). *Application of Maximum Exergy in Structural Dynamic Models* (p. 51). Ph.D. Thesis. Denmark: National Environmental Research Institute.

Nielsen, S.N (1992b). Strategies for structural-dynamical modelling. *Ecological Modelling, 63*, 91–101.

Nihoul, J. C. J. (1984). A non-linear mathematical model for the transport and spreading of oil slicks. *Ecological Modelling, 22*, 325–341.

Nyholm, N., Nielsen, T. K., & Pedersen, K. (1984). Modelling heavy metals transpon in an arctic fjord system polluted from mine tailings. *Ecological Modelling, 22*, 285–324.

O'Connor, D. J., Mancini, J. L., & Guerriero, J. R. (1981). *Evaluation of Factors Influencing the Temporal Variation of Dissolved Oxygen in the New York Bight. PHASE II.* Bronx. New York: Manhattan College.

Odling-Smee, F. J., Laland, K. N., & Feldman, M. W. (2003). *Niche Construction: The Neglected Process in Evolution* (p. 468). Princeton, New Jersey: Princeton University Press.

Odum, E. P. (1953). *Fundamentals of Ecology.* Philadelphia, PA: W.B. Saunders.

Odum, H. T. (1956). Primary production in flowing waters. *Limnology and Oceanography, 1*, 102–117.

Odum, H. T. (1957). Trophic structure and productivity of Silver Springs. *Ecological Monographs, 27*, 55–112.

Odum, E. P. (1959). *Fundamentals of Ecology* (2nd ed). Philadelphia, PA: W.B. Saunders.

Odum, E. P. (1969). The strategy of ecosystem development. *Science, 164*, 262–270.

Odum, E. P. (1971). *Fundamentals of Ecology* (3rd ed). Philadelphia, PA: W.B. Saunders Co.

Odum, H. T. (1983). *System Ecology* (p. 510). New York: Wiley Interscience.

Odum, H. T., & Odum, E. C. (2000). *Modelling for all Scales* (p. 456). San Diego: Academic Press, and CD.

O'Neill, R. V. (1976). Ecosystem persistence and heterotrophic regulation. *Ecology, 57*, 1244–1253.

O'Neill, R. V., Hanes, W. F., Ausmus, B. S., & Reichle, D. E. (1975). A theoretical basis for ecosystem analysis with particular reference to element cycling. In F. G. Howell, J. B. Gentty & M. H. Smith (Eds.), *Mineral Cycling in Southeastern Ecosystems* (pp. 28–40). NTIS pub. CONF-740513.

Onishi, Y., & Wise, S. E. (1982). *Mathematical model, SERA TRA, for sediment-contaminant transpon in rivers and its application to pesticide transpon in Four Mile and Wolf Creeks in Iowa* (p. 56). EPA-60013-82-045, Athens, Georgia.

Orlob, G. T., Hrovat, D., & Harrison, F. (1980). Mathematical model for simulation of the fate of copper in a marine environment. American Chemical Society. *Advances in Chemistry Series, 189,* 195–212.

Parrott, L. (2008). Adaptive Agents. In S. E. Jørgensen & B. Fath (Eds.), *Encyclopedia of Ecology* (vol.1, pp. 47–51). Amsterdam: Elsevier.

Patten, B. C. (1971–1976). *Systems Analysis and Simulation in Ecology* (vols. 1–4). New York: Academic Press.

Patten, B. C. (1978a). Systems approach to the concept of environment. *Ohio Journal of Science, 78,* 206–222.

Patten, B. C. (1978b). Energy environments in ecosystems. In R. A. Fazzolare & C. B. Smith (Eds.), *Energy Use Management* (vol. IV). New York: Pergamon Press.

Patten, B. C. (1981). Environs: The superniches of ecosystems. *American Zoologist, 21,* 845–852.

Patten, B. C. (1982a). Environs: relativistic elementary particles or ecology. *American Naturalist, 119,* 179–219.

Patten, B. C. (1982b). Indirect causality in ecosystem: its significance for environmental protection. In W. T. Mason & S. Iker, (Eds.), *Research on Fish and Wildlife Habitat, Commemorative monograph honoring the first decade of the US Environmental Protection Agency.* EPA-@18-82-022. Washington, D.C: Office of Research and Development, US Environmental Protection Agency.

Patten, B. C. (1985). Energy cycling in ecosystems. *Ecological Modelling, 28,* 7–71.

Peters, R. H. (1983). *The Ecological Implications of Body Size* (p. 329). Cambridge, UK: Cambridge University Press.

Pielou, E. C. (1966). Species-diversity and pattern diversity in the study of ecological succession. *Journal of Theoretical Biology, 10,* 370–383.

Pielou, E. C. (1977). *An Introduction to Mathematical Ecology* (p. 385). New York: Wiley-Interscience.

Prahm, L. P., & Christensen, O. J. (1976). Long-range transmission of sulphur pollutants computed by the pseu<k>spectral model. Danish Meteorological Institute, Air Pollution Section, Lyngbyvej, DK-2100 Copenhagen. In: *Prepared for the ECE Task Force for the preparation of a Co-operative Programme for the Monitoring and Evaluation of the Long-range Transmission of Air Pollutants in Europe, October 1976, Lillestrom, Norway.*

Prigogine, I. (1947). *Etude Thermodynamique des Processus Irreversibles.* Liege: Desoer.

Quinlin, A. V. (1975). *Design and Analysis of Mass Conservative Models of Ecodynamic Systems.* Ph.D. Dissertation. Cambridge, MA: MIT Press.

Rammig, A., & Fahse, L. (2009). Simulating forest succession after blowdown events: The crucial role of space for a realistic management. *Ecological Modelling, 220,* 3555–3564.

Rammig, A., Fahse, L., Bugmann, H., & Bebi, P. (2006). Forest regeneration after disturbance: A modelling study for the Swiss Alps. *Forest Ecology and Management, 222,* 123–136.

Sas, H. (Coordination) (1989). Lake restoration by reduction of nutrient loading. Expectations, experiences, extrapolations. *St. Augustine Academia Verl. Richarz.,* 497.

Scavia, D. (1980). An ecological model of Lake Ontario. *Ecological Modelling, 8,* 49–78.

Schaalje, G. B., Stinner, R. L., & Johnson, D.L (1989). Modelling insect populations affected by pesticides with application to pesticide efficacy trials. *Ecological Modelling, 47,* 223.

Scheffer, M. (1990). *Simple Models as Useful Tools For Ecologists* (p. 192). Amsterdam: Elsevier.

Schoffeniels, E. (1976). *Anti-Chance.* New York: Pergamon Press.

Schrödinger, E. (1944). *What is Life?* (p. 186). Cambridge, UK: Cambridge University Press.

Scotti, M., & Bondavalli, C. R. (2011). *ENA: A free package for Ecosystem Network Analysis in R. Environmental Modelling and Software.*

Secor, D. H., & Houde, E. D. (1995). Temperature effects on the timing of striped bass egg production, larval viability, and recruitment potential in the Patuxent River (Chesapeake Bay). *Estuaries and Coasts, 18,* 527–544.

Seip, K. L. (1978). Mathematical model for uptake of heavy metals in benthic algae. *Ecological Modelling, 6,* 183–198.

Shapiro, J. (1990). Biomanipulation. The next phase-making it stable. *Hydrobiologia, 200/210,* 13–27.

Shevtsov, J., Kazanci, C., & Patten, B. C. (2009). Dynamic environ analysis of compartmental systems: A computational approach. *Ecological Modelling, 220,* 3219–3224.

Shugart, H. H. (1984). *A Theory of Forest Dynamics: The Ecological Implications of Forest Succession Models.* New York: Springer.

Sklar, F. H., Costanza, R., & Day, J. W. (1985). Dynamic spatial simulation modelling of coastal wetland habitat succession. *Ecological Modelling, 29,* 261–281.

Slovic, P., Fischhoff, B., & Lichtenstein, S. (1982). Facts and fears: Understanding perceived risk. In R. C. Schwing & W. A. Albers, Jr. (Eds.), *Societal Risk Assessment: How Safe is Safe Enough?* New York: Plenum Press.

Snape, J. B., Dunn, I. J., Ingham, J., & Presnosil, J. E. (1995). *Dynamics of Environmental Bioporocesses* (p. 492). Weinheim, New York: VCH.

Sommer, U. (1989). Toward a Darwinian ecology of plankton. In U. Sommer (Ed.), *Plankton Ecology: Succession in Plankton Communities.* Berlin: Springer-Verlag.

Starfield, A. M., & Bleloch, A. L. (1986). *Building Models for Conservation and Wildlife Management* (p. 324). New York: Macmillan Publishing Company.

Steele, J. H. (1974). *The Structure of the Marine Ecosystems* (p. 128). Oxford: Blackwell Scientific Publications.

Stenseth, N. C. (1986). Darwinian evolution in ecosystems: a survey of some ideas and difficulties together with some possible solutions. In J. L. Casti & A. Karlqvist (Eds.), *Complexity, Language, and Life: Mathematical Approaches* (pp. 105–129). Berlin: Springer-Verlag.

Stonier, T. (1990). *Information and the Internal Structure of the Universe* (p. 260). Berlin: Springer Verlag.

Straskraba, M. (1979). Natural control mechanisms in models of aquatic ecosystems. *Ecological Modelling, 6,* 305–322.

Straskraba, M. (1980). Cybernetic-categories of ecosystem dynamics. *ISEM Journal, 2,* 81–96.

Suter, G. W. (1993). *Ecological Risk Assessment.* Chelsea, MI: Lewis Publishers.

Swannack, T. M., Grant, W. E., & Forstner, M. R. J. (2009). Projecting population trends of endangered amphibian species in the face of uncertainty: A pattern-oriented approach. *Ecological Modelling, 220*, 148–159.

Tansley, A. G. (1935). The use and abuse of vegetational concepts and terms. *Ecology, 16*, 284–307.

Tetra Tech Inc. (1980). *Methodology for Evaluation of Multiple Power Plant Cooling System Effects, Vol. V: Methodology Application to Prototype-Cayuga Lake*. Lafayette, California: Tetra Tech Inc, For Electric Power Research Institute. Report EPRI EA-1111.

Thomann, R. V. (1984). Physico-chemical and ecological modelling the fate of toxic substances in natural water systems. *Ecological Modelling, 22*, 145–170.

Thomann, R. V., et al. (1974). A food chain model of cadmium in western Lake Erie. *Water Research, 8*, 841–851.

Thomann, R. V., & Fitzpatrick, J. F. (1982). *Calibration and verification of a mathematical model of the eutrophication of the Potomac Estuary*. Mahwah, NJ, to DES, Dist. Col. Washington D.C: Report by Hydroqual, Inc.

Tinkle, I. J. (1967). *The life and demography of the side-blotched lizard* Utastand buriana. *Misc. Publ. Mus. Zool.* University of Michigan Number, pp. 132–182.

Tilman, D., & Kareiva, P. (Eds.). (1997). *Spatial Ecology: The Role of Space in Population Dynamics and Interspecific Interactions*. New Jersey: Princeton University Press.

Turner, M. G., & Gardner, R.H (Eds.). (1991). *Quantitative Methods in Landscape Ecology: An Introduction*. New York: Springer-Verlag.

Turner, M. G., Gardner, R. H., & O'Neill, R. V. (2001). *Landscape Ecology in Theory and Practice: Pattern and Process*. New York: Springer-Verlag.

Uchmanski, J., & Grimm, V. (1996). Individual-based modelling in ecology: what makes the difference? *Trends in Ecology and Evolution, 11*, 437–441.

Uchrin, C. G. (1984). Modelling transport processes and differential accmnulation of persistent toxic organic substances in groundwater systems. *Ecological Modelling., 22*, 135–144.

Ulanowicz, R. E. (1979). Prediction chaos and ecological perspective. In E. A. Halfon (Ed.), *Theoretical Systems Ecology* (pp. 107–117). New York: Academic Press.

Ulanowicz, R. E. (1980). An hypothesis of the development of natural communities. *Journal of Theoretical Biology, 85*, 223–245.

Ulanowicz, R. E. (1982). *NETWORK 4.2b: A package of computer algorithms to analyze ecological flow networks*. Copyright 1982, 1987, 1998, 1999, 2002. UMCEES Ref. No. 82–7 CBL.http://www.cbl.umces.edu/~ulan/ntwk/netwrk.txt

Ulanowicz, R. E. (1986). *Growth and Development: Ecosystem Phenomenology*. New York: Springer-Verlag.

Ulanowicz, R. E. (1995). Ecosystem trophic foundations: Lindeman exonerata. In B. C. Patten & S. E. Jørgensen (Eds.), *Complex Ecology: The Part-Whole Relation in Ecosystems* (pp. 549–560). Englewood Cliffs, New Jersey: Prentice Hall PTR.

Ulanowicz, R. E. (1997). *Ecology, the Ascendant Perspective*. New York: Columbia University Press.

Ulanowicz, R.E (2009). *A Third Window: Natural Life beyond Newton and Darwin*. West Conshohocken, PA: Templeton Foundation Press.

Usher, M. B. (1972). Developments in the Leslie matrix model. In J. N. R. Jeffers (Ed.), *Mathematical Models in Ecology* (pp. 29–60). Oxford: Blackwells.

Vanclay, J. K. (1994). *Modelling Forest Growth and Yield* (p. 312). Wallingford: Cab International.

van den Belt, M. (2004). *Mediated Modelling*. Washington: Island Press.

Verschueren, K. (1983). *Handbook of Environmental Data on Organic Chemicals*. New York: Van Nostrand Reinhold.

Van Donk, E., Gulati, R. D., & Grimm, M. P. (1989). Food web manipulation in Lake Zwemlust: positive and negative effects during the first two years. *Hydrobiological Bulletin, 23,* 19–35.

Vitousek, P. M., Aber, J. D., Howarth, R. W., Likens, G. E., Matson, P. A., Schindler, D. W., Schlesinger, W. H., & Tilman, D. G. (1997). Human alteration of the global nitrogen cycle: sources and consequences. *Ecological Applications, 7,* 737–750.

Volterra, V. (1926). Actuations in the abundance of a species considered mathematically. *Nature, 188,* 558–560.

Wallentin, G., Tappeiner, U., Strobla, J., & Tasserd, E. (2008). Understanding alpine tree line dynamics: An individual-based model. *Ecological Modelling, 218,* 235–246.

Wangersky, P. J., & Cunningham, W. J. (1957). Time lag in population models. *Cold Spring Harbour Symposia Quantitative Biology, 42,* 329–338.

Wheeler, G. L., Rolfe, G. L., & Reinholdt, K. A. (1978). A simulation for lead movement in a watershed. *Ecological Modeling, 5,* 67–76.

Willemsen, J. (1980). Fishery aspects of eutrophication. *Hydrobiological Bulletin, 14,* 12–21.

Wilson, D. S. (1978). Prudent predation: A field test involving three species of tiger beetles. *Oikos, 31,* 128–136.

Wilson, W. (2000). *Simulating Ecological and Evolutionary Systems in C* (p. 300). Cambridge, UK: Cambridge University Press.

WMO (1975) *Intercomparison of Conceptual Models Used in Operational Hydrological Forecasting*. Operational Hydrology Report no. 7, Geneva, Switzerland: World Meteorological Organization.

Wolfram, S. (1984a). Computer software in science and mathematics. *Scientific American, 251,* 140–151.

Wolfram, S. (1984b). Cellular automata as models of complexity. *Nature, 311,* 419–424.

Woodwell, G. M., et al. (1967). DDT residues in an East Coast estuary: A case of biological concentration of a persistent insecticide. *Science, 156,* 821–824.

Wratt, D. S., Hadfield, M. G., Jones, M. T., Johnson, G. M., & McBurney, I. (1992). Power stations, oxides of nitrogen emissions, and photochemical smog: a modelling approach to guide decision makers. *Ecological Modelling, 64,* 185–204.

Wuttke, G., Thober, B., & Lieth, H. (1991). Simulation of nitrate transpon in groundwater with a three-dimensional groundwater model run as a subroutine in an agroecosystem model. *Ecological Modelling, 57,* 263–276.

Yongberg, B. A. (1977). *Application of the Aquatic Model CLEANER to Stratified Reservoir System*. Repon #1. Troy, New York: Center for Ecological Modelling, Rensselaer Polytechnic Institute.

Zhang, J., Gurkan, Z., & Jørgensen, S. E. (2010). Application of eco-energy for assessment of ecosystem health and development of structurally dynamic models. *Ecological Modelling, 221,* 693–702.

Zeigler, B. P. (1976). *Theory of Modelling and Simulation* (pp. 435–621). New York: Wiley.

Index

Note: Page numbers followed by *b* indicate boxes, *f* indicate figures and *t* indicate tables.

Printed and bound by CPI Group (UK) Ltd, Croydon, CR0 4YY

14/10/2024

01774139-0001